CLIMATE CHANGE POLICY

CLIMATE CHANGE POLICY

edited by

MICHAEL BOTHE AND ECKARD REHBINDER

Published, sold and distributed by Eleven International Publishing
P.O. Box 358
3500 AJ Utrecht, the Netherlands
Tel.: +31 30 231 0545
Fax: +31 30 225 8045
info@elevenpub.com
www.elevenpub.com

Printed on acid-free paper.

ISBN 90-77596-04-6

Printed in The Netherlands

Table of Contents

List of Contributors

Buchner, Barbara
Fondazione Eni Enrico Mattei, FEEM, Venice, Italy

Brockhagen, Dietrich
Germanwatch, Berlin, Germany

Caparrós Gass, Alejandro
CIRED, Centre International de Recherche sur l'Environnement et le Développement, Paris, France

Fernández Armenteros, Mercedes
Öko-Institut e.V., Darmstadt, Germany

Jacquemont, Frédéric
FSUR, Forschungsstelle Umweltrecht, University of Frankfurt, Germany

Krohn, Susan Nicole
Bundesumweltministerium, Berlin, Germany

Lefevere, Jürgen
FIELD, Foundation for International Environmental Law and Development, London, UK

Lehmann, Janna
CERDEAU, Centre d'Etude et de Recherche en Droit de l'Environnement, de l'Aménagement et de l'Urbanisme, Paris, France

Massai, Leonardo
FSUR, Forschungsstelle Umweltrecht, University of Frankfurt, Germany

Tazdaït, Tarik
CIRED, Centre International de Recherche sur l'Environnement et le Développement, Paris, France

Introduction

MICHAEL BOTHE AND ECKARD REHBINDER

Climate Change as a Problem of Law and Policy

The International Climate Change Regime and its European Implementation

During the years 1999 to 2003, researchers in six institutions dealing with legal, political and economic aspects of climate change, namely the Forschungsstelle Umweltrecht/Environmental Law Research Centre (ELRC, Frankfurt/M.), the Öko-Institut (Darmstadt), the Centre d'Étude et de Recherche en Droit de l'Environnement, de l'Aménagement et de l'Urbanisme (CERDEAU, Paris), the Centre International de Recherche sur l'Environnement et le Développement (CIRED, Paris), the Fondazione Eni Enrico Mattei (FEEM, Milan/Venice) and the Foundation for International Environmental Law and Development (FIELD, London) joined forces to shed some light on the emerging international regime of climate change and its problems of implementation.

1. The issues of climate change

This international climate change regime is one of the most ambitious undertakings of international governance, highly complex in many respects. The present volume tries to elucidate some key features. The starting point and *raison d'être* of this regime is a natural phenomenon, namely the greenhouse effect. So-called greenhouse gases in the earth's atmosphere account for the effect that the heat derived from solar energy is retained in the earth's atmosphere. It is an effect to which life on earth owes its existence. The regulatory problem addressed by the climate change regime is due to the forecast that a further accumulation of greenhouse gases in the atmosphere will increase the greenhouse effect in a way which is prejudicial. Business as usual forecasts predict a rise in the average global temperature of two degrees by 2025 and four degrees by 2100. This, it is predicted, would negatively affect the biological balance of the earth.

MICHAEL BOTHE AND ECKARD REHBINDER (EDS.), Climate Change Policy, 1–16.

While the greenhouse gas effect as such is an uncontroversial fact of nature, the scope of overall global warming remains uncertain, the distribution of this effect over the earth is still more so, and still more so are the consequences of that effect. The melting of the polar icecaps and the ensuing rise in the sea level are only the most commonly cited effects of climate change. Disastrous weather effects are another perspective. When the first warnings from the scientific community were uttered in the 1970s, the scope of the problem was very controversial. Since then, a tremendous amount of research has been done, providing a clearer picture, but uncertainties remain.

The political reaction to scientific warnings started in the 1980s. In 1985, the United Nation Environment Programme (UNEP) and the World Meteorological Organization (WMO) organized a common conference in Villach. In 1988, the Toronto Conference on the Changing Atmosphere was organized by the Canadian Government. In that year, UNEP and WMO created the Intergovernmental Panel on Climate Change (IPCC) as a scientific, yet official body. The General Assembly of the United Nations addressed the question in Resolutions 43/53, 44/207 and 45/212, the latter Resolution establishing an Intergovernmental Negotiating Committee (INC). In 1990, IPCC rendered its first assessment report. As a result of the efforts of the negotiating committee, the United Nations Framework Convention on Climate Change (UNFCC) could be opened for signature at the Rio Conference in 1992. It must be stressed that this action was based on the precautionary principle, recognized at the same time by the Rio Declaration. The convention is indeed a precautionary measure, based on the rule (which constitutes the core of the precautionary principle) that scientific uncertainty should not bar remedial action.

This definition of the principle is a negative one. Turned positive as a basis for measures to be taken, the precautionary principle is not a hard and fast rule. It is not possible to automatically derive from the principle an indication of specific measures. Thus, the application of the precautionary principle requires political determinations. It is, thus, influenced by political interests. In order to understand the compromise finally reached in the UNFCC of 1992 and later in the Kyoto Protocol (KP), it is necessary to understand the interest at stake, be they real or "only" perceived by relevant actors.

There are, first, the environmental interests pursued by the UNFCCC regime, i.e. the interest to limit the man-made increase of the greenhouse effect and its ensuing negative consequences. The problem of the greenhouse effect is a long-term one. Those really affected are not yet, or only just, born. It is a concern for future generations. But by whom and how strongly could those interests be represented? Much of the pressure to take into account those environmental interests came from civil society. UNCED 1992 was the first big international conference which opened itself systematically to the input from organizations of civil society. As to governments, "green power" was stronger in Europe than elsewhere in the world. Thus, one can acknowledge green preferences of European politicians, expressed by the European Economic Community (as it then still was) and its Member States. Other developed countries (the United States, Japan, Canada, Australia and New Zealand) took a different view. They preferred to leave the solution of the problem to market forces.

Probably the strongest emphasis on environmental interests was put forward by those states which are most probably affected by a sea level rise, the so-called Alliance of Small Island States (AOSIS). The obvious environmental interests of these developing states limited the unity of the G-77 and China Group which otherwise is an important fact in international negotiations.

There were countervailing economic interests: the cost of investment to be made in order to cope with the greenhouse effect, which means high costs threatening certain industries, in particular old, low-efficiency industries in the industrialized countries. It appears that this cost has been of particular concern for the United States. The industrial development interests of developing countries, however, pointed in a similar direction. The fear of the developing countries was that these costs might force them to renounce, or at least restrict, industrialization, a fear which has characterized the stance of developing countries towards environmental policy in general from the early 1970s. The economic interests of fossil fuel producers (OPEC), at least as they currently perceive them, are also adverse to any attempt to limit fuel consumption through climate protection measures.

Combined with the countervailing economic interests, there is also a countervailing social interest: the interest in maintaining a lifestyle based on high energy consumption. This is a politically powerful interest in a great number of industrialized countries. Finally, there is a countervailing political interest, often formulated as a sovereignty interest. It is the perceived threat to state freedom of action which might be the result of the climate change regime. In this perspective, this more general sovereignty aspect joins unilateralist tendencies in the United States' foreign policy.

The UNFCCC is able to accommodate these conflicting interests in a specific way. It is a compromise solution in many respects.

The first aspect of this compromise is the time element: the convention adopts a stepwise approach, which by 1992 had become a standard element in the creation of international environmental regimes. It is a framework convention, which contains only general commitments, thus leaving obligations which really hurt to a later phase. On the other hand, the convention recognizes indeed the existence of the problem, a fact that cannot be underestimated in the face of remaining uncertainties. The first phrase of the preamble reads:

> Acknowledging that change in the Earth's climate and its adverse effects are a common concern of humankind ...

The approach is based on the precautionary principle (Article 3 no. 3) in accordance with the principle of intergenerational equity of Article 3 no. 1. Nevertheless, the convention recognizes, on the other hand, the necessity of development, which should, however, be sustainable (Article 3 para. 4). All this corresponds to the principles adopted at the same time in the Rio Declaration.

As to the specific problem to be solved, the convention adopts what can be called the double track approach: mitigation through stabilization of emissions on the one hand (primary goal) and adaptation to the change, on the other (secondary goal). The former is formulated in Article 2, the latter is reflected in a number of different provisions, e.g. assistance to adaptation by most vulnerable countries (Article 3 para. 2, Article 4 para. 1(e)).

The essential compromise between developed and developing countries lies in a system of differentiated obligations of developed countries on the one hand, and developing countries on the other. On the level of principles, this is formulated in the principle of common but differentiated responsibilities. To quote the preamble:

> Noting that the large share of historical and current global emissions of greenhouse gases has originated in the developed countries, that per capita emissions in developing countries are still relatively low and that the share of global emissions originating in the developing countries grow to meet their social and development needs,
>
> ...
>
> acknowledging that the global nature of climate change calls for the widest possible cooperation by all countries and their participation in an effective and appropriate international response, in accordance with their common but differentiated responsibilities and respective capabilities and their social and economic conditions.

The most important practical difference between developing and developed countries relates to their stabilization duties (for all countries Article 4 para. 1 (d), for developed countries Article 4 para. 2 (a)). Also for developed countries, these stabilization duties are rather soft. In this respect, the convention really adopts a wait-and-see-approach. What is important is the provision for review.

As to implementation and enforcement, the convention calls for the establishment of inventories and reporting; both obligations also being differentiated between developing and developed countries.

A last element of the compromise achieved between developing and developed countries relates to finance. For all practical purposes, the developed countries should bear the actual compliance costs of the developing countries (Article 4 para. 3, 11). It is probably this soft approach which accounts for the smooth ratification process. Within less than two years after the adoption of its text, the UNFCCC entered into force on 21 March 1994 after the deposit of the 50th ratification instrument.

The ensuing development of the regime is characterised by the attempt to achieve stricter stabilization obligations. In 1995, the second IPCC-Assessment Report brought more certainty as to the man-made components of the greenhouse effect and furnished worse predictions for the business-as-usual approach. But it could not remove all uncertainties, in particular those relating to the negative social and economic impact of climate change on specific areas, nor those relating to the actual cost of remedies.

The first Conference of the Parties (COP) took place in Berlin in March/April 1995. In conformity with Article 2 (4) (d), already mentioned, it undertook a review of the existing obligations and held them to be inadequate. A new negotiating group was created with the mandate to elaborate an additional protocol containing concrete stabilization obligations, the so-called Berlin Mandate. That mandate was fulfilled at the Kyoto Conference in 1997 (COP 3). It is the regime created by this Kyoto Protocol which is the object of the research documented in the present publication. The Kyoto Protocol, too, constitutes a compromise, but this time more between various groups of developed countries than between developing and developed countries. The developing countries defended their position as it emerged from the Rio Conference: no strict stabilization obligations for developing countries, not even for threshold countries whose contribution to the greenhouse effect is not negligible. That tough stance of the developing countries furnished an additional reason, even an appearance of justification, for the United States' rejection of the Kyoto Protocol. The United States considers the absence of important developing countries, such as India and Brazil, and in particular of China as unjust and therefore a major flaw of the Kyoto Protocol regime.

The general regulatory approach of the Protocol constitutes a compromise in many respects. The basic obligation of the developed countries is a quantified emission limitation and reduction commitment (QELRC) formulated as an aggregate target, i.e. a stabilization or reduction of the overall greenhouse gas emissions of a particular state, taken as a whole. This method of fixing an absolute target was not uncontroversial from the point of view of the countervailing economic interests. Some would have preferred, and still prefer, a relative target, such as one linked to economic growth factors. On the other hand, the formulation of an aggregate obligation leaves a complete freedom, at least as a matter of principle, as to how the state reaches this overall target. Thus, there is a type of obligation which is not very intrusive, which does not infringe upon claims of state sovereignty. It thus accommodates sovereignty interests.

Like UNFCCC itself, the Kyoto Protocol constitutes a compromise in relation to time. It does not attempt to achieve at once the entire reduction required to solve the problem of the excess greenhouse effect. It only provides for a more modest reduction to be achieved during a first "commitment period", 2008–2012. The full solution of the problem is left to future commitment periods.

It is in a number of important details that further problems involved in this approach appear and where compromises between the developed countries had to be made. As a matter of fact, the concrete shape of the obligations was a matter of horse-trading. The first problem is the determination of the QELRC for each particular country. Concessions had to be made in various respects, leading to a result where certain countries which were reluctant, but necessary to achieve a meaningful participation with the Protocol, were given concessionary QELRCs. This is particularly the case for Russia and Australia. The second serious problem involved in the calculation of the targets is carbon sinks. Recognizing sequestration of greenhouse gases by sinks favours countries having a large potential of afforestation, as growing forests are the most important sink on land.

The next compromise element is flexibilization. Mainly the United States, but not only, insisted on the insertion of a flexibilization mechanism. The basic rational behind the flexibilization mechanism is a process of compensation between a non-reduction in one place and an increased reduction in another. The four flexibilization mechanisms provided by the Protocol are a specific bubble approach, Joint Implementation, Clean Development Mechanism and Emissions Trading. The major argument for flexibilization is economic efficiency. Flexibilization mechanisms allow reductions to be made where it is economically speaking most efficient. Thus, the basic idea is to achieve the overall reduction target at lower cost. That purpose is no doubt laudable, even necessary, given the considerable cost involved in the implementation of the regime. On the other hand, this approach also raises certain doubts. In particular, the Clean Development Mechanism opens up the possibility that developed countries fulfil their reduction obligations by paying for reductions achieved in other countries which do not have quantified reduction obligations of their own without, so to say, making a significant effort at home. For many, this approach seems to be unreasonable, and there have therefore been attempts to put a cap on flexibilization, meaning that the use of flexibilization mechanisms can only be supplementary to a serious stabilization effort made by each developed country.

Similar considerations apply to sinks, in particular in combination with Joint Implementation. This allows a country having relatively high emissions to reduce its burden by financing afforestation in another country. The calculation of individual stabilization targets for states as a result of aggregate emissions and the use of sinks and in combination flexibilization mechanisms posed difficult questions of calculation, monitoring compliance and last but not least sanctions.

When the basic elements of the compromise were adopted in Kyoto, it was not possible to deal with all these technical details. Thus, in a way, Kyoto was still characterized by a wait-and-see approach. The real meaning of Kyoto could only become clear when these relevant details were settled. Thus, there developed a widespread reluctance to ratify the Kyoto Protocol until these details became clear. COP 6, which took place at The Hague in 2000, brought a final show-down on these matters. It was not possible to solve the outstanding issues during that conference. COP 6 had to be adjourned and resumed in Bonn in 2001. It was at this conference that the difficulties were settled as a matter of principle, while the finalization took place at COP 7 in Marrakech in 2001. The result are the so-called Marrakech Accords, a huge volume of instruments most of which will have to be formally adopted by the first Conference of the Parties to the UNFCC serving as a meeting of the Parties of the Kyoto Protocol (COP/MOP).

The Clinton administration had taken steps in the direction of a ratification of the Protocol by the United States, but was not successful against an hostile Senate. In 2000, the Bush administration announced its intention not to ratify. The other countries decided to go ahead nevertheless, although the United States is by far the biggest emitter of greenhouse gases. But the US decision fundamentally changed the negotiating environment. The negotiations now practically took place between the rest of the industrialized countries. *Vis-à-vis* the EU which pushed for a strong Protocol, the relative bargaining power of some hesitant countries grew. The position of Russia became particularly strong as the entry into force of the Protocol then depended on the ratification by Russia. Entry into force requires ratification by a number of countries which, taken together, account for at least 55% of the world's emissions. Russia accounts for 17%. After the non-ratification by the United States, it was not possible to reach the necessary 55% without Russia. The Russian decision to ratify finally came late in 2004.[1] The Protocol enters into force on 16 February 2005.

As the consequence of the process of progressive concretization of the UNFCC and Kyoto Protocol regime, a most ambitious and complex system emerges. Four categories of problems can be distinguished:

1. The calculation of QELRCs: the stabilization obligation is calculated in a unit which makes the different greenhouse gas emissions comparable. To arrive at a national emission figure, sequestration of greenhouse gases in

[1] The contributions contained in this volume were concluded before this event.

sinks has to be deducted. The method of calculating this sequestration is the next problem as there is no practical means to actually measure it.

2. Flexibilization: flexibilization means that reductions achieved in one place may be credited to a QELRC of another, or, vice versa, that an emission taking place in one place is debited to the account of another state. This means, first of all, that the transactions in question have to be documented, registered and monitored. The calculation problems already mentioned are complicated by that fact.

3. Implementation: as the QELRC is expressed in an aggregate quantity of emissions originating in one state, the state has a complete freedom as to the choice of the means how to achieve the required stabilization of those aggregate emissions. These measures are different in relation to different greenhouse gases. In relation to CO_2, the basic principle is reducing the combustion of fossil fuels, which can be achieved in a number of ways, such as more energy-efficient combustion engines, energy saving through measures such as modification of the construction and situation of buildings, and the reduction of combustion processes by change in consumption habits, such as reducing the speed of cars or the use of air-conditioning systems. To achieve these results, very different regulatory approaches are possible, such as command and control approaches in relation to certain points of a chain of causation leading to the use of combustion processes, economic incentives or disincentives such as an energy tax or social incentives such as premiums on good environmental behaviour. The appropriate choice of a mix of implementation measures is the major challenge of a national (or, as the case may be, EU/EC) policy to achieve compliance with the QELRC. In the light of the costs involved, a cost-efficient choice of instruments is vital. The problem is further complicated by the fact that these national measures of implementation may also be subject to other international legal regimes. An example is, for instance, land-use measures to develop sinks which may be in contradiction to an international legal regime such as the Biodiversity Convention which may require that nature is left in the existing state and not be modified through measures of afforestation.

4. Compliance control: information concerning national compliance, i.e. relating to the aggregate figure of emissions as calculated in conformity with the rules just described, has to be collected by a body established under the Convention and/or the Protocol. As in many other environmental regimes, the Protocol relies on national reporting. For this purpose, the Kyoto Protocol develops the relevant provisions of the Convention. There are a number of procedural obligations of the developed country parties to ensure compliance with the Protocol. They include the development of inventories of all greenhouse gas emissions in

the base year 1990 and the following years (Article 3 (4), 10 (a) KP), the establishment by 2007 of a national system for estimating greenhouse gas emissions from sources and their removal by sinks (Article 5 KP) and reporting on compliance with quantitative emission reduction obligations (Article 7 KP). These obligations have to be concretized by COP/MOP decisions. They are reinforced by the verification procedure set forth in Articles 3, 4, and 7 KP – a task to be undertaken by the technical body established by the Convention, namely the review teams of the Subsidiary Body for Scientific and Technical Advice (SBSTA). All that is a very complex challenge. The reporting is complex and ambitious, and so is the scrutiny of those reports.

Following the modern trend towards non-conflictual compliance regimes, the compliance provisions of the Protocol have been supplemented, through a text adopted at COP 7 in Marrakech, by a compliance mechanism which contains important procedural safeguards, including an appeals procedure, as required by Article 18 KP.

5. Sanctions: reactions of the system in cases of non-compliance are essential. As in other environmental regimes, the system of carrots and sticks is applied. The sticks, so far, are rather weak. In case of non-compliance, a Party may no longer be eligible to participate in the flexible mechanisms. Emissions exceeding the assigned amount, multiplied by a factor of 1.3, will be debited to the next reduction period.

Article 3 I KP allows each Annex I State to achieve its reduction obligation (assigned amounts) not only individually but also collectively. Article 4 specifies this empowerment by setting forth the methods of compliance and responsibility for non-compliance. States joining in this mechanism can add up their respective reduction and stabilization commitments and redistribute them internally in a different way. These provisions are not limited to supranational organizations; they could also be used by any two or more Annex I Parties, for instance as a substitute for emissions trading between these Parties. However, they are primarily designed to afford regional economic integration organizations such as the EU flexibility in reallocating their reduction burdens internally according to criteria other than the flat rate reduction obligations provided by the KP. Consequently, Article 4 VI specifically addresses the situation that Annex I Parties acting jointly and together with an organization of regional economic integration are both Parties to the KP. In case of non-compliance with the collective (aggregated) reduction obligation each Member State is individually and jointly with this organization responsible for meeting its reduction obligation as agreed upon within the organization.

In the course of the negotiations on the KP, the EU Council concluded a burden-sharing agreement under the EU bubble taking into account aspects of cost-effectiveness, national starting points (base year emissions), economic structure and growth expectations, energy efficiency and capabilities to reduce emissions. Consequently, Germany, Austria and the United Kingdom are to reduce their greenhouse gas emissions in the first commitment period by 21, 13 and 12.5% respectively, Belgium, Italy and the Netherlands by 7.5, 6.5 and 6% respectively. France and Finland have only to stabilize their emissions. By contrast, Ireland, Spain, Greece and Portugal are entitled to increase their emissions by 13, 15, 25 and 27% respectively. This agreement was adopted in a binding form of a Council decision on the occasion of EU ratification of the KP.

The development since the conclusion of the burden-sharing agreement has shown that a number of Member States miscalculated their reduction capabilities. While Sweden, Finland, France and the United Kingdom have overcomplied or at least complied with their obligations and Germany is close to complying, other countries such as Spain, Denmark, Austria and Belgium are far behind meeting their obligations; as a matter of fact in view of the actual development of emissions in these countries, only drastic reduction measures or extensive acquisition of emission allowances within the emissions trading system to be instituted by the EU or from joint implementation or clean development mechanism projects could ensure compliance in the future.

Implementation of, and compliance with the KP requires a number of steps to be taken by the EU and/or at national level. The EU has already undertaken some efforts to comply with the procedural obligations under the UNFCCC and the Kyoto Protocol. Directives 93/389/EC and 99/296/EC, as amended by Decision 280/2004/EC, already oblige Member States to establish inventories and to set up a reporting and monitoring mechanism for GHG. As regards the program obligation under Article 10 KP, the Commission adopted in July 2000 a first EU climate change program (ECCP) which identified potential sources of emission reduction. In 2002, the second phase of the ECCP started, aiming at the implementation of the proposals of the first phase. Finally, two core instruments of EU legislation concerning climate change were adopted after lengthy debates between the Council and the Commission as well as between the Council and the European Parliament: Council Directive 2003/96/EC restructuring the Community framework for the taxation of energy product and electricity and the European Allowance Trading Directive 2003/87/EC (EATD).

The EATD introduces a EU-wide trading system which as a first step is limited to CO_2 and designated categories of facilities in electricity generating and industry. The system is operated by the Member States under supervision of the Commission. Essential decisions such as the determination of the respective burdens of the emissions trading and other sectors of the economy, the allocation of emission allowances and the granting of greenhouse gas permits are taken by the Member States according to criteria and requirements set by the directive. With respect to the allocation of burdens between the emissions trading and other sectors of the national economy, the directive requires consideration of the respective share in emissions without, however, making strict proportionality mandatory. Allocation of allowances must be based on a national allocation plan for which certain criteria, including the applicability of EC rules on state aids, are set forth and which have to be approved by the Commission. Early reduction action and new entrants must be considered in a way essentially to be determined by the Member States. Even the base year on which allocation must be based is – in contrast to pertinent EP proposals – not set forth in the directive. At least 95% of the initial allocation will be free of charge, based on existing emissions (grand-fathering), while in the second commitment period this is reduced to 90%. This means that no Member State is required to auction off allowances, but can do so to a limited extent. There are provisions for temporary exclusion of individual facilities (Article 27) and pools of facilities (Article 28) from the system and for unilateral inclusion of further greenhouse gases and categories of sources in the second period (Article 24). Links with joint implementation and the clean development mechanism as well as the relation to emissions trading between signatory states provided under Article 17 KP were left to be decided at a later stage. The "linking" directive 2004/101/EC was finally adopted after lengthy negotiations on 24 October 2004.[2] Credits obtained from JI and CDM may be used for emissions trading, with the exception of credits from nuclear projects and sinks.

As regards compliance, a comprehensive, two-tier compliance regime will be established. The directive sets forth monitoring, reporting and verification obligations for enterprises (Articles 14, 15) on which required reporting by Member States must be based (Article 21). The new decision 280/2004/EC on the EU's reporting and monitoring mechanism also applies to emissions trading. Moreover, Member States are required to lay down rules on sanctions applicable to infringements and ensure that they are implemented. These sanctions must be effective, proportionate and dissuasive; there is a minimum penalty set forth by the directive. When a Member State does not comply with its obligations under the burden sharing decision as well as under the emissions trading directive the general infringement procedure under Article 234 EC Treaty is available. This undoubtedly is a cumbersome procedure, yet its advantage is that court decisions on infringement are binding on Member States and in case of non-compliance with the judgement penalties can be imposed.

---

[2] The contributions contained in this volume were concluded before this event. They, thus, refer only to the draft linking directive.

2. The contributions of this publication

The papers presented in this volume highlight a number of fundamental issues of the climate change regime as it has evolved and try to provide for their better understanding. In relation to the international regime, they address three layers of problems: the political forces at stake, the principles and the regulatory approaches.

A key to understanding the current shape of the regime, and its potential for further development, is the clash and the alliances between the various interests at stake in the negotiations leading to the UNFCCC, then to the Kyoto Protocol and finally to the complex system of norms resulting from the post-Kyoto negotiations. The introductory paper by *Barbara Buchner* concentrates on the latter phase, which is also a key to a possible development of the regime during the negotiation of the second reduction period. The paper gives a detailed description of political and economic problems, the political forces involved and the course of the negotiations up to COP 9 in Milan.

The following contributions address a selection of key problems of the Kyoto Protocol.

A key principle of the UNFCCC regime is equity, analysed in the paper by *Barbara Buchner* and *Janna Lehmann*. Equity has different connotations. In the UNFCCC, it is reflected in the principle of common but differentiated responsibilities, intergenerational equity and sustainable development, and the polluter-pays principle, but also in the differentiation of obligations between the developed countries. The paper shows that equity constitutes a major principle of interpretation although it is not a directly applicable norm. The paper also shows that equity is an element influencing the efficiency of international agreements, though the equity approaches adopted in the climate change regime do not necessarily promote the most efficient solutions. The paper concludes that a simultaneous consideration of equity and efficiency in climate policy improves the overall performance of the regime.

The regulatory approaches of the climate change regime, as is shown in the paper by *Janna Lehmann*, constitute the result of a history of regulatory experiments made in the earlier treaty regimes concerning air pollution, i.e. the 1979 Convention on Long-Range Air Pollution in Europe and the 1985 Vienna Convention on Ozone Depleting Substances, supplemented by the Montreal Protocol of 1987.

The following papers address a number of specific regulatory instruments of the UNFCCC/KP regime.

A distinctive feature of the Kyoto Protocol is the flexibilization of obligations through the use of economic instruments. Among them, probably the most controversial and in many ways puzzling one is emissions trading.

In his paper, *Jürgen Lefevere* discusses the background of the concept of emissions trading, and how it was introduced into the Kyoto Protocol. He provides an overview of the concepts, including its ethical dimensions.

The following paper by *Barbara Buchner*, *Alejandro Caparrós* and *Tarik Tazdaït* on green technology transfers is concerned with another of the flexibilization mechanisms of the Kyoto Protocol, namely the Clean Development Mechanism. Based on the analytical framework of game theory, it concludes that the welfare benefit for developing countries ensuing from technology transfers made under this mechanism is more than doubtful. The picture changes, however, if certain ancillary benefits are taken into account.

A general issue raised in the calculation of QELRCs and in the flexibilization mechanism is carbon sequestration in sinks. This raises general questions of forest conservation and/or management which are also relevant under other international regimes, in particular the Biodiversity Convention. The paper by *Alejandro Caparrós* and *Frédéric Jacquemont* shows that the economic incentives for carbon sequestration provided by the Marrakech Accords may have negative impacts on biodiversity in the case of afforestation and reforestation programmes. Thus, policies duly taking into account both regimes are needed.

The major natural sink, namely ocean sequestration, has been neglected by the express provisions of the UNFCCC and the KP. The paper by *Susan Krohn* first shows the technical possibilities of ocean sequestration and then evaluates these measures in the light of various applicable international regimes (the Biodiversity Convention, UNCLOS, the London Dumping Convention, regional seas treaties) and general principles of international environmental law (common heritage of mankind, intergenerational equity).

The last two papers, thus, point to a general problem of the current international legal order, namely its fragmentation and the ensuing issue of harmonization between overlapping regulatory regimes. In a problem area which cuts across as many natural phenomena and divers human activities as does the problem of climate change, this is of particular importance.

The second part of the book deals with specific regulatory problems which climate change raises for the EU.

Both the Member States and the EU itself try to promote activities and behaviour which can lead to a stabilization or reduction of GHG emissions. But this raises the problem of undesirable effects which positive financial incentives may have on competition and, thus, on the proper functioning of the internal market. The paper by *Mercedes Fernández* shows how the Community norms concerning state aid address this problem, trying to strike an appropriate balance between the concern for undistorted competition and the environmental interest to promote climate-friendly patterns of energy production and consumption.

The paper by *Jürgen Lefevere* addresses another economic instrument, namely emissions trading. It discusses the background of the Emission Allowance Trading Directive, some of its key design issues and also the possible link to other economic instruments (Joint Implementation, Clean Development Mechanism).

Emissions trading raises in particular the problem of the initial allocation of allowance, which is highly controversial in economic theory and in the political arena. The paper by *Dietrich Brockhagen* provides an economic analysis of this issue.

The UNFCCC and the Kyoto Protocol create a highly complex multi-level system of governance. As to the EU, this is further complicated by the fact that both the EU and the Member States are Parties to these treaties and that, in addition, they have availed themselves of the possibility admitted by the Protocol to jointly fulfil their obligations in a differentiated way. This is the so-called EU bubble. It poses difficult problems for compliance control, as the paper by *Frédéric Jacquemont* shows.

The so-called EU bubble is limited to the "old" Member States of the EU. But the "new" Members, too, are Parties to Protocol. Their interactions are determined both by the Kyoto Protocol and EU law – a complicated relationship analysed in the paper by *Mercedes Fernández* and *Leonardo Massai*.

3. Perspectives

At a time when the Kyoto Protocol is about to enter into force (December 2004) a debate of its perspectives is appropriate. The challenge is diverse. It concerns both the application of the existing law, i.e. the implementation of the climate preservation regime established by the Protocol and its rule of application, and the further development of the law, i.e. negotiations for the second commitment period which should start in 2013.

As to the application of existing law, this is, so to say, the hour of truth: now, it has to be tested whether the grand design is going to work.

The Protocol sets up a very ambitious system of international governance. It is designed to implement fundamental principles of the international order down to the practical detail. It tries to find a balance between the respect of the sovereignty of States and the effective solution of a problem which by its very nature is universal. It tries to achieve intra-generational equity by applying the principle of common but differentiated responsibilities. It intends to attain intergenerational equity by a set of rules which are based on the principle of sustainable development and the precautionary principle.

The Protocol, unlike many international declarations and also some treaties, does not just recognize these principles; it breaks them down to the nitty-gritty technical detail. It does so, *inter alia*, through novel regulatory instruments, flexible mechanisms, so-called economic instruments of environmental policy which go in many respects beyond the legal tools for the protection of the environment so far tested at the national level. These instruments are analysed in the present volume.

They have to face a double challenge: efficiency and uniform application.

Efficiency: the use of economic instruments, i.e. of market mechanisms, is not per se a guarantee of an optimal solution of problems. The market mechanisms are created by law, the law has been developed through a process of political negotiations, where the word "horse trading" was sometimes a fitting description. The market signals may, thus, be distorted by that law. As a consequence, the optimal allocation which the market is expected to achieve does not necessarily take place. This artificial market has rather high transaction costs which may also lead to sub-optimal results. Despite the high degree of theoretical analysis, in particular in economic theory, which has accompanied the formation of this law and which is in part reflected in the present volume, the ambitious regulatory approach is an experiment. The results have to be carefully observed.

Uniform application: the mechanisms of the Kyoto Protocol depend on national implementation which is subject to international scrutiny. The regulatory problem has to be addressed by a multi-level approach, which has been described in a number of contributions in this volume. There has to be a division of responsibilities between the international and the national level, the latter being divided, for the European region, between the EU and its Member States. Whether this division is able to bring about cost-effective and uniform application of the rules remains to be seen. The latter is both a matter of efficiency and a requirement of justice: free-riding has to be avoided, and equality before the law must be secured.

As to the development of the law: it is not in doubt that the relative stabilization of the greenhouse gas content of the atmosphere which might be achieved during the first commitment period 2008–2012 will not be enough to save the world from the deleterious consequences of global warming. The debate on the next commitment period has started at COP 10 in Buenos Aires. Two problems must inspire the negotiation process. There is, first, the need to take into account the results of the "test phase" which is only starting. The negotiations have to be a "lessons learned" exercise. This is somewhat problematic as the negotiations will have to take place while the test is still going on.

The second problem, however related to the first, is the lack of participation with the Kyoto regime. This concerns, first of all, the world's biggest producer of greenhouse gas emissions, the United States, and a few other industrial States like Australia. It is obvious that the problem of global warming cannot adequately be addressed without the participation of the United States. This is both a matter of efficiency and of distributive justice. What is needed for the purpose of bringing the United States back into the boat is both a process of political persuasion among the industrial countries and a further serious debate on the economics of the climate problem. Some of the analyses provided in this volume may also serve as a point of departure for this debate.

The other participation problem is the exemption of the major new industrialized countries from quantitative emission reduction or limitation obligations. This concerns, in particular, China, India and Brazil. It has to be asked whether the equitable *raison d'être* of the principle of common but differentiated responsibilities is not overstressed if these major contributors to current emissions are free from these obligations, even if it is true that their contribution to the creation of the problem has not been significant.

PART I

International Negotiations and Implementation of the Kyoto Protocol

Chapter 1

BARBARA BUCHNER

The Dynamics of the Climate Negotiations: A Focus on the Developments and Outcomes from The Hague to Delhi

1 Introduction: A Brief Glance Into the Past

As early as the 19th century, French scientist Jean-Baptiste Fourier described the existence of an atmospheric effect which – similar to a greenhouse – enables the earth to maintain an average temperature of approximately 15°C instead of –19°C, which otherwise would be the case. Indeed, scientists had anticipated that adding carbon dioxide to the atmosphere could change the climate of the earth, but initially they expected that oceans would harmlessly absorb the CO_2 caused by the industrialisation. From the 1890s to 1940, scientific efforts regarding greenhouse effects increased due to growth in the global average surface air temperature by 0.25°C, but diminished in the next three decades when a world-wide cooling of 0.2°C took place. In its beginnings, research on the greenhouse effects was directly linked to changes in temperature. Around 1970, global temperatures started to rise again and accordingly interest in climate change re-surfaced. In the 1970s, the US Department of Energy intensified its research and published a number of studies, which augmented the concerns about future global warming. As a consequence of this growing awareness on environmental concerns in general, in 1972 environmental issues were for the first time placed on the international political agenda: the United Nations Conference on the Human Environment (UNCHE) took place in Stockholm and is generally regarded as the landmark event marking the emergence of environmentalism and the starting point of current international environmental negotiations.

However, the first instance when climate change was discussed internationally can be traced back to the year 1979: the First World Climate Conference was organised by the World Meteorological Organisation (WMO) in Geneva. At this

MICHAEL BOTHE AND ECKARD REHBINDER (EDS.), Climate Change Policy, 19–43.

conference, climate change was for the first time recognised as a serious problem and increased atmospheric concentrations of carbon dioxide resulting from the burning of fossil fuels, deforestation, and changes in land use were identified as the principal cause of global warming. In addition the declaration of this first major international meeting in the context of climate change urged governments from all over the world 'to foresee and prevent potential man-made changes in climate that might be adverse to the well-being of humanity.'

This event initiated a series of international conferences on climate change. In 1985, the first major scientific international conference on the greenhouse effect was held at Villach, Austria, jointly organised by WMO, UNEP (United Nations Environmental Programme) and ICSU (International Council of Scientific Unions). This conference finally established greenhouse warming as an international concern, warning that greenhouse gases will 'in the first half of the next century, cause a rise of global mean temperature which is greater than any in man's history'. At this event, researchers predicted possible implications such as sea level rises by up to a metre and reported that gases other than CO_2, such as methane, ozone, CFCs and nitrous oxide, also contribute to warming.

In 1988, global warming attracted world-wide attention when scientists at Congressional hearings in Washington DC blamed a major US drought on the changing climate. In the same year, the Toronto Conference on the Changing Atmosphere took place where scientists, politicians and officials from 48 countries as well as the UN called for a 20 per cent reduction of global CO_2 emissions by 2005 as compared to 1988 levels. The approach taken in Toronto, i.e. setting a near term target relating to the reduction of the emissions volume, proved to be very influential in the international negotiations on the development of a multilateral framework to address climate change that followed.

Through pressure from the scientific community which emphasised the need for more scientific research into the causes and effects of climate change, the WMO and UNEP subsequently set up the Intergovernmental Panel on Climate Change (IPCC), a global network of over 2000 scientists, to analyse and report on scientific findings. In particular, the IPCC was given the mandate to make an inventory of current scientific knowledge of our climate system, to identify the effects of climate change and to suggest the possible response strategies.

Increasing scientific evidence of the likelihood of global climate change in the 1980s led to an increasing awareness that human activities induce a growth in the atmospheric concentrations of greenhouse gases. The year 1990 can be considered as a crucial year towards a binding Convention on Climate Change. On the one hand, representatives of 137 countries and the European Community attended the Second World Climate Conference in Geneva, calling in particular for the drafting of an international Convention on Climate Change by June 1992. At the same time, the first assessment report of the IPCC drew further attention to climate change, finding that the planet warmed by 0.5°C in the past century. The IPCC warned that only strong measures to halt rising greenhouse gas emissions would prevent serious global warming. While the Second World Climate Conference provided the political pressure, the first IPCC report contained the scientific basis for initiating the process of the UN negotiations on a climate convention. Indeed, the supreme decision-making body of the UN, the General Assembly, decided in December to enter into negotiations aimed at such a framework convention. In particular, a resolution was adopted establishing the Intergovernmental Negotiating Committee for a Framework Convention on Climate Change (INC/FCCC). Its mandate consisted of preparing an effective Framework Convention on Climate Change, supported by UNEP and WMO.

In June 1992, the 'Earth Summit', the UN Conference on Environment and Development (UNCED), was held in Rio de Janeiro, Brazil. At this conference, after two years of intense work, the UN Framework Convention on Climate Change (UNFCCC) was adopted and opened for signature. The Convention was signed by 155 countries, agreeing thereby to prevent 'dangerous' warming from greenhouse gases, and setting an initial target of reducing emissions from industrialised countries to 1990 levels by the year 2000[1]. At the Rio Earth Summit, the need for industrialised countries to somehow control their emissions of greenhouse gases – although in a non-binding fashion – was thus recognised.

In 1994, the Convention entered into force with the ultimate objective to induce

> (...) stabilisation of greenhouse gas concentrations in the atmosphere at a level that would prevent dangerous anthropogenic interference with the climate system (...) (UNFCCC, Art. 2),

and emphasising that

> (...) the Parties should protect the climate system (...) on the basis of equity and in accordance with their common but differentiated responsibilities and respective capabilities. (UNFCCC, Art. 3).

The UNFCCC builds on the principle of precaution (Art. 3) and places the main responsibility for taking action on the industrialised countries, dividing the world along the classic UN line of developed and developing countries[2]. The treaty is ambitious in requiring, for the first time in history, targets for GHG emissions in the atmosphere. However, it only sets a vague stabilisation target. It does not impose specific levels of concentrations to be achieved in order to avoid

---

[1] More states have signed it since then. As of 24 May 2004, 166 countries signed the Convention and 189 ratified it. For a detailed and up-dated list of signatories see http://unfccc.int/resource/conv/ratlist.pdf.

[2] The developed countries are further separated into the Annex I, consisting of industrialised countries including those countries with economies in transition and the Annex II, comprising only those countries that were OECD members when the UNFCCC was established in 1992.

dangerous climate change, nor does it specify policies, imply commitments or mandate which remedies governments have to pursue in order to achieve a stabilisation of GHGs[3]. Instead, the UNFCCC was primarily meant to form the basis for an international agreement that can ensure emissions reductions in the future by providing the framework for future negotiations, expecting that both the basic science and a political consensus on global warming would develop over the following years[4]. Most importantly, the UNFCCC established the institutions for the future work, in particular the Conference of the Parties (COP). The COP represents the highest body of the UNFCCC, comprising all Parties, and is supported by two 'subsidiary bodies', SBI (Subsidiary Body on Implementation) and SBSTA (Subsidiary Body on Scientific and Technological Advice). By ratifying the UNFCCC, countries do not only acknowledge that climate change is a real problem, but they also accept to start working on this problem through the UNFCCC institutions.

The application of the Convention represented a crucial step towards future negotiations. Indeed, the year after in 1995, the COP met for the first time (in Berlin) in order to assess progress towards the promises made at the Earth Summit[5]. At this first Conference of Parties to the Climate Convention (COP-1), the countries noted that progress since Rio was very modest. Parties adopted the Berlin Mandate in which they agreed that the commitments embedded in the Convention were inadequate in order to reach the Convention's objectives. The need for negotiations about real emissions cuts was emphasised and consequently, a negotiation process aiming at the adoption of a protocol or another legal instrument to strengthen GHG emission reductions in Annex I countries after 2000 was launched. In particular, the process required Annex I Parties to elaborate policies and measures (PAMs) to limit their GHG emissions and to set quantified limitations and reduction objectives with specified time-frames, such as 2005, 2010 and 2020[6]. The Berlin Mandate thus exempted developing countries from mandatory emission reductions. Nonetheless, the developing countries certainly participated in the negotiations. COP-1 also established an

---

[3] The only commitments that the Parties have are to ensure that the work of implementing measures is initiated.

[4] For more details on the UNFCCC see for example Anderson (2000), Grubb *et al.* (1999) and Torvanger *et al.* (2001).

[5] As indicated above, the COP represents the assembly of all the countries that have ratified the UNFCCC and currently convenes once a year to report on the progress of the Convention.

[6] Policies are procedures developed and implemented by one or several governments with the objective of mitigating climate change through the use of technologies and measures, whereas measures are actions that can be taken by one or several governments, also together with the private sector, to accelerate the use of technologies or other practices that reduce GHG emissions.

open-ended *Ad-Hoc Group on the Berlin Mandate* (AGBM) in which all Parties were invited to negotiate a protocol on the future commitments, which was to be finalised in 1997 at COP-3.

In July 1996, the Second Conference of the Parties (COP-2) met in Geneva and sent out some important political statements. Shortly before the meeting, the IPCC issued its Second Assessment Report (SAR). The SAR's strong conclusions were endorsed by the Ministerial Geneva Declaration, which consequently called for legally binding commitments and significant GHG reductions. A number of ministers showed concern about the discernible human influence on the global climate, as concluded by the SAR, and agreed that its scientific elements need to be considered in decision-making. However, due to sharp differences between delegations, the declaration was only 'noted' by the COP-2. The conclusions drawn by the IPCC motivated also the US to significantly change its position, supporting for the first time an instrument containing legally binding reduction commitments to fulfil the Berlin Mandate. Yet, the US decision to support the binding emission reduction commitments was linked to a strong preference for a tradable permit system, which raised new complexities for the negotiations. Therefore, notwithstanding the important positive elements of the COP-2, the number of disagreements during the conference indicated that the future negotiations would have to cope with many obstacles.

2 The Background: The Kyoto Protocol and its Key Issues

After a number of meetings in the AGBM, the Third Conference of the Parties (COP-3) adopted in December 1997 the Kyoto Protocol, establishing for the first time in history legally binding emission reductions targets for the industrialised world. The agreement represents without any doubt a milestone in international climate policy. This section will now briefly analyse its evolution and its main components.

2.1 The Negotiating Process: Key Players and their Strategies

The negotiations were mainly influenced by four countries or groups of countries: the European Union (EU); the so-called Umbrella group, comprising the United States, Japan, Canada, Australia, New Zealand, Norway, Iceland, Russia and Ukraine; and the G-77/China group consisting of developing countries including China[7]. Obviously, negotiations between these groups implied a lot of different

[7] Main negotiating group of developing countries within the UN system. Despite its name, it represents more than 130 countries.

viewpoints and priorities, not only among the groups but also within them. A further crucial negotiating group within the UNFCCC process is the Alliance of Small Island States (AOSIS), which is a group of 42 countries that are among those most threatened by rising sea levels and other impacts of climate change.

The European Union took the role of a 'green consciousness' in the negotiation process, emphasising from the beginning the necessity of substantial, legally binding emission reduction targets to enable real emission reductions and strongly criticising attempts to build 'loopholes' into the architecture of the protocol. Consisting of many environmentally sensitive member states that have called for action on climate change ever since the issue emerged politically, the EU had a strong sense of global and environmental responsibility. Notwithstanding differences in the development between the North and the South, the EU thus has been at the forefront of stronger climate change control. The ambitious negotiating position of the EU represented indeed the main motivation behind the numerical targets of the Kyoto Protocol. However, AOSIS was decisive in proving the first step into this direction. Indeed, AOSIS countries worked intensively to achieve the wording of the objective of the Climate Convention, namely to stabilise concentrations of greenhouse gases in the atmosphere at levels that would not present a danger to the global climate system. Toward this end, AOSIS submitted a draft Protocol to the first Conference of the Parties to the Climate Convention held in Berlin in April of 1995. This draft Protocol established a goal of a 20 per cent reduction of greenhouse gasses by the year 2005. The 'AOSIS Protocol', became a centrepiece for continued international negotiations on climate policy known, as noted above, as 'the Berlin Mandate' and thus remained the basis and the framework for the negotiations.

On the other hand, the US provided a major influence on the institutional approach to implementation, in particular with respect to the shaping of emissions trading. In contrast to Europe, the US has been more hesitant about actions on climate change, emphasising instead the economic consequences of mitigation activities. Apparently contrary to its strong position in the scientific analysis of the issue, demonstrated for example by the US leadership on the way to the establishment of the IPCC, the US initially only asked for emission limits and its acceptance of binding commitments was achieved only by admitting a higher degree of flexibility in the Protocol. In general, the US – supported by other members of the Umbrella group, in particular Australia and Canada who also have economies characterised by high energy-intensity and population growth – advocated a maximum of flexibility in achieving emission reductions in order to weaken the domestic impact of the reduction targets, favouring, therefore, a flexibility mechanism and carbon sequestration in the form of carbon sinks[8]. The inclusion of 'negative emissions', allowing for sinks from the land-use and forestry sector was insisted upon by the US.

Japan had an ambiguous role in the negotiation process, strongly dependent on the fact that they hosted the conference at which the final aspects of the Protocol were negotiated. Being strongly related to the US in an institutional sense, Japan tried to find a mediating position by following the European lead in adopting strong emission reduction targets. Motivated by hopes to strengthen and revive energy conservation through emission constraints, Japan supported the EU notwithstanding the concerns of its industrial sector and scant worries about climate change impacts.

The G-77/China group represents a number of countries with a large range of very different characteristics. Apart from AOSIS, a group of mostly very poor countries, G-77 comprises the Organisation of the Petroleum Exporting Countries (OPEC) and a number of other developing countries, including big countries such as India and Brazil, supported by China. According to their differing interests, this group followed two main objectives during the negotiations. On the one hand, it supported the EU strategy by pushing for high emissions targets, while it also forcefully vetoed, on the other hand, the broad inclusion of voluntary commitments for developing countries in the Kyoto Protocol. In particular, the key players of this group demonstrated decisive opposition to the article of voluntary commitments and they succeeded in dropping this component by inducing a debate on the pace at which emissions trading is supposed to come into effect. However, diverging interests within this group led to a number of strange situations in the negotiating process, most impressively illustrated by the process leading to the creation of the Clean Development Mechanism (CDM). Notwithstanding the strong resistance of China and India, Brazil pushed – with support by the US – for this type of hybrid institution, bringing together certified joint implementation and emissions trading with voluntary developing country participation. Several developing countries declared their voluntary participation in this mechanism, and after the US recognised it as the 'politically correct avenue' for involving some of the key developing countries, the CDM became one of the cornerstones of the Kyoto Protocol.

Intense negotiations among these key players led to the finalisation of the Protocol's text after delegates worked throughout the last nights of the COP-3 in the final session. In particular, the EU lowered its resistance against the inclusion of six gases, sinks, and emissions trading and agreed to a broader differentiation of targets after other delegations conceded to include policies and measures according to 'national circumstances'. At about 1:00 p.m. on 11 December 1997, the final COP Plenary convened and adopted the Kyoto Protocol.

---

[8] Sinks are ecosystems such as forests, soil and oceans which absorb carbon dioxide, reducing thus a country's net CO_2 emissions in proportion to the existence of large ecosystems.

2.2 The Contents: Main Components of the Kyoto Protocol

The Kyoto Protocol sets binding emissions reduction targets for industrialised countries and requires that by the year 2005 'demonstrable progress' must be made in achieving those targets[9]. By 2012 the overall GHG emissions of industrialised countries should decline to an average 5.2 per cent below their 1990 levels. For the first time in history, industrialised countries accepted legally binding targets for emissions of a range of greenhouse gases. In addition, by signing the Protocol, industrialised countries appear to admit to bear the main responsibility for the prevailing concentration of greenhouse gases. Under the Protocol, economies in transition, including Russia and several states from the former Soviet Union, are allowed to catch up with their economic development after experiencing serious economic recessions in the early 1990s before they must comply with strict targets. The choice of the base year of 1990 allows these countries to dispose of notably large amounts of emission credits due to their currently lower emission levels compared to 1990. Developing countries, including the large economies of India and China, were exempted from binding reduction targets under the Kyoto Protocol.

The Protocol establishes also a series of flexibility measures, which enable countries to meet their targets by cooperating on emission reductions across country borders and by vaguely establishing carbon sinks such as certain forestry and land-use activities to soak up emissions. In particular, the following three flexibility mechanisms are introduced: international emissions trading (IET, also referred to simply as 'emissions trading'), joint implementation (JI) of GHG mitigation or sequestration projects between developed countries, and the CDM to set incentives for joint projects in abating GHG emissions between Annex I (developed country) and Non-Annex I (mostly developing country) Parties to the Convention. These mechanisms are meant to assist countries in meeting their emission targets in a flexible, cost-effective way.

In addition, an entry-into-force provision was established: a precondition for the Protocol to enter into force is that at least 55 Parties to the Convention, representing at the same time at least 55 per cent of 1990 carbon dioxide emissions of Annex I Parties, must have ratified the treaty.

The main components of the Kyoto Protocol are as follows:

---

[9] According to the principle of common but differentiated responsibilities which has been set in the UNFCCC, the developing countries, including the large economies of India and China, were exempted from binding reduction targets in the first commitment period. Indeed, the per capita emissions of these countries are much lower than the industrialised countries' emissions and at the moment their priorities are rather centred on the maintenance of development than on environmental issues.

- Emission limitation/reduction requirements for industrialised countries: the industrialised countries agree to reduce their aggregate GHG emissions by 5.2 per cent in the so-called first commitment period between 2008 and 2012 in comparison to the base year of 1990. The specific targets are defined in Annex B to the Kyoto Protocol, which represents an up-dated version of the industrialised countries defined in Annex I to the Convention, including, in addition to the original OECD countries, economies in transition to a market economy. The emission reduction targets are differentiated between countries and vary from +10 per cent for Iceland, +/–0 for Russia, –8 per cent for the EU, –7 per cent for the US and –6 per cent for Japan.
- Emission limitations can be jointly managed through a 'bubble' or group of countries, which has to achieve the overall target of all its participating countries, with the possibility to distribute the burden within the bubble. The EU has formed such a bubble.
- Emission targets can be achieved through the implementation of three flexibility mechanisms (emissions trading, Joint Implementation and the CDM), which are intended to increase the cost-effectiveness of climate policies by reducing the countries' abatement costs.
- Sequestration of CO_2 in forests and sinks is included in the Kyoto Protocol as one of the options to comply with the reduction targets, although the extent of the use of this option was left vague.
- No emission reduction commitments for developing countries in the first commitment period.
- Coverage of six greenhouse gases: carbon dioxide (CO_2), methane (CH_4), nitrous oxide (N_2O), hydrofluorocarbons (HFC), perfluorocarbons (PFC), and sulphur hexafluoride (SF_6).
- An entry into force provision requiring ratification by at least 55 Parties to the Convention, representing at the same time at least 55 per cent of 1990 carbon dioxide emissions of Annex I Parties.

In the Kyoto Protocol, the targets, methods and timetables for global action against climate change were set. However, the precise rules were missing and left for further negotiations. The Kyoto Protocol thus left a lot of questions unresolved, with the common understanding that these challenging issues would require further negotiations. In particular, the rules for the flexible mechanisms still needed to be negotiated, the accounting for sinks and sources of GHGs from land-use changes and forestry had to be specified and questions regarding compliance issues, the funding mechanisms and capacity building in developing countries needed to be resolved.

3 From Kyoto to Delhi

3.1 Setting the Scene: The Developments up to the COP-6

a) The Political Controversies after Kyoto

In November 1998, the Fourth Conference of the Parties (COP-4) convened in Buenos Aires, Argentina. Characterised by a lack of ambition and fatigue after Kyoto, COP-4 failed to resolve the disputes over the rules for the Kyoto Protocol. Nevertheless, the conference represented a further step on the way towards effective climate change control: the Parties adopted the Buenos Aires Plan of Action (BAPA), aimed at finalising the negotiations on the remaining points by COP-6 in The Hague, the Netherlands, in November 2000. By this decision, the Parties demonstrated their determination to show substantial progress on the outstanding issues in order to prepare for the future entry into force of the Kyoto Protocol, strengthening at the same time the implementation of the Convention. In addition, after a lack of attention paid to developing countries in the COP-3, this conference achieved a number of positive developments for the G-77/China group, related above all to technology and finance issues. The strategy of making concessions to the developing world illustrated the long-term objective of diminishing the dis-equilibrium between the viewpoints of developed countries and developing countries with regard to developing countries' voluntary commitments. COP-4 demonstrated also the conflicting interests within developing countries, when Argentina announced its intentions of a voluntary commitment at COP-5, strongly opposed by the other group members. Nonetheless, it was BAPA, which triggered a further important event during the COP-4, namely the US signature of the Kyoto Protocol.

In 1999, COP-5 was held in Bonn, Germany, aimed at continuing the work towards the fulfilment of the BAPA. The conference revitalised the process after the more hesitant approach in Buenos Aires, demonstrated by the 'unexpected mood of optimism' that was created after the delegates completed their work ahead of schedule. COP-5, therefore, sent a clear and powerful sign to the world, demonstrating that work on climate change control was being taken seriously.

After the COP-5, a number of formal and informal consultations were held in order to enable the best preparations for the subsequent climate talks, which were expected to demonstrate the progress of international climate policy since its beginnings. In particular, these numerous meetings were supposed to help lay the foundations for an agreement at COP-6.

b) The Open Issues in November 2000

After the first euphoria over the adoption of the Kyoto Protocol had passed, the situation changed dramatically as soon as the various countries recognised the scope of the Protocol and the level of detail that needed to be elaborated upon. Almost all the definitions and exact formulations of the details were postponed to later meetings. Only afterwards were the dimensions of the delayed work recognised. In particular, the complexity of the mechanisms and the dimensions of the potential 'loopholes' as well as the extent of the concession regarding 'hot air' (i.e. the concern that countries with economies in transition will sell large amounts of emission credits due to emission reductions acquired because of the economic recession in the early 1990s) had been under-estimated. Therefore, after three years of negotiations and preparations for the COP-6, a large number of controversial key issues required more specific discussions. The main questions were related to the key issues on the BAPA and regarded the following issues:

- **Mechanisms**: as discussed above, the Kyoto Protocol established three mechanisms meant to assist countries in meeting their emission targets in a flexible, cost-effective way: an emission trading system, joint implementation (JI) and the Clean Development Mechanism (CDM). At the COP 6, operational details governing the use of the mechanisms had to be elaborated and a compromise between the countries' differing views should have been found. Key issues in this context included: the determination of the roles and responsibilities of various institutions; the 'hot air' debate; decisions regarding carbon 'sink' projects; and the 'supplementarity' issue (i.e. whether or not there should be a limit in the form of a ceiling on the quantity of emissions reductions that can be met through the use of the mechanisms). Further questions to be addressed included decisions concerning the baselines for measuring the emissions-reduction contribution of specific projects, accounting rules for allocating credits and a definition of criteria for the eligibility of projects.
- **Carbon sinks**: forestry and other forms of land use have the potential of sequestering atmospheric carbon. For this reason, the Kyoto Protocol vaguely promised to provide emission reduction credits for some land use, land-use change, and forestry (LULUCF) activities. The question that should have been answered at the COP-6 was twofold: 1) which types of LULUCF activities can be accounted for; and 2) how much credit countries could get for their forestry and land-use activities and consequently in which dimension these activities could lessen the adverse effects of emission reduction measures. The problem is that plants and soils can act as carbon sinks, but there is no scientific certainty about how much carbon is being removed from the atmosphere, or whether the removal from the atmosphere is permanent. Therefore, the use of carbon sinks in meeting emission targets was not only controversial but also complex and questions concerning an accurate definition of 'sinks', their consequences and possibilities to measure them needed to be clarified.

- **Compliance**: the Kyoto Protocol identified a compliance system as a key element of the agreement. This system should be able to determine whether countries have met their commitments under the Protocol at the end of the commitment period and the eventual consequences of non-compliance. The aim should be a compliance system that sets incentives for the positive performance of the participants. Decisions should have been made at COP-6 regarding the enforcement, organisational details, monitoring and verification and non-compliance.
- **Developing country issues**: developing countries, including the large economies of India and China, were exempted from binding reduction targets in the first commitment period based on the principle of common but differentiated responsibilities. The per capita emissions of developing countries are indeed much lower than the industrialised countries' emissions and industrialised countries are the main contributors to the prevailing atmospheric concentration of greenhouse gases. A very important task for COP-6 was to strengthen developing countries' capacities to cope with their efforts to combat climate change. Key issues included capacity building and transfer of technology, particularly in relation to financial and technical assistance from Annex I Parties.
- **Adverse effects**: developing countries are suffering adverse effects both from climate change itself and some developing countries claim that they will suffer from the adverse impact of response measures undertaken to combat climate change. Under the UNFCCC, Parties already agreed to address the needs and concerns arising from these adverse effects. Details on assistance to least developed countries (LDCs), small island developing states, countries with low-lying coastal areas and other vulnerable countries were still needed to be elaborated upon. At the COP-6, characteristics of this assistance and particularly the question of compensation should have been discussed.

3.2 The Sixth Conference of the Parties (COP-6) – The Hague, The Netherlands

COP-6 has been considered as the most important negotiation on global climate change since the Kyoto Protocol was agreed to in 1997. The COP-6 was intended to determine specific rules and operational details that describe how the commitments on reducing emissions of greenhouse gases under the 1997 Kyoto Protocol can be achieved and in which way the countries' mitigation efforts will be measured. The talks were thus primarily supposed to clarify the details of vague language in the Protocol before its final implementation and ratification. A further aim was to reach agreement on actions to strengthen implementation of the UNFCCC itself as well as to clarify the key issues on the BAPA (see above).

a) The Negotiations and Main Outcomes

The key negotiations took place between the United States supported by Canada, Australia, Japan and New Zealand (the 'Umbrella Group'), the European Union guided by France (which had the presidency of the EU) and the G-77 group of developing countries (however, China and AOSIS took separate positions from the G-77 group)[10].

During the first days of the negotiations, progress was made with respect to the definition of the mechanisms and the completion of other text. Yet already at the end of the first week, frustration was expressed due to the slow progress and the amount of remaining political and technical issues for negotiation. The last days were therefore focused on several proposals by the President of the COP-6, the Dutch Minister of Environment Mr. Jan Pronk who tried to find a compromise on a number of important issues by first identifying four 'clusters' of key political issues with the aim of facilitating progress on many disputed political and technical issues. However, no consensus could be reached on the core issues within the clusters and the negotiations appeared stalled. In an attempt to advance the negotiations, President Pronk distributed a note containing his proposals on key issues, the so-called 'Pronk document'. Further intense talks on these proposals did not achieve a breakthrough and so the final effort to seek a consensus concluded without agreement. The final negotiations ended on Saturday, 25 November 2000. The formal Plenary decided to suspend COP-6 and to continue talks in 2001.The failure of COP-6 is mainly the result of four key concerns for which compromises could not be reached between the participating negotiators:

- Land use, land-use change and forestry (LULUCF); in particular sinks
- Supplementarity and the flexible mechanisms
- The nature of the compliance regime
- Financing mechanisms

b) A Brief Analysis of COP-6

The difficulty of the COP-6 consisted in finding the right balance between promoting ratification of the Kyoto Protocol while at the same time maintaining

---

10 Indeed, most of AOSIS states are a member of G-77 and the same is true for China. Actually, the group is called 'G-77 and China'.

its environmental integrity. This conflict was best illustrated during the very controversial discussions regarding 'sinks'. A main cause of the climate talks' failure was the fundamental differences between the EU and the US (backed by the other members of the Umbrella Group) regarding the question of sinks. But sinks were not the only problem; there were also various technical matters on which text had been developed, but that were characterised by highly divergent political points of view. From the beginning of the conference, it was obvious that the negotiators had the enormous task of finding common ground on the emission reduction options for all the various countries. Nevertheless the breakdown of the climate talks created disappointment within the participants and observers, in particular due to the high expectations that were set for this conference. Notwithstanding progress and the approximations on difficult key issues, most importantly the greater acknowledgement of climate change, the conference presented a missed opportunity given the fact that the chances of a deal were more favourable at that time than ever before[11].

The Kyoto Protocol was to have been put into effect at The Hague in that most details were to have been clarified at COP-6. The consequence of the failed climate talks was the delay of the Kyoto Protocol's ratification, implying also that some countries started worrying about the possibility to reach their emission reduction targets on time.

3.3 The Resumed Sixth Conference of the Parties (COP-6*bis*) – Bonn, Germany

As a consequence of the failure at COP-6, the next round of negotiations became very important for the success of the Kyoto Protocol. Several meetings were convened after the failure in The Hague in an effort to get negotiations back on track. However, the resumed COP-6 was delayed to July 2001 due to the US strategy to require more learning-time for the new administration. However, after this date had been fixed, in March 2001 the new US President George W. Bush announced his opposition and consequently the US rejection of the Kyoto Protocol. In particular, President Bush declared the Protocol as 'fatally flawed' due to the harm that this climate treaty would incur on the US economy and the exemptions of the developing countries from fully participating in the agreement. Notwithstanding this step backward in international climate policy, the remaining Annex I Parties decided to continue the Kyoto process and moved forward with the resumed COP-6 in Bonn in July 2001 in order to complete the negotiations on the operational details of the Kyoto Protocol and to bring the talks initiated at The Hague to a successful end.

---

[11] For a thorough discussion of the COP-6 see Buchner (2001).

a) The Negotiations and Main Outcomes

During a number of meetings after COP-6, some progress was achieved on the sticky points that had caused the failure of the talks in The Hague. Nonetheless, most of the open issues were forwarded to the resumed COP-6, and in particular decisions on financial issues, compliance, LULUCF and supplementary in the use of the mechanisms were needed. In Bonn, the negotiations proceeded between the same Parties as in the previous meetings, although the US kept a very low profile and chose not to intervene in Kyoto Protocol issues unless it created a precedent for international law or affected key US interests.

The negotiations were characterised by the participants' strong will to strive for a break-through by achieving a political agreement on the key issues. However, after the first week of negotiations, the process seemed to be slowed down and President Pronk tried one more time to revive the talks by providing a compromise proposal. Although a number of countries signalled agreement to the proposal, disagreements on the section on compliance emerged. After intense consultations, agreement was achieved by changing the proposed provisions on compliance and leaving the remaining proposals on the political issues. The COP formally adopted the political decision called 'Bonn Agreement'. In addition, political declarations by developed countries pledged additional funding for climate change activities in developing countries. However, notwithstanding this political success, the work on several key issues could not be completed but was forwarded to subsequent COP. Most importantly, decisions on mechanisms, compliance and LULUCF needed further negotiations. Summarising, the Bonn agreement resolved key political issues in those decision texts, but did not finalize the actual decision texts.

b) A Brief Analysis of COP-6*bis*

Notwithstanding the withdrawal of the US from the Kyoto Protocol, the other Annex I Parties continued the process and indeed reached a political compromise on the climate treaty at the resumed COP-6 in Bonn, July 2001.

The outcome of the Bonn negotiations, the so-called 'Bonn Agreement', has been interpreted as a commitment to the approach embodied in the Kyoto Protocol and represents the basis for its implementation. However, a number of concessions were made to the key players in order to keep the process alive, and in this way the text of the Protocol was substantially watered down. For example, both Russia and Canada obtained a larger number of credits for their forestry (carbon sinks) than was offered in earlier negotiations and requirements for supplementarity, a previously high EU priority, was reduced in significance. Several elements were included which reduced the Protocol's environmental integrity – which had been the main obstacle to finding an agreement with the US in The Hague – from the level set out in previously rejected proposals. In addition to the sacrifices to keep the negotiations successful, a number of key issues could not be concluded (most notably, compliance) contributing to concerns that the Bonn Agreement neither assures the Kyoto Protocol's ratification nor its environmental effectiveness.

Still, in the light of the low expectations prevailing after the failure in The Hague and the US decision to defect from Kyoto, the outcome of the resumed COP-6 represents a considerable success. In particular, the willingness to continue efforts in multilateral cooperation on climate change control were demonstrated, indicating that the Kyoto Protocol plays the role of a first step into the right direction of more courageous future climate policy.

3.4 The Seventh Conference of the Parties (COP-7) – Marrakech, Marocco

Notwithstanding the political deal reached in Bonn, a number of 'technical' issues related to the structure of the Kyoto Protocol remained unresolved and draft decisions were redirected to the next meeting. COP-7 in November 2001, therefore, represented a crucial event for ensuring the Protocol's timely entry into force. The conference sought to close three years of negotiations on the goals set out in Buenos Aires, and to complete the tasks left unfinished at the two last meetings.

a) The Negotiations and Main Outcomes

In Marrakech, the delegates continued the work on the outstanding technical issues on the basis of the Bonn Agreement, attempting to resolve, among others, details on mechanisms, the compliance system, LULUCF, and the Protocol Articles 5 (methodological issues), 7 (communication of information) and 8 (review of information). On these issues, and on the input to the World Summit on Sustainable Development (WSSD) to be held in September 2002, a package deal was proposed towards the end of the second week. However, notwithstanding the positive response from a number of negotiating groups (for example the EU and G-77/China), key members of the Umbrella Group (Canada, Australia, Japan, New Zealand and Russia) opposed the deal due to controversies on eligibility requirements and banking under the mechanisms[12]. After further intense negotiations, consensus was found on these points and both compliance eligibility requirements and limited banking of units by sinks under the CDM were agreed to. The Annex I countries (except the US) agreed therefore to a package deal which ensured the implementation of the Buenos Aires Plan of Action. Being aware of the necessity to successfully close this conference in order to support the Kyoto process, the key Umbrella countries that were needed to

[12] Note that Norway and Iceland agreed with the compromise.

satisfy the 55 per cent clause succeeded in further weakening their contributions[13].

b) A Brief Analysis of COP-7

Given the considerable concessions to the demands of the Umbrella Group, and the corresponding undermining of the Kyoto Protocol's environmental integrity, the final deal left many observers and negotiators disappointed. However, through the political conclusions achieved in Bonn, decisions on a number of critical issues had been forwarded to COP-7 and therefore the stakes in Marrakech were considerably high. The threat that, after years of negotiations, the climate talks could still end without consensus (implying a failure of international efforts in climate change control), characterised the atmosphere of the conference. Therefore, emphasising that the final deal included not only concessions but also considerable progress on various issues as the CDM and compliance, the agreement was welcomed as a realistic first step in climate change control, being both politically and economically feasible. The Marrakech Accords were thus received as a milestone in international climate policy due to the broad consensus on the important role of the Kyoto Protocol in serving as a precedent for multilateral action against climate change.

The Marrakech Accords to the Bonn Agreement have politically opened up the path for the ratification of the Kyoto Protocol. Shortly after COP-7, the WSSD took place in Johannesburg from 26 August to 4 September 2002. This conference (which was the follow-up of Rio's Earth Summit in 1992), happened at a very crucial point in the determination of the Kyoto Protocol's future. Indeed, several declarations and discussions about the Kyoto Protocol's future were made at the WSSD, which can be seen as a crucial milestone in the Kyoto process[14]. During this conference, China declared its ratification of the Protocol while both Canada and Russia announced their intents to ratify the Protocol very soon. The World Summit thus demonstrated that multilateral support for the Kyoto process was still intact.

3.5 The Eighth Session of the Conference of Parties (COP-8) – New Delhi, India

COP-8 took place in New Delhi from 23 October to 1 November 2002 and marked the beginning of a new phase of negotiations. After the last conference

---

[13] For example, Russia managed to almost double its sinks allowances.

[14] For an overview on the key outcomes of the Johannesburg Summit see for example United Nations (2002).

had completed three years of negotiations on the operational details of the Kyoto Protocol, the COP-8 started discussion on the implementation of the Marrakech Accords and UNFCCC issues which had been left unresolved at previous meetings. India wanted to use this meeting mainly to address the implementation of issues agreed thus far and to look at developing country matters – in particular adaptation – and wanted to desperately prevent looking ahead to anything to do with developing country commitments after 2012.

a) The Negotiations and Main Outcomes

The negotiations in New Delhi were characterised by progress on a number of issues previously left aside due to the pressing negotiations under BAPA. The biggest issues were the continuing work on sinks in the CDM, the finishing of the work on the Non-Annex I national communication guidelines, the Canadian proposal on clean energy exports, and the Delhi Declaration. In addition, decisions and conclusions were taken on the following topics: guidelines for Non-Annex I national communications; issues under financial mechanisms; research and systematic observation; cooperation with relevant international organisations; and methodological issues. Agreement on a number of technical issues relating to the implementation of existing commitments was also reached. COP-8 adopted the Delhi Declaration calling for the timely ratification of the Kyoto Protocol and urging countries that had not yet ratified to do so in a timely manner.

b) A Brief Analysis of COP-8

By adopting the Delhi Declaration, COP-8 further strengthened the process towards the Kyoto Protocol's ratification. Nonetheless, the outcome of the COP-8 reflects a number of compromises, which were needed to conclude a set of the climate negotiations and to prevent international climate policy from being a complete failure. In particular, the issue of future development of the climate change regime was central to COP-8. However, in this context, the negotiations unravelled a lot of dichotomies, most importantly between developed and developing countries, which were expressed through different views on the Kyoto Protocol and the UNFCCC, on adaptation and mitigation and on environment and development. As a consequence, expectations at COP-8 related to offering clearer long-term visions on climate change control or even explicit future steps beyond the first commitment period were not satisfied. The Indian hosts had far-reaching ambitions, tempting to bring climate policy closer to developing countries by unifying climate-change control with sustainable development. Above all, negotiators had hoped to weaken the usual division between developed and developing country positions, highlighting the variety of interests of Non-Annex I Parties. The breakdown of this dichotomy would have enabled a more ambitious declaration, capable of initiating a dialogue on broadening the commitment after 2012. However, the few supportive voices within the group of developing countries were overruled by the more powerful players, notably the oil producers, who emphasised the need for a declaration concentrating on adaptation[15]. Although mitigation was acknowledged as continuing to have high priority, the developing countries' concerns about adaptation and their sensitivities in relation to accepting commitments led to a focus on adaptation in the Delhi Declaration.

The final outcome of the COP-8 was characterised by a shift in the negotiating Parties' positions. In particular, the former 'Green Group' consisting of the EU and the developing countries, which had loosely co-operated throughout previous COPs, experienced a number of difficulties and divisions[16]. This development was supported by a further remarkable position change: the US modified its standpoint regarding the developing countries, and instead of continuing to require their substantial participation, it backed them in the short term by opposing the 'unreasonable' demands by, in particular, the EU, Canada and Japan, which would imply an economic slowdown. Although having promised in 2001 to not obstruct the negotiations on the Kyoto Protocol, the US thus seemed to have changed its mind and sided with Saudi Arabia and other OPEC countries in reducing the substance of various decisions.

Notwithstanding the missed opportunity to start a dialogue with the developing countries, observers report that informal discussions among the developed and developing countries were initiated at COP-8.

3.6 The Ninth Session of the Conference of Parties (COP-9) – Milan, Italy

The COP-9 was held in Milan, Italy, from 1–12 December 2003. Over 5000 participants attended the session. The negotiations were aimed at strengthening and building on the original UNFCCC treaty and at completing some technical issues related to the UNFCCC and the Kyoto Protocol. However, only modest results were expected given the inertia of the political situation. Indeed, although the international infrastructure for the Kyoto Protocol's entry into force is to a large degree in place, the implementation of the Protocol has been delayed while awaiting that Russia ratifies the agreement. As a consequence, also the formal negotiations on major next steps in the international climate effort and on broader post-2012 perspectives are behind time.

Given this situation, a major focus at COP-9 was devoted to a large number of high quality side events which were meant to stimulate the discussion both on

---

15 In fact, among the developing countries only AOSIS provided support.

16 For a discussion of the COP-8 see Ott (2002).

topics which were already on the agenda and also on issues which have proved too 'hot' for official talks. In this context, the conference had to show whether the relationship between the industrialised and the developing world could be improved in order to lay the foundation for a more ambitious future approach to global climate-change control.

a) The Negotiations and Main Outcomes

Given the continued uncertainty on the Kyoto Protocol's fate, the formal agenda of COP-9 was very light. Nonetheless, the negotiations reached consensus on some decisions, in particular concerning the technical rules for sinks projects in the CDM for which COP-9 will be remembered as the 'forest COP.' Additional outcomes were decisions on guidelines for the operation of two funds to assist developing countries: the Special Climate Change Fund and the Least Developed Countries Fund. The official negotiations produced thus modest progress on a handful of largely technical issues; still they remained essentially deadlocked on issues touching on the broader question of next major steps in the international climate effort.

b) A Brief Analysis of COP-9

The COP-9 was characterised as a new type of meeting in the climate negotiation process. On one hand, the official negotiations aimed at strengthening the UNFCCC treaty and the Kyoto Protocol produced modest progress on some technical issues. However, the negotiations basically remained deadlocked when broader aspects related to next major steps in international climate policy were addressed. On the other hand, though, there was a so-called 'second face' of COP-9, reflected by over 100 side events organised by numerous observer constituencies, such as e.g. developed and developing country delegations, intergovernmental and non-governmental organisations, research institutions and the private sector. These side events addressed both pertinent issues related to the official negotiations, and explored a number of highly important aspects which did not yet receive a lot of attention in the negotiations given that they have proved simply too 'hot' to be handled by the COP. In short, the side events tackled issues related to meeting the UNFCCC's ultimate objective and demonstrated that efforts to address the adverse effects of climate change are already underway.

While resolving differences in the official negotiations remains complex, the significant number of side events at COP-9 signals a change towards a more positive outlook for future COP sessions. Indeed, regardless of the lack of significant progress, vigorous efforts to address the adverse effects of climate change are already underway. In that sense, it has been demonstrated that economic and interdisciplinary research can provide important inputs for the design of effective climate change control and is gaining momentum in the policy process.

4 Conclusions and the Outlook for the Future of Climate Policy

During the last several decades, climate change has evolved as a major topic in international policy discussions. The adoption of the Kyoto Protocol, establishing for the first time in history binding emissions reduction requirements for industrialised countries, represents a milestone in the international efforts to control climate change. Notwithstanding its limitations, the Kyoto Protocol constitutes one of the best-written texts within the history of environmental agreements since it links environmental, economic and legal components. However, being the outcome of last-minute negotiations motivated by the high expectations of the world as well as by the ambitious target to accomplish a historic agreement, a number of issues in the Protocol were not sufficiently elaborated or were left deliberately vague and therefore their definitions posed problems and disputes. A further difficulty consisted in the complexity and technicality of the issues, expressed in the enormous volume of texts that had to be agreed. As a consequence, the negotiations on the rules that were needed to prepare for the Kyoto Protocol's entry into force turned out to be very difficult. In addition, also the positions of some of the various groups of states involved in the negotiations changed during the negotiating process, due to political interests or further scientific information. In particular, the decision of the by far biggest producer of carbon dioxide emissions, the US, to not ratify the Kyoto Protocol induced a number of consequences for the Kyoto process, increasing above all the bargaining power of the remaining players – including Russia – due to the fact that the rules for entry into force of the Kyoto Protocol require 55 Parties to the Convention to ratify the Protocol, including Annex I Parties accounting for 55 per cent of that group's carbon dioxide emissions in 1990.

This 55 per cent clause and the pressing need to conclude the negotiations on the Kyoto Protocol in order to enable its timely entry into force have led to a number of concessions at the last COPs. Indeed, starting in The Hague, the Kyoto Protocol seems to have been watered down during the negotiations, inducing a much lower degree of environmental effectiveness than initially expected.

Nonetheless, the Kyoto Protocol still represents remarkable progress in international climate policy and there is common consensus that its major value is to start action against climate change and to make a signal to the public. Its implementation could indeed trigger a chain reaction, creating incentives for further activities. Indicators of the science, public and business sectors show that there is hope for a common climate control consciousness, strengthened by the series of major floods and droughts around the world, which have reinforced fears that global warming is also raising the risk of extreme weather events.

After COP-7, the international infrastructure for the Kyoto Protocol's entry into force is to a large degree in place. However, the Protocol can only become operational if Russia ratifies. Although repeatedly emphasising its intention to do so, Russia has long been uncertain regarding its final strategy. Only recently signs indicate that Russia is finally very close to the ratification of the Kyoto Protocol. Indeed, on 30 September 2004, the Russian government approved the Kyoto Protocol on climate change and sent it to parliament for ratification (BBC News, 30 September 2004). The decision by the parliament, whose approval is necessary for the treaty's ratification, is expected to take place within the next weeks and is in general considered as a mere formality. The Russian ratification would open the way for the Kyoto Protocol's entry-into force, sending thus a strong signal for international climate policy.

In accordance with these improved signals for the Kyoto Protocol's entry into force, 'Beyond Kyoto' issues are receiving increased attention and the post-Kyoto strategy is becoming an essential part of the international climate agenda. Post-2012 perspectives need to be discussed and compromises among the diverse countries have to be found. In this context, future climate talks will need to tackle an issue which represented a key area of disagreement in New Delhi: the formal engagement by the Parties on international climate-change control after the Kyoto Protocol's first commitment period. Delegates at COP-10 in Buenos Aires, will have the difficult task of improving the relationship between industrialised and developing countries, which suffered a setback since COP-8. Indeed, COP-9 has further highlighted the division between developed and developing countries, notwithstanding some progress observable in innovative approaches discussed in the side events. Still, communications between the EU and the developing countries require improvement, acknowledging that the question of equity needs to be tackled not only on the scientific, but also on the political level. Only after such a debate has been seriously started, a common base for negotiations between developed and developing countries on the future process of climate-change control can be created.

The last round of climate negotiations in Milan has ended on a positive note, emphasising the role of the UNFCCC's constituencies as an important component of the process to deliver an equitable global climate change regime, a point clearly made by their demands for strong climate action, dedicated leadership, information sharing and future thinking. Indeed, while resolving differences in the official negotiations remains complex, the significant number of side events signals a change towards a more positive outlook for future COP sessions.

As of 5 October 2004, 84 Parties had signed the Kyoto Protocol and 126 had ratified it, representing the countries that contributed 44.2 per cent of 1990 emissions. In addition, in particular Russia's recent decision to ratify the Kyoto Protocol generates optimism with regard to the treaty's entry into force. While this is encouraging, it nevertheless should not be forgotten that the Protocol itself – and in particular in its current form – will not affect climatic development in any observable manner. There is, therefore, consensus that the international process on climate-change control needs to be kept alive, considering the Kyoto Protocol is only an important first step, which should induce a chain reaction to a more ambitious future climate strategy. Indeed, studies and opinions coincide in concluding that climate policy requires global emission reductions in order to find a long-term solution to combat climate change. In particular, without the involvement of the key CO_2 emitters – above all the US and the large developing countries – global emission reductions will remain small. Given the circumstances in the developing countries, binding commitments are certainly not a strategy in the short term. However, in the longer term, their involvement in reducing global emissions will be a crucial component to a successful approach to climate-change control, as is the need to re-involve the US. As a consequence, any type of effective strategy to tackle the global problem of climate change needs to be based on a global approach, involving as many countries as possible, and accounting in particular for the large emitters of greenhouse gases.

Academic work has begun to identify possible future architectures, taking into account near-term and long-term incentives. Indeed, the progress achieved in national climate policy, as for example the internal emissions trading scheme of the EU or the various state-level initiatives in the US, as well as the emergence of numerous bilateral cooperative efforts focusing on climate-change control suggest that – despite the slow process of putting the Kyoto Protocol into force – the long-term prospects for climate protection look reasonable. Now the gap between science and politics needs to be bridged. Since no universal solution to climate change is available, the focus must be on the design of policy mixes[17]. Indeed, to strengthen the development of public awareness or consciousness of the problem, it is important to start with every available tool that could help to combat climate change. Climate-change control may depend to a degree on the expectations that the business community has in playing a role and on its engagement as an engine behind climate action. Further, the plan to use a wide range of instruments could be successful in the long run. After the foundations have been laid with this strategy, there is hope that the incentives and spillover effects of the various measures can induce further action to protect the climate. Even more important than the broad use of alternative instruments is a second main focus of a future strategy: the need to be equipped with reliable institutions that can control the process of emission reductions. This requirement is crucial because – due to the global problem of climate change – the long-term objective in

[17] The series of climate negotiations linked to the Kyoto Protocol and the scientific research undertaken in parallel have highlighted a number of key issues for the future climate policy. In particular, a crucial element of future climate strategies relates to technology and technological transformation. In addition, high priority in the design of the 'next' climate regime should be devoted to adaptation. Recent research suggests also that regional agreements/fragmented regimes can sometimes induce stronger incentives than a global approach and can therefore be seen as a first step in direction of effective climate change control.

climate-change control consists of implementing global governance. In this context, institutions play a key role since the incentives for participation, compliance and enforcement are strictly linked.

Past experiences have shown that economics can provide parts of the puzzle necessary to combat global warming. Nevertheless, since economists are not acting independently of policy processes, a broader viewpoint needs to be adopted in order to implement a successful strategy for controlling climate change, incorporating elements from a variety of disciplines. In particular, consensus has started to emerge that the future of climate policy should be seen as part of a strategy directed towards the broader aim of sustainable development. It may well be that future consensus on how to protect the global climate system differs considerably from the present form as proposed in the Kyoto Protocol. What matters is whether a substantial long-term approach to respond to the increasing threat of climate change can be found that takes into account the need for a broad strategy and strong institutions.

References

Anderson, J.W. (2000). 'The U.N. Framework Convention on Climate Change", in R.J. Kopp and J.B. Thatcher (eds.), *The Weathervane Guide to Climate Policy: An RFF Reader*. RFF: Washington, D.C., 3–4.

Buchner, B. (2001). 'What Really Happened in The Hague? Report on the COP6, Part 1'. *FEEM Working Paper*, 38:01. Milan: FEEM.

Grubb, M., C. Vrolijk and D. Brack (1999). *The Kyoto Protocol: A Guide and Assessment*. London: The Royal Institute of International Affairs.

International Institute for Sustainable Development (IISD), (1995). 'Summary of the First Conference of the Parties for the Framework Convention on Climate Change', *Earth Negotiations Bulletin*, 12:21. http://www.iisd.ca/linkages/vol12/

International Institute for Sustainable Development (IISD), (1996). 'Summary of the Second Conference of the Parties for the Framework Convention on Climate Change', *Earth Negotiations Bulletin*, 12:38. http://www.iisd.ca/linkages/vol12/

International Institute for Sustainable Development (IISD) (1997). 'Summary of the Third Conference of the Parties for the Framework Convention on Climate Change', *Earth Negotiations Bulletin*, 12:76. http://www.iisd.ca/climate/kyoto/coverage.html

International Institute for Sustainable Development (IISD) (1998). 'Summary of the Fourth Conference of the Parties for the Framework Convention on Climate Change', *Earth Negotiations Bulletin*, 12:97. http://www.iisd.ca/linkages/climate/ba/

International Institute for Sustainable Development (IISD) (1999). 'Summary of the Fifth Conference of the Parties for the Framework Convention on Climate Change', *Earth Negotiations Bulletin*, 12:123. http://www.iisd.ca/linkages/climate/cop5/

International Institute for Sustainable Development (IISD) (2000). 'Summary of the Sixth Conference of the Parties for the Framework Convention on Climate Change', *Earth Negotiations Bulletin*, 12:163. http://www.iisd.ca/linkages/climate/cop6/

International Institute for Sustainable Development (IISD) (2001a). 'Summary of the Resumed Sixth Conference of the Parties for the Framework Convention on Climate Change', *Earth Negotiations Bulletin*, 12:176. http://www.iisd.ca/linkages/climate/cop6bis/

International Institute for Sustainable Development (IISD) (2001b). 'Summary of the Seventh Conference of the Parties for the Framework Convention on Climate Change', *Earth Negotiations Bulletin*, 12:189. http://www.iisd.ca/linkages/climate/cop7/

International Institute for Sustainable Development (IISD) (2002a). 'Summary of the Eighth Conference of the Parties for the Framework Convention on Climate Change', *Earth Negotiations Bulletin*, 12:209. http://www.iisd.ca/linkages/climate/cop8/

International Institute for Sustainable Development (IISD) (2002b). 'Summary of the World Summit on Sustainable Development', *Earth Negotiations Bulletin*, 22:51. http://www.iisd.ca/linkages/2002/wssd/

International Institute for Sustainable Development (IISD) (2003a). 'Summary of the Ninth Conference of the Parties for the Framework Convention on Climate Change', *Earth Negotiations Bulletin*, 12:231. http://www.iisd.ca/linkages/climate/cop9/

International Institute for Sustainable Development (IISD) (2003b). 'Summary of the Side Events from COP-9', *Earth Negotiations Bulletin On the Side*, 13:12. http://www.iisd.ca/linkages/climate/cop9/enbots/

IPCC (1990) *First Assessment Report*, Cambridge: Cambridge University Press.

IPCC (1995) *Second Assessment Report*, Cambridge: Cambridge University Press.

IPCC (2001) *Third Assessment Report*, Cambridge: Cambridge University Press.

Kopp, R.J. and J.B. Thatcher (eds.) (2000). *The Weathervane Guide to Climate Policy: An RFF Reader*. Washington, D.C.: Resources for the Future.

Ott, H.E. (2002). 'Warning Signs from Delhi. Troubled Waters Ahead for Global Climate Policy', Wuppertal Institute for Climate, Environment and Energy.

Torvanger, A., K.H. Alfsen, H.H. Kolshus and L. Sygna (2001). 'The State of Climate Research and Climate Policy', *CICERO Report*, 2001:2.

United Nations (2002). *Key Outcomes of the Summit*, Johannesburg Summit 2002.

Chapter 2

BARBARA BUCHNER AND JANNA LEHMANN

Equity Principles to Enhance the Effectiveness of Climate Policy: An Economic and Legal Perspective

1 Introduction

The international equity debate on mitigating the risks of global climate change can find its origins in the 1992 UN Framework Convention on Climate Change (UNFCCC). Article 3 of the UNFCCC requires that the Parties engage in the protection of the climate system with "common but differentiated responsibilities". This phrase marks the real beginning of the search for proposals to differentiate emission reduction commitments, both in the international and intergenerational realms.

A wide variety of equity criteria and principles have been investigated and used for developing proposals to share the mitigation burden in a "fair" way. However, notwithstanding strong efforts in research and diplomacy, no definitive guideline has yet been chosen reflecting true equity in the context of international climate policy. Instead, the ongoing debate about the timing of emission reduction commitments has gained new intensity due to the crucial question of the involvement of developing countries. In order to achieve, in the long term, global participation to curb climate change, the adequate incorporation of "fairness" or "equity" considerations plays a crucial role.

Since the time that negotiations on climate change control commenced, the definition of an internationally equitable distribution of the burden of reducing greenhouse gases has been a important factor. The reason for its importance is twofold: on the one hand, countries differ strongly in their vulnerability to damages arising from global warming, while on the other, there are significant differences in their abilities to bear mitigation costs.

Acknowledging different national circumstances, it is obvious that different countries have differing viewpoints with respect to the adequacy of mitigation requirements. In particular, the developing world asks the industrialised coun-

MICHAEL BOTHE AND ECKARD REHBINDER (EDS.), Climate Change Policy, 45–72.

tries to bear their historical responsibility with respect to the build-up of greenhouse gas (GHG) emissions in the atmosphere and to act based on this. Although developing countries are increasing their share of the overall global GHG emissions, their emissions intensities are still far below the ones in the developed world. In order to continue their development paths, developing countries also have to cope with other basic problems and thus their priorities do not necessarily rank environmental protection amongst those of importance. As a result, climate change control should consist of initial activities by the industrialised world bearing the main responsibility for the presently accumulated emissions situation.

However, in the long run, the threat of global warming can only be effectively prevented, if responsibility is globally shared. All countries contribute to the problem of climate change and the sources of climate change risk (e.g. GHG emissions, land-use changes) are globally distributed; therefore, all countries must ultimately implement mitigation activities. In particular, climate change control should be allocated according to the capacity of each country.

Three main fairness principles have characterised the debate: need; capacity; and guilt/responsibility. According to the principle of guilt/responsibility, the costs of reducing GHG emissions should be related in some proportion to the degree by which the various actors are responsible for the actual emissions. Since some countries have been causing large quantities of GHG emissions for years and others almost none, the idea of historical responsibility needs to be considered. In contrast, both the principles of need and capacity focus more on the impact of abatement measures on each actor and its abilities: while the first would allocate the abatement costs according to the country's legitimate need for economic development, the second would distribute them in accordance with the country's ability or capacity to tackle climate change issues[1]. It has been shown that burden sharing is primarily a matter of allocating the costs of reducing GHG emissions; however, at the same time, the notion of "need" demonstrates that the benefits also have to be distributed in an equal manner[2].

In international law, the idea of equity exists in various ways: equity under the concept of "common but differentiated responsibility"; and under the polluter-pays principle. The polluter-pays principle, which is mentioned expressly in the UNFCCC, tries to encompass the three main fairness principles (need, capacity, responsibility) in juridical terms. The principle, although it is mentioned in the (not legally binding) Rio Declaration of 1992, is not directly applied as such in international law. Some governments have translated this principle into legal

---

1 For a discussion of these principles, see, for example, Ringius, Torvanger and Underdahl (2001), Ringius and Torvanger (2000) or Jansen *et al.* (20001b).

2 This principle is mostly translated into the argument for giving credits for past achievements in terms of emissions reductions or for guaranteeing compensation to developing countries compensation for their losses.

terms by introducing it in national legislation. However, the idea of "polluter pays" remains a national legal principle and not an international one. On a regional level, only European Community (EC) law has given a juridical shape to the principle, but not in such an extent as in national law.

In order to cope with the increasing weight that equity plays in climate change control, this chapter is aimed at clarifying the role of equity in both economic and legal terms. It demonstrates how equity is taken into account in legal and economic provisions together with the implications of different ways of dealing with "fairness". In this fashion, the real importance of equity considerations can be highlighted and commonly used arguments in the debate can be verified or falsified.

The chapter is structured as follows: in the first part, we will examine the equity debate from a legal point of view (Section 2). Then, by investigating equity criteria and recently suggested equity proposals in climate policy, we will look at economic considerations surrounding this debate (Section 3). Finally, Section 4 will draw some conclusions with respect to the insights that the equity debate provides for the effective future design of climate policy.

2 Equity Principles from a Legal Point of View

In this section, we will give a closer look at the equity debate from a legal point of view. After having examined some general aspects arising in the context of equity and environmental justice (2.1), we will analyse the concept of common but differentiated responsibility (2.2). As a third step, the polluter-pays principle and its relation to the concept of common but differentiated responsibility will be examined (2.3).

2.1 What are "Equity" and "Environmental Justice"?

According to a fundamental principle of international law, all States are equal. No State's sovereignty is worth more than another State's and each voice is counted equally. Each State has a right to be treated in the same manner and each State has the same duties. Thus, all States are obliged in the same way to protect and preserve the environment.

However, the idea of environmental justice has revealed concepts of equity, which modify this rigid principle. Indeed, environmental justice seeks to ensure that the benefits of environmental resources, the costs associated with protecting them, and any degradation that occurs are equitably shared. Principles of environmental justice are also invoked in order to combat "environmental racism", which includes acts by industrialised countries, such as establishing polluting industries and dumping toxic waste in poorer countries.

Equity is not equality. Rather, it translates the idea of justice and fairness. It is a principle of justice outside common or positive law, which is used to correct laws or to disregard them in particular circumstances. The principles of equity exert an element of balance; however, their content is very broad and not well-defined.

National laws rarely endorse principles of equity, because equity suggests a correction or non-application of substantive law. International law, however, being more flexible, recognises equity and applies it to rules considered as unjust and too rigid in specific cases. Thus, the Statute of the International Court of Justice (ICJ), annexed to the Charter of the United Nations, enables the ICJ to use equity by admitting the option to judge *ex aequo et bono*[3], if the parties to the litigation agree, instead of referring to a Convention or a substantive rule (Article 38, para. 2). This possibility is also envisaged in the Statute of the ICJ where a gap in rules exists. Generally, equity is a "quality of law" which imbues all the rules of international law. For this reason, it significantly underlies any interpretation of an international legal rule[4].

The ICJ explained the concept of equity in its decision "Continental Shelf", (Tunisia v Libya), 1982, holding that the juridical concept of equity is a general principle *directly* applicable as law. As such, equity can be applied also in every decision and not only in cases, where the court judge *ex aequo et bono*. This means, agreement of the parties to the dispute is not necessary.

2.2 Common but Differentiated Responsibility

Characteristics of the Concept

Common but differentiated responsibility is a concept and not a principle. Every juridical system is based on concepts (abstract expressions of social objectives). In contrast to principles, concepts cannot be directly applied; they underlie all rules forming a legal system. However, they can play a significant role in the development of law[5].

The concept of common but differentiated responsibility is more of a moral and political concept than a legal one. Thus, an international court may not condemn a State for simply not taking action to reduce pollution (responsibility), but may condemn a State, which has damaged a neighbouring State's territory by air pollution (liability). Hence, liability is the consequence either of damage

---

[3] The court can propose this possibility to the parties, but the judges are not obliged to do so.

[4] Dinh, Dailler and Pellet (1999), 232.

[5] Kiss and Beurier (2000), 290.

caused to somebody, or of a violation of law. The concept of common but differentiated "responsibility" corresponds to the concept of responsibility.

The acceptance by an industrialised country of its differentiated responsibility should result in increased efforts to enhance sustainable development. There are connections between fulfilment by industrialised countries of their responsibilities and reciprocal compliance by other groups of States[6].

Background

Financial mechanisms: the concept of common but differentiated responsibility is reflected particularly in the financial mechanisms of an increasing number of environmental agreements. In these agreements, compliance by developing Parties is linked to the provision of funds and technology by developed country Parties.

Prior to the Rio Conference, the Vienna Convention for the Protection of the Ozone Layer and its Montreal Protocol called on State Parties to provide subsidies, aid and credit, in addition to alternative technology and products to substitute ozone-depleting substances. The London Amendment of 1990 established a financial mechanism to further assist Parties in complying with the control measures. The mechanisms' Executive Committee adopted specific operational policies and guidelines to ensure that the funds are used to achieve the objectives of the agreement.

The following Conventions all foresee technology transfer or financial assistance between Parties as being important components of their regimes:

→ Basel Convention on the Control of Transboundary Movements of Hazardous Wastes and Their Disposal, 1989 (Article 10 (2) (1989))
→ Montreal Protocol on the Protection of the Ozone Layer (1987), amended in 1992
→ Convention on Biological Diversity (Articles 16, 20 and 21) (1992)
→ UNFCCC (1992)
→ UN Convention to Combat Desertification (UNCCD) (1994)

Obligations: the Montreal Protocol created a special status for developing countries by granting them a moratorium under certain conditions. The 1994 Protocol for Further Reduction of Sulphur Emissions to the 1979 Convention on Long-Range Transboundary Air Pollution makes a similar distinction by imposing different rates of reduction for the annual sulphur emissions of different countries[7].

---

6 Gündling (1998), 199 *et seq.*
7 Gündling (1998), 296.

The responsibilities of the contracting Parties are even more differentiated in the UNCCD, adopted in 1994. The Convention made a distinction not only between the rights and duties of developed and developing countries, but also between different regions suffering from desertification. Specific annexes to the Convention are applied to different regions with precise rules varying from one area to another: Africa, Latin America and the Caribbean, and the Northern Mediterranean.

Rio Declaration: the concept of common but differentiated responsibility was expressly mentioned for the first time in an international instrument in Principle 7 of the Rio Declaration (1992) which states:

> States shall co-operate in a spirit of global partnership to conserve, protect and restore the health and integrity of the Earth's ecosystem. In view of the different contributions to global environmental degradation, States have common but differentiated responsibilities. The developed countries acknowledge the responsibility that they bear in the international pursuit of sustainable development in view of the pressures their societies place on the global environment and of the technologies and financial resources they command.

Principle 8 adds that States "*should reduce and eliminate unsustainable patterns of production and consumption and promote appropriate demographic policies*". This is addressed implicitly to both developing countries and developed countries. The concept of common but differentiated responsibility seems at first glance to be contrary to the guiding principle of international law that declares that all States are equal. Nevertheless, in practice, the former concept is applied in international law and by international institutions. In fact, this concept has been incorporated within many international environmental conventions adopted since the end of the 1980s.

"Common but Differentiated Responsibility" Under the UNFCCC

In climate change policy, two key considerations exist in regard to the concept of common but differentiated responsibility: equality of results and equality of efforts. Equality of results means that all countries will eventually reach the same level, whereas equality of efforts requires all countries to take equal efforts. Both results and efforts can be defined in a variety of ways as is apparent in the climate change regime[8].

Indeed, the UNFCCC is a typical example of the advancement of the principle of equity. Like other conventions, it envisages technology transfer and financial assistance, but it extends the idea of differentiation to the level of

---

[8] Lammi and Tynkkynen (2001).

emission reduction obligations by making a distinction between three categories of states:

> 1. The developed country Parties are to take the lead in combating climate change and the adverse effects thereof (Article 3 (1)). They shall provide new and additional financial resources to meet the agreed full costs incurred by developing country Parties in complying with their obligations (Article 4 (3)). They also shall assist the developing country Parties that are particularly vulnerable to the effects of climate change in meeting costs of adaptation to those adverse effects (Article 4 (4). They shall facilitate the transfer of environmentally sound technology and know-how to developing country Parties (Article 4 (5)).
>
> 2. Most Eastern and Central European countries are considered to be undergoing a process of political and economic transition and are granted some flexibility to enhance their ability to address climate change (Article 4(6)).
>
> 3. Developing countries are to receive financial assistance and benefits from the transfer of technology. They have more time to make their initial communication on the measures they have taken to implement the Convention and the least developed Parties may make their reports at their discretion.

In 1990, at the Second World Climate Conference (SWCC), the 137 States attending the meeting agreed that:

> Recognising further that the principle of equity and the common but differentiated responsibility of countries should be the basis of any global response to climate change, developed countries must take the lead. They must all commit themselves to reduce their major contribution to the global net emissions and enter into strengthened co-operation with developing countries to enable them to adequately address climate change without hindering their national development goals and objectives.

The leadership of the industrialised countries is expressed implicitly in terms of their obligation to assist the developing countries[9].

The UNFCCC and the Secretariats of the Convention on Biological Diversity, the Convention on Wetlands (Ramsar Convention, 1972), and the UNCCD are linked and their respective secretaries work together[10]. They all support the concept of common but differentiated responsibility, so that it can be considered as a basic concept of the entire climate protection system.

---

[9] Gupta (1998), 182.

[10] See the reports submitted to the 2nd meeting of the Subsidiary Body for Scientific and Technological Advice (SBSTA) at COP7, Tuesday, 30 October 2001, agenda item 7, "Cooperation with relevant international organizations", (FCCC/SBSTA/2001/MISC.7; FCCC/SBSTA/2001/MISC.8; FCCC/SBSTA/2001/MISC.9).

Above all, the UNFCCC relies on the transfer of technologies. In the different obligations concerning reporting duties under the UNFCCC, technology transfer is used as a means to bring about more equity by applying the concept of common but differentiated responsibility.

The UNFCCC Secretariat has already focussed on the development and transfer of technologies, in particular on the development of technology information. It has commissioned a technical paper, which served as the basis for discussion in the UNFCCC's Subsidiary Body for Scientific and Technological Advice (SBSTA) and acted as the foundation of a draft decision on the subject adopted in the Marrakesh accords (2001). The technology information system has its own web-site[11].

The European Union, during the COP7 SBSTA meeting on 30 October 2001, expressed its interest in having the Expert Group on Technology Transfer playing a major role in the follow-up on this issue. The EU invited the UNFCCC Secretariat to prepare a summary for the 16th Session of SBSTA regarding:

- what role Parties should play in supporting the technology information system?
- what the result will be of the testing phase of the technology information system in relation to the content as well as the performance to the system?[12]

Members of the Umbrella Group[13] pointed out that funding will be necessary to support the technology information system and asked about the future of the Expert Group on Technology Transfer in this regard[14]. The United States suggested that the private sector should be more efficient in promoting technology transfer.

---

[11] FCCC/SBSTA/2001/L.7. The page is only accessible with a password, in order to protect the information.

[12] Statement by Belgium on behalf of the European Community and its Member States, Marrakech, 29 October–9 November 2001, SBSTA 15 – Agenda Item 5 "Development and Transfer of Technologies".

[13] The Umbrella Group represents one of the predominant negotiating groups that have emerged during the climate talks regarding the Kyoto Protocol. Originally, the group consisted of the United States, Australia, Canada, Japan and New Zealand. In March 2001, the US announced its withdrawal from the Kyoto Protocol.

[14] Opinion expressed by Canada.

2.3 The Polluter-Pays Principle and its Relationship to the Concept of Common but Differentiated Responsibility

Characteristics of the Principle

The polluter-pays principle was elaborated by the Organisation for Economic and Cultural Development (OECD) as an economic principle aiming to assign funds to the safeguarding of the environment and measures of control as well as to encourage the reasonable use of environmental resources[15]. The principle appears also in some international treaties as well as in the Treaty of Maastricht on the European Union (1992) (Article 130 R, al.2) and the Treaty of Amsterdam (1997) (Article 174, al.2)[16].

One of its most simplified formulations can be found in the Convention on the Marine Environmental Protection of the North-East Atlantic 1992 (OSPAR Convention). According to Article 2(2)(b) of this instrument, the polluter-pays principle means that the costs of the prevention of pollution, of control measures and of reduction must be paid by the polluter.

The formulation within the Rio Declaration is more abstract. Principle 16 states:

> National authorities should endeavour to promote the internationalisation of environmental costs and the use of economic instruments, taking into account the approach that the polluter should, in principle, bear the cost of pollution, with due regard to the public interest and without distorting international trade and investment.

Although many international treaties relating to environmental protection mention the polluter-pays principle, they do not determine its content. The careful formulation of Principle 16 of the Rio Declaration presents doubts about the character of the polluter-pays principle, which mainly seems to be an economic objective rather than being an international legal rule. In the final analysis, this principle can be applied in an area where constraints imposed by national environmental legislation on the economic actors are the same for everyone. This is true for the European Union and should be the situation within other free trade areas as well. However, the question may arise, whether or not the globalisation of free trade will lead to an approximation of international environmental legislation[17].

---

[15] Mentioned for the first time in a recommendation of the Council on 26 May 1972 (C(72) 128 (final), OECD, 1972.

[16] However, the EC directive concerning air pollution does not mention the polluter-pays principle.

[17] Kiss and Beurier (2000), 288, 298.

In essence, the polluter-pays principle is a measurement of economic integration (see below, point 2) with a general function of simultaneously being retributive[18], preventive[19] and curative[20].

Community law: in a statement regarding the EC's adoption of the polluter-pays principle, Jans[21] says that:

> The action of the European Community is based on the polluter-pays principle which is one of the cornerstones of the Community environmental policy even before it was incorporated into the Treaty. The polluter-pays principle is an essential element of environmental policy contained in Article 130 r, para. 2 of the EC Treaty and now also incorporated in Article 174 of the Amsterdam Treaty.

Many proposals and action plans mention the principle, and so it would appear that the polluter-pays principle is already a principle legally applied in the EC legislation. However a more profound study shows that this is not necessarily the case.

The polluter-pays principle is to be implemented via secondary legislation, and civil liability for remediation costs is a concrete application of the principle. Moreover, since the polluter-pays principle is formulated as a fundamental principle, it is not possible to deduce its exact character. The details of the EC legal regime, however important, cannot be deduced from the polluter-pays principle alone. For instance, it is unclear whether a polluter can be liable for damages also *without* fault and on an unlimited scale. Finally, each Member State will add another feature to the polluter-pays principle, if it is implemented in the national legislation.

Community practice is still a long way from enforcing the polluter-pays principle. In fact, liability on the part of individual EC citizens on the basis of primary law under the polluter-pays principle is not envisaged under Article 174 of the EC Treaty. For the moment, there are as yet no EU rules on civil liability, even if they are under advanced discussion[22].

Concerning liability rules relating to the Member States, the positive and the customary laws of the EC do not contain provisions that could be taken as a basis for liability on the part of the Member States for transboundary harm among them. In fact, there is a duty to pay compensation to another Member State in cases of transboundary harm. There are also no EC rules for cases where harm

---

[18] Redistributing in the sense of the internationalisation of the social costs of pollution, by equalizing the benefits of the polluter and the costs of pollution supported by the society.

[19] Preventing by dissuading the polluter from polluting.

[20] The "penalty" paid by the polluter should serve to repair the damages caused by the pollution.

[21] Jans (2000), 37 *et seq.*

[22] "White Paper on Civil Liability", Brussels: European Commission, 9 February 2000.

occurs because of a violation of an EU obligation, or where the harm occurs because of an authorisation of a dangerous installation or an authorisation of transboundary movements of hazardous waste originating from their territory[23].

International law: there is no international legal text giving an obligatory character to the polluter-pays principle. International conventions and declarations only advise States to implement the principle in their legislation. In this way, the polluter-pays principle would need to initially become national law, if it is to be applied on a national level. In fact, in international texts, States often abstain from giving an obligatory character to the principle. There are not enough States that have incorporated the principle in their legislation to presume a "long state practice" and thus, despite the assertion of several authors[24], there is insufficient justification for claiming the polluter-pays principle as customary international law[25],[26].

The Supportive Relationship between the Polluter-Pays Principle and the Concept of "Common but Differentiated Responsibility"

The polluter-pays principle is mainly understood as a principle applied by the State to private actors situated within its territory, to make them pay for the degrading effects of their activities. Its importance on the legal relationship between States is therefore low.

However, nothing hinders the polluter-pays principle being considered not as a principle with direct application, but as a "principle of *interpretation* of rules" between States. The duty of a State which pollutes is to be seen in light of the idea that those States who pollute more, should pay more in relation to other States, and, in general, to the international community. They should pay for their emitted pollution, be it in the form of compensation or in the form of reducing their emissions. Interpreted in this way, the polluter-pays principle is related to the concept of common but differentiated responsibility, in terms of its role as a form of clarification and as a guiding principle. In a broader sense, an extension of the polluter-pays principle also implies that those States that have the capacity to pay more, should pay more. Consequently, the obligation is not only to pay in respect of one's pollution, but also to pay in respect of one's ability to pay. Here we find the link between the polluter-pays principle and the concept of common but differentiated responsibility. Every interpretation of international environmental

---

[23] Wolfrum and Langenfeld (1999), 143.

[24] De Sadeleer (1999), 53 *et seq.*; Sands (1995), 213.

[25] Concerning the conditions for the development of international customary law, see Dinh, Daillier and Pellet (1999), 317 *et seq.*

[26] Some authors defend another opinion, e.g. de Sadeleer (1999), 53 *et seq.*; Sands (1995), 213.

law should be done in light of the polluter-pays principle in conjunction with the concept of common but differentiated responsibility. The polluter-pays principle could also be understood as a solidification of the concept of common but differentiated responsibility. The vague formulation of Principle 16 of the Rio Declaration regarding the polluter-pays principle provides the possibility for such an interpretation. The concept of common but differentiated responsibility asserts the laying of responsibility on those States that polluted first. In this regard, it is an application of the polluter-pays principle extended to past acts or omissions.

Contradictions Between the Polluter-Pays Principle and the Concept of Common but Differentiated Responsibility

Where the polluter-pays principle is applied to private actors, it can be seen as an obstacle to the application of the concept of common but differentiated responsibility[27]. Indeed, one may wonder whether financial assistance granted to developing countries to help them reduce pollution complies with this principle or at least whether or not it is counter-productive[28]. Under the polluter-pays principle, developing countries should make enterprises pay for their pollution, or stop them from polluting. However, at the same time, enterprises receive direct or indirect support from industrialised countries in order to combat pollution.

Thus, State support removes the punitive character of the polluter-pays principle, whose objective is to penalise the polluting enterprise by making them pay. State financing of enterprises may help them control pollution, but it is not necessarily a strong incentive to do so. To take away money from one side, and to give money on the other side, may lead to a neutralisation of both activities[29].

In the case of the application of the principle as a tool for the interpretation of rules, instead of as a principle of direct application, this danger does not occur.

---

[27] Mickwitz (1998), 46 *et seq.*

[28] De Sadeleer (1999), 67 *et seq.*; Kiss and Beurier (2000), 290.

[29] In this line of thought, the OECD and the EEC justified the application of the polluter-pays principle in view of its role in forbidding State aid for anti-pollution investments. In fact, the polluter-pays principle is of particular relevance with respect to new EC guidelines on State aid for environmental protection (OJ 1994 C 72/3). According to the Commission, the application of the EC Treaty rules on State aid (Arts. 87–89) must reflect the role economic instruments can play in environmental policy. This means taking into account a broader range of financial measures in this area. State aid control and environmental policy must, in the Commission's view, also support one another in insuring stricter application of the polluter-pays principle.

3 Equity Principles from an Economic Point of View

This section is devoted to the economic analysis of the equity debate that has accompanied climate change policy since its beginnings. Firstly, we will clarify the main motivations underlying the debate and classify them within a scheme of equity criteria. We will then present some of the equity proposals that have been suggested in climate policy and highlight their underlying rationales. As a final step, insights will be drawn with respect to the interrelations between economy and equity focussing again on the field of climate policy.

3.1 General Equity Considerations in Environmental Economics with a Focus on Climate Policy

Global climate policy has always been accompanied by efforts directed towards the definition of a "fair" distribution of abatement requirements. The need to set reduction targets and thus to determine the burden for individual countries constitutes one of the major obstacles to reaching a comprehensive agreement on climate change control. Depending on the initial allocation of resources and other characteristics of the different regions, differing strategies contributing to climate change control have been proposed by the different regions.

Notwithstanding great efforts to search for equitable guidelines, no definitive answer has been found regarding what constitutes an adequate commitment in terms of a fair distribution of the GHG mitigation burden. Instead, the main conclusion drawn in literature emphasises that no single principle is alone capable of resolving this issue[30]. In order to successfully find a satisfactory approach for the greatest possible number of countries, a strategy based on several equity criteria incorporating at least one of the three main equity principles needs to be implemented. As a consequence, we will now focus on equity criteria that have been identified in economic theory.

Equity proposals can usually be classified by distinguishing whether the applied equity criterion has been chosen according to the initial allocation of emissions ("allocation-based equity criteria"), according to the final outcome of the implementation of the policy instruments ("outcome-based equity criteria"), or according to the process by which the criterion has been chosen ("process-based equity criteria")[31].

[30] For a deeper discussion, see, for example, Cazorla and Toman (2000), Ringius, Torvanger and Underdal (2000), Ringius and Torvanger (2000) and Jansen *et al.* (2001b).

[31] For further explanations regarding this distinction see, among others, Rose, Stevens, Edmonds and Wise (1998) and Schmidt and Koschel (1998).

Tables 1–3 below summarise the main features of these three different groups of equity proposals and describe the ways in which they are usually implemented. Note should be taken that "allocation-based equity criteria" are implemented with reference to the abatement cost function. They are the dominating concepts used and examined in the literature (Eyckmans, 2000b; Schmidt and Koschel, 1998), because they can be easily applied without specifying the welfare function for each country. Nevertheless, a number of alternative equity formulations are possible and have emerged, mainly related to a redistribution of total welfare. For example, Tol (2001) analyses the impacts of three equity concepts based on welfare distribution. The first one relates to Kant with a "Rawlsian touch" ("Do not do to others what you do not want to happen to you", whereby the "others" are the least well-off regions – thus "act as if the impact on the worst-off country is your own"). The second one can be seen as a principle based on Varian's no-envy criterion (for all regions, at all times, the sum of costs of emissions reductions and the costs of climate change should be equal; income distribution should be at the same level that it would have been without climate policy). The third one maximises a global welfare function, which explicitly includes an inequality aversion.

The reasons why most empirical studies focus on cost-related equity concepts are their simple implementation and the possibility of comparing the results across studies. Indeed, criteria based on welfare distribution depend on the specification of the welfare function. Existing specifications largely differ across models. In some models, the welfare function is not even defined. By contrast, the specification of abatement costs, and in particular of marginal abatement costs, is subject to much lower variability across models.

Different interpretations have been made from these considerations and they have often been the roots of misunderstandings between developed and developing countries. Drawing upon the various equity criteria, many countries have developed proposals to differentiate GHG emission reductions. In the next section, we will give some examples of equity proposals made during the Kyoto Protocol negotiations.

3.2 From Theory to Practice: A Discussion of Equity Proposals in Climate Policy

Since the debate about the adequacy of scope and timing of emission reduction commitments has been ongoing for some time, it becomes more and more obvious that the definition of "fairness" or "equity" in the context of climate change control is not a straightforward task. There are different ways of translating the equity principles and criteria and a number of proposals regarding what could constitute fairness in GHG mitigation efforts exist. Corresponding to

Table 1. Allocation-based equity criteria

Equity principle	Definition	Implied burden-sharing rule
Egalitarian	All people have an equal right to pollute and to be protected from pollution.	Equal emissions reductions (abatement costs) per capita (in proportion to population or historic responsibilities). Implementation criterion: Equal per capita abatement costs
Ability to pay	Abatement costs should vary directly with economic circumstances and national well-being.	Equal emissions reductions (abatement costs) per unit GDP Implementation criterion: Equal abatement costs per unit of GDP
Sovereignty[32]	All nations have an equal right to pollute and to be protected from pollution.	Grandfathering (equal emissions reductions or abatement costs in proportion to emissions) Implementation criterion: Equal average abatement costs

Source: Adapted from Cazorla and Toman (2000), Tol (2001), Rose and Stevens (1993), Rose, Stevens, Edmonds and Wise (1998) and Schmidt and Koschel (1998)

the wide variety of equity criteria, a range of possible burden-sharing rules has emerged[33]. In particular, during the Ad Hoc Group on the Berlin Mandate (AGBM) work, which was initiated at the first Conference of the Parties (COP1)

---

[32] Closely related to the equity principle of sovereignty is the "polluter-pays principle" which also says that the abatement burden has to be allocated corresponding to emissions (which may include historical emissions). As in the case of sovereignty, equal emissions reductions (abatement costs) in proportion to emission levels are required. Since interpretation of this principle almost coincides with that of the principle of sovereignty, only rarely is a distinction made between them in the literature (see, for example, Cazorla and Toman, 2000). Due to the similarities, we also decided not to take it explicitly into account, but to deal with it implicitly through the sovereignty equity concept.

[33] For further details see, for example, Ringius, Torvanger and Underdahl (2001), Jansen et al. (2001b), Cazorla and Toman (2000), Tol (2001), Rose and Stevens (1993), Rose, Stevens, Edmonds and Wise (1998) and Schmidt and Koschel (1998).

Table 2. Outcome-based equity criteria

Equity principle	Definition	Implied burden-sharing rule
Horizontal	All nations have the right to be treated equally both concerning emission rights and burden sharing responsibilities.	Welfare changes across nations such that welfare costs or net abatement costs as a proportion of GDP or of population are the same in each country. Implementation criterion: Equal welfare costs per unit of GDP or per capita
Vertical	Welfare gains should vary inversely with national economic well-being; welfare losses should vary directly with GDP The greater the ability to pay, the greater the economic burden.	Emissions reductions such that net abatement costs grow with GDP. Implementation criterion: Equal abatement costs per unit of DP.
Compensation (Pareto rule)	"Winners" should compensate "losers" so that both are better off after mitigation.	Distribute abatement costs so that no nation suffers a net loss of welfare. Implementation criterion: Strong profitability

Source: Adapted from Cazorla and Toman (2000), Tol (2001), Rose and Stevens (1993), Rose, Stevens, Edmonds and Wise (1998) and Schmidt and Koschel (1998).

to the UNFCCC, various Parties submitted burden-sharing proposals[34]. This process induced research efforts into equity rules, which consequently lead to a number of suggestions. We will give an overview below of the most innovative approaches that have been used in different proposals.

[34] For a detailed and comprehensive discussion of the single proposals, see, for example, Jansen *et al.* (2001b), Ringius, Torvanger and Underdahl (2000) or Ringius, Torvanger and Underdahl (2001).

Table 3. Process-based equity criteria

Equity principle	Definition	Implied burden-sharing rule
Rawls' max-min	The welfare of the worst-off nation should be maximised, thus maximise the net benefit to the poorest nations.	Distribute largest proportion of net welfare change to the poorest nations; majority of emissions reductions (abatement costs) imposed on wealthier nations.
Market justice	The market is "fair", thus make greater use of markets.	Distribute emissions reductions to highest bidder; lowest net abatement costs by using flexible mechanisms (ET).
Consensus	The international negotiation process is fair, thus seek a political solution promoting stability.	Distribute abatement costs (power weighted) so the majority of nations are satisfied.
Sovereign argaining	Principles of fairness emerge endogenously as a result of multistage negotiations.	Distribute abatement costs according to equity principles that result from international bargaining and negotiation over time.
Kantian allocation rule[35]	Each country chooses an abatement level at least as large as the uniform abatement level it would like all countries to undertake.	Differentiate emissions reductions by country's preferred world abatement, possibly in tiers or groups.

Source: Adapted from Cazorla and Toman (2000), Tol (2001), Rose and Stevens (1993), Rose, Stevens, Edmonds and Wise (1998) and Schmidt and Koschel (1998).

Convergence Approach

Under this approach, emission allowances are required to converge to equal per capita levels over time. Basically, the atmosphere is seen as a global commons to which each human being is equally entitled. This consequently calls for an equitable allocation of emission allowances[36]. The differentiation of future commitments results as a compromise between theory and reality, achieved after

[35] According to Rose *et al.* (1998) this rule can be considered roughly equal to the principle of sovereignty plus elements of the principle of consensus.
[36] Berk and den Elzen (2001).

a transition period, which enables the convergence of the actual status quo emissions to equal per capita emissions.

In particular, there are two different types of convergence methodologies: linear convergence and non-linear convergence. The latter mode has been introduced by the Global Commons Institute as the *"Contradiction & Convergence"* approach. Under this methodology, developed countries would reduce emissions over time in proportion to their population, and developing countries would increase emissions according to their population, applying thus a non-linear (exponential) way of convergence[37].

During the AGBM negotiations, proposals by France, Switzerland and the EU followed the convergence approach.

Historical Responsibility Approach

The responsibility approach has its roots in the polluter-pays principle in that it entails that the larger a country's emissions, the larger its share of the emission reduction burden. Developing countries support this approach, emphasising that a fair differentiation of (future) commitments has to be based on historical responsibilities.

A famous example focusing on this approach is the Brazilian proposal, which was made during the negotiation of the Kyoto Protocol in order to allocate permits according to relative historical contributions to global warming, quantified by the cumulative historical emissions. After having analysed the proposal, The Netherlands National Institute for Public Health and the Environment (RIVM) presented a revised version of the Brazilian proposal, giving suggestions for elements that could be improved[38].

Triptych Approach

This approach combines different rules for the differentiation of future commitments and applies them to differing sectors of the economy. An example of this approach is the EU's triptych proposal, which was developed for the differentiation of emission reduction and stabilisation targets in order to meet the EU's commitments under the Kyoto Protocol. In particular, the proposed measure has a bottom-up, sector-oriented nature and is characterised by the division of the economy into three sectors. The main motivation behind this proposal was the need for a method that would take into account the differences in emission-producing activities across the EU's Member States[39].

---

[37] For further details, see Global Commons Institute (1997).

[38] For further details, see Berk and den Elzen (1998) or den Elzen *et al.* (1999).

[39] For further details, see Blok, Phylipsen, and Bode (1997) and Phylipsen *et al.* (1998).

Menu Approach

This mode of dealing with the differentiation of future commitments has been brought up through two Japanese proposals. Both of them share the common feature that a country may choose one of several options (based on different equity criteria) in order to minimise their emission reduction commitments. In this way, a type of "menu" is offered and countries are free to implement their favourite option.

Multi-criteria Approach

Another group of proposals incorporates an approach based on a combination of various burden-sharing criteria. The main advantage of this multi-criteria option is its capacity to address the needs of countries with differing emissions, population and economic development structures. Examples for this approach are the Norwegian and Icelandic proposals.

Multi-stage Approach

Basing their approach on the objective of "increasing participation", the RIVM introduced a multi-stage option characterised by a gradual increase in the number of countries involved and their level of commitment according to participation and burden-sharing rules. Based on income and/or emission thresholds, various stages of participation are foreseen which enable a gradual transition over time towards global emission reduction efforts[40].

Multi-sector Convergence Approach

This framework has been jointly developed by the ECN (Netherlands Energy Research Foundation) and CICERO (Center for International Climate and Environmental Research – Oslo). Its main objective is to facilitate international negotiations on national emission mitigation targets after the first budget period set out in the Kyoto Protocol. The multi-sector convergence framework can be described as a sector-oriented approach with global coverage, which requires that the amount of per capita emission assignments will ultimately converge at the same level for all countries. Building on some features of the region-oriented triptych approach, it goes further by facilitating a global perspective on the issues and involving a greater variety of sectors. This latter characteristic renders a more flexible approach since country-specific circumstances can be included. In particular, additional allowances may be conceded to countries facing specific circumstances that warrant higher emission needs than countries with more

---

[40] For a discussion, see, for example, Berk and den Elzen (1998), den Elzen *et al.* (1999) and Berk and den Elzen (2001).

favourable specific emission mitigating circumstances, all other factors being the same[41].

Due to the variety of potential approaches and implementation possibilities for fairness, the development of climate policy has been characterised by a number of differing proposals regarding the equity debate and, in particular, burden-sharing rules. Not very surprisingly, States have proposed differentiation metrics based above all on their own self-interests. However, although research has attempted to combine the most successful elements of different proposals, no definitive rule or principle for sharing the mitigation burden has yet been adopted in climate policy. Different pre-conditions and characteristics of the countries, strong and diverse self-interests, incentives to free ride, as well as the special features of climate change, render the approval and acceptance of equity criteria difficult. Nonetheless, it is clear that the issue of equity will constantly gain weight because of the future involvement of developing countries.

This analysis has discussed only a selection of the commitment regimes proposed during the Kyoto Protocol negotiations[42]. Nonetheless, we can conclude that the majority of equity proposals made in the last few years focuses above all on the fair distribution of the burden of addressing climate change in terms of abatement costs. However, in order to account for all aspects needed for a more equitable climate policy, costs arising from the damages of climate change also need to be included. This is particularly true when designing schemes of compensation or of adaptation assistance. In a comprehensive approach to equity in climate policy, burdens thus have to be defined in a broader way, taking into account damage costs.

3.3 Economic Insights Related to/from Equity Considerations in Climate Policy

The IPCC Third Assessment Report (2001) has removed the remaining doubts concerning the existence of global warming and has, moreover, confirmed that global warming is also caused by human activities. Climate change has thus become the most important and pressing current global environmental problem. In order to cope with the consequences induced by climate change, the problem needs to be tackled on a global scale, involving as many countries as possible. Global participation is also supported by the economic argument of cost-

---

[41] For a comprehensive presentation of this approach, see Jansen *et al.* (2001a).

[42] For a more detailed discussion, see, for example, Ringius, Torvanger and Underdahl (2000) who also take three further proposal groups into account focusing on: (i) fossil fuel dependency (e.g. Australia and Iran), (ii) GDP per capita (e.g. Poland *et al.*, Estonia, Poland and Russia, Korea) and (iii) cost-effectiveness (e.g. New Zealand).

effectiveness. Since no supranational authority exists which could impose an effective international climate policy, international environmental agreements such as the Kyoto Protocol are used to enable emission reductions. Ultimately, the most important way of strengthening international efforts to curb global warming consists of implementing these agreements.

However, the insights gained by economic theory demonstrate that it is very difficult to achieve broad participation in international environmental agreements[43], which means that the likelihood of an effective global climate agreement is not very high. This argument is also emphasised by the recent developments in climate policy, which demonstrate that in fact the opposite is happening, i.e. the regionalisation of climate policy seems to be taking place. The recent series of climate negotiations was supposed to clarify the details of the vague language of the Kyoto Protocol in order to prepare for its implementation and entry into force[44]. However, the negotiations did not proceed as expected. Instead of finding consensus among the Annex I Parties in order to jointly implement the Kyoto Protocol, different negotiating groups of countries emerged.

These events confirm the theoretic findings that point at the strong free-riding incentives prevailing in the context of global environmental goods. As a consequence, a major research question is related to the search for other reasons that could justify climate change control and thus increase the number of countries signing and ratifying international agreements. Recent literature analyses the incentives underlying the emergence of international environmental cooperation and the formation of climate coalitions within the general framework of non-cooperative games. This literature highlights that "self-enforcing agreements", i.e. agreements based on profitable and stable coalitions, may emerge at the equilibrium[45]. However, in most studies, the size of stable coalitions remains limited regarding any functional specification of State welfare functions[46]. Hence, the need to develop strategies that enhance the incentives to sign a climate agreement by making it profitable for relevant countries to become engaged and by offsetting their incentives to free-ride.

In this context, the equity debate has demonstrated its importance for economic considerations since one of the currently proposed ideas in the debate on climate change policy is that the development of more equitable agreements could be a way of increasing consensus and thus encouraging more accessions to

---

[43] Carraro and Siniscalco (1993).

[44] The recent series of negotiations started in November 2000 in The Hague, The Netherlands, at the first part of UNFCCC COP 6. The negotiations of COP 6 were completed in Bonn, Germany in July 2001, culminating in the well-known Bonn Agreement. The final negotiations that were needed to make the Kyoto Protocol "ratifiable" took place at COP 7 in Marrakech, Morocco, from 29 October to 10 November 2001.

[45] Carraro and Siniscalco (1993) and Barrett (1994).

[46] Hoel (1991), Barrett (1992), Carraro and Siniscalco (1992) and Heal (1994).

the Kyoto Protocol. This idea is at the heart of many of the recommendations contained in the recent 2001 IPCC Summary for Policymakers from Working Group 3. As a consequence of the insight that both efficiency and equity issues regarding different policy proposals need to be addressed in order to adequately implement the Kyoto Protocol, recent research focuses on the links between these two aspects: on the one hand, an equitable policy strategy induces more countries to commit to emissions control, thus enhancing its effectiveness; on the other hand, a cost-effective policy reduces conflict on distributional issues. The main policy implication that can be drawn from this literature is that strategies in climate change policy should balance equity, efficiency and political feasibility criteria in order to be both practical and effective. Policy integration is crucial, and in order to increase the probability of the implementation of such integrated approaches, equity aspects play an important role[47].

A particularly interesting study by Reiner and Jacoby (1997) investigates the implications of various Annex I country differentiation proposals with respect to welfare, equity and policy. The main finding is that the introduction of emissions trading can yield various benefits, both for overall welfare and equity considerations. In particular, potential conflicts within the developed world can be reduced while at the same time the complicated negotiations over burden-sharing agreements can be avoided. The analysis illustrates the difficulty in finding a consensus among States on what constitutes equitable burdens since the exclusive application of any of the equity proposals being considered induces problems. Moreover, the study highlights the argument that equity is often just used as an excuse for minimising domestic abatement costs. If this were true, "fairness would have little influence on a decision to undertake international commitments, and instead costs of compliance and willingness to pay will determine differences in burdens rather than more formal burden-sharing formulae[48]."

The main recommendation of this study consists, thus, of a primary focus on a trading regime that would both reduce overall mitigation costs and provide incentives to entice developing countries to enter into the emissions reductions regime. As a secondary step, the design of an equitable differentiation system to accomplish meaningful reductions at lower costs needs to be discussed on the global stage.

---

47 See, for example, Yohe, Montgomery and Balisteri (2000), Tol (2001), Rose and Stevens (1993), Rose and Stevens (1998), Rose, Stevens, Edmonds and Wise (1998), Manne and Richels (1995), Blanchard, Criqui, Trommetter and Viguier (2001) and Cazorla and Toman (2000). For a collection of papers dealing with the links between efficiency and equity of climate policy, see Carraro (2000).

48 Reiner and Jacoby (1997), 18.

Building on these results, in a recent research study we tried to get a deeper insight into the opportunities that are provided by a more equitable distribution of burdens[49]. The goal was to check empirically whether a more equitable differentiation of commitments improves the economic effects arising from an agreement's implementation. If this conjecture were true, i.e. if equity would indeed induce more countries to sign a climate agreement, we could conclude that equity indeed enhances efficiency, because a larger number of relevant signatories (possibly the big emitters) obviously implies a larger amount of emission abatement.

Using Nordhaus' RICE model (Nordhaus and Yang, 1996) in a game-theoretic framework in order to represent the interactions between economic and climate variables, we have analysed the incentive structure of different types of climate agreements and assessed whether increased equity enhances the likelihood that more countries – particularly, the largest emitters – would sign and ratify a climate agreement.

As a first step, we focused on the Kyoto Protocol. After having identified the burden-sharing rules in the Kyoto Protocol, we analysed the profitability of the Protocol and its stability, namely the inevitable incentives it provides to free ride. The conclusion is that almost all Annex 1 countries lose by signing the agreement and that more than one of these countries have an incentive to free ride, i.e. the net benefit from letting the other countries reduce emissions is larger than the net benefit from reducing emissions. These net benefits take into account the avoided damages from climate change at least as far as they are represented in RICE.

Secondly, we analysed the conjecture that a more equitable *ex-post* distribution of the burden of reducing emissions could enhance the incentives for more countries – particularly large emitters – to accept an emission reduction scheme defined within an international climate agreement. Our optimisation experiments only partly support this conjecture. Even though more equitable burden sharing rules provide better incentives to sign and ratify a climate agreement than the burden-sharing rules in the Kyoto Protocol, a stable agreement cannot generally be achieved, i.e. equity seems to enhance the profitability of climate agreements, but it does not offset the incentives to free ride.

Thirdly, we verified whether a transfer mechanism exists that could help broaden an initial stable, but partial, coalition achieved by agreeing on an equitable burden-sharing scheme. Our results suggest that transfers can indeed help broaden a given coalition. However, the coalition could not be achieved at least with the three equity rules considered in this paper (equal average abatement costs, equal per capita abatement costs, and equal abatement costs per unit of GDP).

---

[49] For a comprehensive presentation of the approach and its results, see Bosello, Buchner, Carraro and Raggi (2001).

The only strategy which we found could achieve a global agreement without free-riding incentives is a policy mix in which global emission trading is coupled with a transfer mechanism designed to offset ex-post incentives to free ride. This policy mix can achieve a stable global agreement whatever the initial stable coalition.

As a consequence, our results seem to suggest that an excessive focus on equity rules is not fruitful. In particular, it is more effective to minimise overall abatement costs via emission trading and then use the resulting surplus to provide incentives for free-riding countries to join the initial coalition. Our results therefore support the insights achieved in other research studies, as e.g. Reiner and Jacoby (1997). However, due to the limited consideration of burden-sharing criteria and the specifications of the RICE model, these results have to be considered as preliminary and they need to be checked with respect to their robustness.

4 Conclusions: Legal and Economic Conclusions on the Role that Equity Plays in Climate Policy

The global public goods nature of the climate change problem and the powerful argument of cost-effectiveness emphasise that international cooperation on emission controls is crucial in order to curb global warming. As a consequence of strong free-riding incentives, additional reasons have to be established which induce such cooperation to occur. Recent research results seem to suggest that equity constitutes one means of influencing the efficiency of environmental agreements. However, since the different approaches to the differentiation of emission reduction commitments are related to specific consequences for each State, equity cannot be characterised as the best strategy to improve the economic implications of climate change agreements. Instead, economic research studies suggest that other strategies exist, that are more adequate in dealing with the issue, e.g. it would be better to directly apply economic measures in order to avoid distortions.

Although research, both theoretical and empirical, is still missing, the insights gained in this analysis provide useful indications on how equity aspects influence the economic consequences of environmental agreements. In particular, they demonstrate that a simultaneous consideration of efficiency and equity in climate policy improves its overall performance. In addition to verifying the obtained results with the help of various modelling frameworks, future research should concentrate on how to incorporate equity aspects into future strategies in order to maximise the effectiveness of international climate policy.

The importance gained in recent years regarding equity issues in the climate debate is also highlighted by their inclusion in international law. Equity may be applicable as a rule, but the importance of the principle is generally in its application as a rule of *interpretation* of the provisions concerning climate change policy. International law rules cannot be interpreted without taking this principle into account. Therefore, it becomes obvious that equity plays an important role in the climate change context in order to promote environmental justice.

The concept of common but differentiated responsibility is interconnected with the principle of equity, but is more precise than the latter. As the above analysis has demonstrated, this is realized in rules on technology transfer and in some differentiated obligations in the UNFCCC and the Kyoto Protocol.

The polluter-pays principle, which could be considered as a way to give shape to the idea of environmental justice, is, in our opinion, not yet a principle of international law, although some authors take a different view. In international law, the principle is mostly used as an underlying pattern and as an argument in the debate regarding burden sharing and compensation of harms caused by climate change. It has nevertheless more of a political and moral character than a juridical one. One can consider that in the evolution of international law, the polluter-pays principle has become a "principle of *interpretation* of rules", and that there has been a concretisation of the concept of common but differentiated responsibility, which underlies the interpretation of international environmental rules. Based on this perspective, it has the same meaning for international climate change legislation as the principle of equity: it serves as a basis for rule interpretation.

Economic considerations related to the role of equity thus play an important role in identifying a successful approach to international climate policy and, in combination with the results of legal analysis, it becomes clear that equity constitutes a crucial factor in the theory and practice of climate change control. Notwithstanding the broad array of equity concepts, the equity principles discussed by economists appear to be quite specific and well-defined. In addition, economic considerations related to the role of equity play an important role in identifying a successful approach to international climate policy. The equity principles recognised under international law are much less specific, but are still crucial to the shaping of international accords.

The concept of common but differentiated responsibility is found throughout the Kyoto Protocol regime. Therefore, we can draw the conclusion that both economics and law support the premise that equity constitutes a crucial factor in the theory and practice of climate change control.

References

Babiker, M.H. (1997). *The CO_2 Abatement Game: Costs, Incentives and the Stability of a Sub-Global Coalition*. Washington: Economic Working Paper Archive Washington University.

Banuri, T., Göran-Mäler, K., Grubb, H., Jacobson, H.K. and Yamin, F. (1992). 'Equité et considérations sociales', in *Le Changement Climatique, Contribution du Groupe de travail III au Deuxième Rapport d'évaluation du Groupe d'experts intergouvernemental sur l'évolution du climat*. UNEP, 75–120.

Barrett, S. (1992). *Conventions on Climate Change: Economic Aspects of Negotiations*. Paris: OECD.

Barrett, S. (1994). 'Self-enforcing International Environmental Agreements'. *Oxford Economic Papers*, 46:878–894.

Berk, M. and den Elzen, M.G.J. (2001). *Options for Differentiation of Future Commitments in Cimate Policy: Insights from the FAIR Model*. Bilthoven, The Netherlands: National Institute of Public Health and the Environment (RIVM).

Berk, M.M. and den Elzen, M.G.J. (1998). The Brazilian Proposal Evaluated'. *CHANGE*, 44:19–23.

Blanchard, O., Criqui, P., Trommetter, M. and Viguier, L. (2001). 'Equity and Efficiency in Climate Change Negotiations: A Scenario for World Emissions Entitlements by 2030'. *Cahier de Recherche*, 26. Institut d'Economie et de Politique de l'Energie.

Blok, K., Phylipsen, G.J.M. and Bode, J.W. (1997). *The Triptique Approach: Burden Differentiation of CO_2 Emission Reduction Among European Union Member States*. Utrecht, The Netherlands: Department of Science, Technology and Society, Utrecht University.

Bosello, F., Buchner, B., Carraio, C. and Raggi, D. (2001). 'Can Equity Enhance Efficiency? Lessons from the Kyoto Protocol'. *FEEM Working Papers*, 49. Venice: FEEM.

Carraro, C. (ed.) (2000). *Efficiency and Equity of Climate Change Policy*. Dordrecht: Kluwer Academic Publishers.

Carraro, C. (ed.) (1994). *Trade, Innovation, Environment*. Dordrecht: Kluwer Academic Publishers.

Carraro, C. and Siniscalco, D. (1992). 'The International Protection of the Environment'. *Journal of Public Economics*, 52:309–328.

Carraro, C. and Siniscalco, D. (1993). 'Strategies for the International Protection of the Environment'. *Journal of Public Economics*, 52:309–328.

Cazorla, M. and Toman, M. (2000). 'International Equity and Climate Change Policy'. *RFF Climate Issue Brief*, 27. Washington.

Den Elzen, M., Berk, M., Schaeffer, M. Olivier, J., Hendriks, C. and Metz, B. (1999). The Brazilian Proposal and Other Options for International Burden Sharing: An evaluation of methodological and policy aspects using the FAIR model'. *RIVM Report No. 728001011*. Bilthoven, The Netherlands: National Institute of Public Health and the Environment (RIVM).

De Sadeleer, N. (1999). *Les principes du pollueur-payeur, de prévention et de précaution*. Brussels: Bruylant.

Dinh, N.Q., Daillier, P. and Pellet, A. (1999). *Droit International Public*. Paris: Pedone.

Eyckmans, J. and Tulkens, H. (2001). 'Simulating Coalitionally Stable Burden Sharing Agreements for the Climate Change Problem'. *CORE Discussion Paper 9926 and CLIMNEG Working Paper 18*.

Global Commons Institute (1997). *Contraction and Convergence: A Global Solution to a Global Problem*. London, UK: Global Commons Institute.

Gründling, L. (1998). *International Environmental Law: Atmosphere, Freshwater and Soil*. UNITAR, UNEP, IUNC.

Gupta, J. (1998). 'Leadership in the Climate Regime: Inspiring the commitment of the developing countries in the Post-Kyoto phase'. *RECIEL*, 7:180–190.

Heal, G. (1994). 'The Formation of Environmental Coalitions', in C. Carraro (ed.), *Trade, Innovation, Environment*. Dordrecht: Kluwer Academic Publishers.

Hoel, M. (1991). 'Global Environmental Problems: The Effects of Unilateral Actions Taken by One Country'. *Journal of Environmental Economics and Management*, 20/1:55–70.

Intergovernmental Panel on Climate Change (IPCC) (2001). *Third Assessment Report and Summary for Policymakers*. http://www.ipcc.int.

Jans, J.H. (2000). *European Environmental Law*. Groningen: Europe Law Publishing.

Jansen, J.C., Battjes, J.J., Sijm, J.P.M., Volkers, C.H. and Ybema, J.R. (2001a). 'The Multi-Sector Convergence Approach. A Flexible Framework for Negotiating Global Rules for National Greenhouse Gas Emissions Mitigation Targets'. *CICERO Working Paper*, 4.

Jansen, J.C., Battjes, J.J., Ormel, F.T., Sijm, J.P.M., Volkers, C.H. and Ybema, J.R. (2001b). 'Sharing the Burden of Greenhouse Gas Mitigation'. CICERO-ECN.

Kiss, A. and Beurier, J.P. (2000). *Droit international de l'environnement*. Paris: Pedone.

Kiss, A. and Shelton, D. (1999). *Developments and Trends in International Environmental Law*. UNITAR, UNEP, IUNC.

Kiss, A. (1997). *Introduction au droit international de l'environnement*. UNITAR, UNEP.

Lammi, H. and Tynkkynen, O. (2001). *The Whole Climate – climate equity and its implications for the North*. Finland: Friends of Earth.

Manne, A. and Richels, R. (1995). 'The Greenhouse Debate: Economic Efficiency, Burden Sharing and Hedging Strategies'. *The Energy Journal*, 16/4:1–37.

Mickwitz, P. (1998). *Implementation of Key Environmental Principles*. Nordic Councils of Ministers.

Müller, B. (2001). 'Varieties of Distributive Justice in Climate Change. An Editorial Comment'. *Climatic Change*, 48:273–288.

Müller, B. (2001). *Fair Compromise in a Morally Complex World*. Paper presented at the PEW Center, Equity and Global Climate Change Conference. Washington D.C., 17–18 April 2001.

Nordhaus, W.D. and Yang, Z. (1996). 'A Regional General-Equilibrium Model of Alternative Climate-Change Strategies'. *The American Economic Review*, 86/4.

Phylipsen, G.J.M., Bode, J. W., Blok, K., Merkus, H. and Metz, B. (1998). A Triptich Sectoral Approach to Burden Sharing; Greenhouse Gases in the European Bubble'. *Energy Policy*, 26:929–943.

Reiner, M.D. and Jacoby, H.D (1997). 'Annex I Differentiation proposals: Implications for Welfare, Equity and Policy'. MIT Joint Program on the Science and Policy of Global Change, *Report No. 27*, Cambridge, Massachusetts.

Ringius, L. and Torvanger, A. (2000). 'Burden Differentiation: Criteria for Evaluation and Development of Burden Sharing Rules'. *CICERO Working Paper*, 1.

Ringius, L., Torvanger, A. and Underdahl, A. (2000). 'Burden Differentiation: Fairness Principles and Proposals'. *CICERO Working Paper*, 13.

Rose, A. (1999). 'Burden-Sharing and Climate Change Policy beyond Kyoto: Implications for Developing Countries'. *Environment and Development Economics*, 3:392–98.

Rose, A. and Stevens, B. (1993). 'The Efficiency and Equity of Marketable Permits for CO2 Emissions'. *Resource and Energy Economics*, 15/1:117–146.

Rose, A. and Stevens, B. (1998). 'A Dynamic Analysis of Fairness in Global Warming Policy: Kyoto, Buenos Aires, and Beyond'. *Journal of Applied Economics*, 1: 329–362.

Rose, A., Stevens, B., Edmonds, J. and Wise, M. (1998). 'International Equity and Differentiation in Global Warming Policy: An Application to Tradable Emission Permits'. *Environmental and Resource Economics,* 12(1): 25–51.

Sands, P. (2000). *Principles of International Environmental Law.* Manchester: Manchester University Press.

Schelling, T. (2000). 'Intergenerational and International Discounting'. *Risk Analysis,* 20/6.

Schmidt, T.F.N. and Koschel, H. (1998). 'Climate Change Policy and Burden Sharing in the European Union – Applying Alternative Equity Rules to a CGE-framework'. *ZEW Discussion Paper,* 98/12.

Tol, R.S.J. (1998). 'The Optimal Timing of Greenhouse Gas Emission Abatement, Individual Rationality and Intergenerational Equity'. *FEEM Working Paper,* 3.

Tol, R.S.J. (2001). 'Equitable Cost-Benefit Analysis of Climate Change Policies'. *Ecological Economics,* 36: 71–85.

Torvanger, A. and Ringius, L. (2000). 'Burden Differentiation: Criteria for Evaluation and Development of Burden Sharing Rules'. *CICERO Working Paper,* 1.

Wigger, B.U. (2001). 'Pareto-Improving Intergenerational Transfers'. *Oxford Economic Papers,* 53: 260–280.

Wolfrum, R. and Langenfeld, C. (1999). *Environmental Protection by Means of International Liability Law.* Berlin: Erich Schmidt Verlag.

Yohe, G. W., Montgomery, D. and Balisteri, E. (2000). 'Equity and the Kyoto Protocol: Measuring the Distribution Effects of Alternative Emissions Trading Regimes'. *Global Environmental Change,* 10: 121–132.

Chapter 3

Janna Lehmann

A Comparative Analysis of the Long-Range Transboundary Air Pollution, Ozone Layer Protection and Climate Change Regimes

In order to facilitate understanding of the implementation of the atmospheric protection system, it is beneficial to compare the international legal regimes concerning air pollution, ozone layer protection and climate change. In fact, their common concern for the atmosphere, the similarities in terms of air pollutants and their historical interaction, justify a systemic and comparative approach.

The ozone layer protection regime relies on elements of regulation contained in previous experiences gained through the implementation of the Long-Range-Transboundary Air Pollution (LRTAP) regime, and the climate change regime relies on experiences gained in the ozone layer regime. The United Nations Framework Convention on Climate Change (UNFCCC) even mentions *expressis verbis* the ozone layer protection regime.

In terms of evaluation and long-term risk, ozone layer depletion and climate change are comparable. The net effect of ozone depleting substances (ODS) on the density of the ozone layer is only measurable on a long-term basis because of the influence of natural occurrences such as volcanic eruptions, weather patterns and the strength of the sun's radiation. Complete recovery from the ozone damage through application of the 1985 Vienna Convention and its Montreal Protocol is not expected until as late as 2045.

In respect of contributing factors, air pollution and climate change have similar sources. The burning of fossil fuels is a primary source of pollution, acid rain, and climate change by producing greenhouse gases (GHGs). By contrast, damage to the ozone layer is provoked by use of halogen compounds (most of which are hydrochlorofluorocarbons (HCFCs), although there are other classes

Michael Bothe and Eckard Rehbinder (ed.), Climate Change Policy, 73–102.

of compounds in this category). All of these are chemical substances created by man that do not normally exist in nature[1].

Scientific uncertainty, the need for long-term measurements and the necessity to fulfil environmental obligations in favour of the next generations are common characteristics of three ecological problems: ozone layer depletion, the greenhouse effect, and transfrontier air pollution[2].

These three environmental problems were known long ago. The effects of CO_2 as a GHG were discussed as long ago as the turn of the twentieth century. In 1896 the Swedish chemist Svante Arrhenius identified CO_2 as a regulator of warmth and in 1908 made the calculation that a doubling of CO_2 concentration in the atmosphere would lead to a rise of temperatures by 4–5.5°C.

The first research on chlorofluorocarbons (CFC) was lead in 1969 by J. Lovelock. Later, M. Molino and F. Sherwood Rowland published a report in 1974 on the correlation between CFCs and the destruction of the ozone layer. The phenomenon of acid rain was observed in the 19th century.

After a comparison of the legal structure and the scope of application of the three regimes, and a brief historical overview, this chapter will review the common guiding basic principles of the regimes.

I Comparison of the Legal Structure and of the Scope of the Three Regimes

The natural factors of the three ecological issues are narrowly linked. Thus, it is necessary to make clear the different fields of application of the three regimes and to give a background of the existing legal structure of each (1). The general commitments and substantive obligations at the heart of these structures differ in their technical regulations (2).

1 Background and History of Long-Range Transboundary Air Pollution (LRTAP), Ozone Depletion and Climate Change Issues

This section will briefly present the legal frameworks of the three regimes, reviewing their scope of application, and giving some background information on the ecological problems that they address.

---

[1] Note the severe impact of volcanic gases especially chlorine, which is a notable example of a non-man-made emissions.

[2] Trunko (1987) and Sonnemann (1986); on ozone 65 *et seq.*, on climate change 86 *et seq.*

1.1 Legal Framework

The three regimes practically adopt a very similar approach. They have first established framework treaties (although in the beginning they were not called that way) with fairly general obligations (either because the political consensus for stricter obligations was not mature enough at that stage, or because of a restricted scientific knowledge base), to be later developed through the adoption of protocols or amendments. The three regimes all took comparable routes. The LRTAP convention was developed through a series of protocols, relating in particular to control measures for specific substances. The Vienna Convention for the Protection of the Ozone Layer was first elaborated through the Montreal Protocol, under which then a series of amendments and adjustments were introduced, relating in particular to control measures. Finally, the UNFCCC has been elaborated by the Kyoto Protocol, which introduced precise quantified reduction commitments for the period 2008–2012.

a) The LRTAP Convention was adopted as a reaction to a phenomenon which at the end of the 1970s and beginning of the 1980s[3] was discussed in such terms as "acid rain" and "forest dieback"[4]. The 1979 Convention, containing rather vague and general substantive obligations, turned out to be a framework for a complicated legal regime that was elaborated upon more and more by Protocols:

1979: Convention on Long-range Transboundary Air Pollution (LRTAP)

1984: Protocol to the 1979 Convention on LRTAP on the Long Term Financing of the Co-operative Programme for Monitoring and Evaluation of the Long-range Transmission of Air Pollution in Europe (EMEP) (entry into force: 28 January 1988)

1985: Protocol to the LRTAP Convention on the Reduction of Sulphur Emissions or their Transboundary Fluxes by at least 30 per cent (entry into force: 2 September 1987)

1988: Protocol to the LRTAP concerning the Control of Emissions of Nitrogen Oxides or their Transboundary Fluxes (entry into force: 14 February 1991)

1991: Protocol to the LRTAP concerning the Control of Emissions of Volatile

---

[3] In reality, forest dieback was known for more than two centuries. Already in the 19th century, scientific articles were published on damage to forests caused by acid rain. A publication by Heß (1878) outlined the damage caused by coal fires and industrial smoke to forests. In the 1940s, Swedish researchers started monitoring freshwater acidity levels.

[4] For a detailed history on the Convention and the Protocols, see Hanf (2000), 30, 42.

Organic Compounds or their Transboundary Fluxes (entry into force: 5 August 1998)

1994: Second Protocol to the LRTAP on the Reduction of Sulphur Emissions (entry into force: 5 August 1998)

1998: Protocol to the LRTAP on Persistent Organic Pollutants (entry into force: 23 October 2003)

1998: Protocol to the LRTAP on Heavy Metals (entry into force: 29 December 2003)

1999: Protocol to Abate Acidification, Entrophication and Ground-level Ozone (not yet in force)

b) Ozone Depletion. The first legal steps concerning the ozone layer protection regime were undertaken on a national level.

In 1978 Canada forbade the non-essential use of HCFCs as propellants in aerosol spray cans.

In 1979 Sweden followed with a similar regulation, then Norway in 1981. Later the Netherlands, Germany and the EEC introduced regulations; however, instead of a prohibition of the use of HCFC, industry was only obliged to display warnings.

In 1985 the Vienna Convention on the Protection of the Ozone Layer was adopted and entered into force on 22 September 1988. It, too, contains only vague and general substantive provisions. Precise phasing-out obligations were added by the Montreal Protocol, signed in 1987 and further elaborated by the amendments of 1990 (London), 1992 (Copenhagen), 1997 (Montreal) and 1999 (Beijing)[5].

c) Climate Change. The development of the climate change regime is more varied than that of the ozone layer protection and the LRTAP regimes.

Despite the fact that knowledge of the impacts of CO_2 on the climate had existed for many years, it has taken a long time for an international climate

[5] For more details on the development of the Ozone Layer Protection Regime, see Sand (1999), 209–211.

change regime to be developed[6]. Public discussion of the subject began only in the 1970s. In 1979 the climate change problem was included for the first time in the agenda of the Geneva Environment Conference. Ten years later, an official declaration on the subject was made in Toronto at the World Conference on Changing Atmosphere (1989), resulting in the 'Toronto Target', a voluntary objective to reduce CO_2 emissions by about 25% by 2005. Later, although many UN resolutions were adopted, the subject was still too controversial for the adoption of a binding text.

In 1990 the General Assembly of the United Nations created the Intergovernmental Negotiating Committee for a Framework Convention on Climate Change (INF/FCCC). The UN Framework Convention on Climate Change (UNFCCC) was adopted in May 1992 and entered into force in March 1994[7]. Precise greenhouse gas reduction or limitation obligations were added by the Kyoto Protocol adopted in 1997, which finally entered into force on 16 February 2005. It has been elaborated by secondary norms agreed upon at the Marrakech Conference in 2001 which remain to be formally adopted by the first Meeting of the Parties to the Kyoto Protocol.

1.2 Scope and Purpose of the Three Regimes

a) The LRTAP Convention is in particular concerned with substances which cause acid rain: sulphur dioxide, nitrogen oxides, sulphur, stratospheric ozone, and various heavy metals, such as lead, cadmium and nickel (waste from burning fossil energy).

The Convention defines 'air pollution' as every introduction by man of substances or energy having deleterious effects on nature and harming living resources and ecosystems and that interfere with amenities and other legitimate uses of the environment (Article 1). 'Long-range transboundary air pollution' means air pollution whose physical origin is situated wholly or in part within the area under the national jurisdiction of one State and which has adverse effects in the area under the jurisdiction of another State at such a distance that it is not generally possible to distinguish the contribution of individual emission sources or groups of sources (Article 2).

b) The Vienna Convention on the Protection of the Ozone Layer is concerned with the destruction of the ozone layer ("the layer of atmospheric ozone above the

[6] First publication of Svante Arrhenius on the influence of CO_2, mentioning that the doubling of concentration of CO_2 in the atmosphere leads to a temperature rise of 4° to 5.5°, were published between 1896 and 1908.

[7] In Rio, 155 States signed the Convention. For the current situation regarding ratification: http://www.iisd.ca (Earth Negotiation Bulletin) or www.unfccc.int (UNFCCC). See the introductory article by Bothe and Rehbinder, *supra*.

planetary boundary layer") by ozone-depleting substances, ODS. Ozone itself is not taken into consideration in the Convention. Ozone as a pollutant at ground level is treated under the LRTAP Convention under its Nitrogen Oxides Protocol and its VOC Protocol. Ozone in the stratosphere[8] (where it is vital as a protective layer) is treated under the Vienna Convention on the Protection of the Ozone Layer (Vienna Convention 1985).

The Vienna Convention 1985 states that the ozone layer shall be protected from 'adverse effects' which means changes in the physical environment or biota, including changes in climate, which have significant deleterious effects on human health or on the composition, resilience and productivity of natural and managed ecosystems, or on materials useful to mankind (Article 1). To achieve this level of protection, alternative substances shall be used.

c) The Climate Change Convention specifically excludes greenhouse gases "controlled by the Montreal Protocol" (Article 4). GHG are defined as "... those gaseous constituents of the atmosphere, both natural and anthropogenic, that absorb and re-emit infrared radiation" (Article 1).

1.3 Background to the Environmental Issues

a) The concept of LRTAP was developed as a result of the phenomenon of acid rain destroying forests, plantations, buildings, and poisoning/contaminating water and having a negative effect on human health[9]. Acid rain is due to air-polluting substances such as sulphur dioxide, sulphate, nitrogen oxides, stratospheric[10] ozone, and heavy metals such as lead, cadmium and nickel. They are a by-product of burning fossil fuels, released into the atmosphere as gases or aerosols and are bound in the upper atmosphere with water under the influence of the sun's radiation. They return to earth with rain, snow and mist, or are deposited as dry deposition on plants, the ground, and buildings and on water surfaces. Global weather patterns can transport the substances over thousands of kilometres.

---

8 The second layer of the atmosphere extends from about 12 to 48 km above the earth's surface. It contains small amounts of gaseous ozone (O_3), which filters out about 99 percent of the incoming harmful ultraviolet (UV) radiation. http://www.fluorocarbons.org/g-info/glossary/glossary.htm

9 Damage was first noticed by the Scandinavians who then pushed forward the creation of the LRTAP regime. For more historical details, see Hanf (2000), 25–30.

10 The stratosphere is the air level higher than 12 km above the earth.

Rain is 'acid', when it has a pH of less than 6.5. The compounds that make the rain 'acid' are sulphuric acid[11] (derived from sulphur trioxide and water), and nitric acid[12] (derived from nitrogen oxides and water). The main polluting sources are power plants, automobile emissions, heating plants, and, less importantly, industrial emissions, and household emissions[13].

b) Ozone is found naturally, but it is distributed in very different concentrations in the atmosphere. The main concentration is found in the stratosphere (90%), where it acts as a protective layer against sun radiation. Too much ozone in the upper stratosphere and in the troposphere[14] acts as a GHG. Excessive ozone concentrations in the near ground troposphere cause 'summer-smog'.

The natural concentration of ozone in the troposphere is very low. At this level, ozone emerges as a reaction between nitrogen oxides (from burning fossil fuel), oxide and sunlight. Very little of this ozone arrives at upper levels as there is an interface between the levels where hardly any ozone can pass through[15]. The problem of ozone in the troposphere is regulated in the LRTAP Nitrogen Oxides Protocol and VOC Protocol[16].

In the stratosphere, ozone is vital for conditions on earth as it filters out the ultraviolet rays that would otherwise cause damage to skin and genetic material, kill new generations of water-born animals and plants[17], and hinder the growth of plants. It also increases the temperature on the ground. Anthropogenic chemical substances destroy ozone in the stratosphere. In regard to the actual damage, there are no absolute or precise scientific data – a situation similar to that in the climate change discussion.

---

[11] Sulphuric acid is derived from sulphur trioxide SO_3 + water H_2O. Sulphurous acid is derived from sulphur dioxide SO_2 + water H_2O ($H_2O + SO_2$) and is a weaker acid. For a glossary of all technical and chemical terms concerning ozone and acid rain, see http://www.fluorocarbons.org/g-info/glossary/glossary.htm.

[12] HNO_3.

[13] Hanf (2000), 22 *et seq.*

[14] The troposphere is the lowest layer of the atmosphere and contains about 95 percent of the mass of air in the earth's atmosphere. The troposphere extends from the earth's surface up to about 10 to 15 km. All weather processes take place in the troposphere. Ozone that is formed in the troposphere plays a significant role in causing both the greenhouse effect and urban smog. http://www.fluorocarbons.org/g-info/glossary/glossary.htm.

[15] The interchange of air between the levels hardly exists because of the different sun radiation levels. In the stratosphere, the sun's radiation is very strong, thus chemical reactions take place which cannot occur in the troposphere where the radiation levels are much lower.

[16] Ground level ozone does not travel itself because of its high reactivity, but ozone precursors do.

[17] 70% of the oxygen produced on earth is delivered by vegetable plankton.

The family of chlorofluorocarbons (CFCs) is one of the main groups of damaging substances affecting ozone concentrations[18]. However, in the press and in public information materials, these substances are referred to as 'HCFCs', when speaking generally about ODS. CFCs travel up through the troposphere without reacting with other substances and it is only in the stratosphere, with the powerful radiation from the sun, that chemical reactions lead to the destruction of ozone molecules. Some ODS are very long lasting: some exist for up to 50 000 years; but the most common ones last for up to 70–80 years. Thus, the actual effect can only be ascertained after comparing measurements taken over many years. Additionally, natural influences such as certain weather conditions can play an important role in destroying ozone in the stratosphere[19].

c) The greenhouse effect (a rising of the average temperature of the earth) is due substantially to a higher concentration of so-called greenhouse gases in the atmosphere. As CO_2, water vapour, methane[20], CFC[21] and nitrous oxide N_2O[22] (laughing gas) absorb sunlight, they are transformed into heat. Like ozone, CO_2, which is responsible for 60% of the greenhouse effect, exists normally in nature and is not a toxic substance. It is only its growing concentration in the atmosphere that renders it harmful.

Unfortunately, trapping CO_2 emissions from fossil fuel burning plants is very expensive, pointing to the need to adopt measures which result in the suspension or delay of CO_2 production, rather than its physical removal.

2 General Commitments under the Three Regimes and Substantive Environmental Obligations

The main commitments set out in environmental agreements generally concern the protection of a special natural milieu and/or the regulation of pollutants. The formulations may be more or less specific. The LRTAP regime deals with specific pollutants in different ways (2.1), the ozone layer regime elaborates a very complicated set of regulations for phasing out production and consumption of ODS (2.2), whereas the climate change regime promotes the reduction of GHG emissions (2.3).

18 The other substances are organic halogen compounds.

19 Information Brochure 'Ozone', Ministry of Environment of Nord-Rhein-Westfalen, 34.

20 Among other sources, methane is produced from the digestive systems of cows, in rice plantation and in the extraction of natural gas.

21 This is still used in high quantities in many developing countries and is already addressed under the ozone regime.

22 Laughing gas is produced mainly by fertilizers used in agriculture.

2.1 General Commitments under the LRTAP Convention

The commitments are laid down in Article 2 of the Convention. The LRTAP Convention is very brief on its general obligations, stating that "the Contracting Parties, taking due account of the facts and problem involved, are determined to protect man and his environment against air pollution and shall endeavour to limit and, as far as possible, gradually reduce and prevent air pollution including long-range transboundary air pollution."

The LRTAP legal framework is very technical and detailed. As described above, there are already eight Protocols to the LRTAP Convention. It is possible to distinguish several types of substantive environmental protection commitments in the Convention and its protocols. They are described below[23] in the order of their development:

1. Source-oriented (general): the Convention generally obligates Parties to use the best available technology, which is economically feasible, and low or non-waste technology (Article 6).

2. Emission-oriented: the first Sulphur Protocol of 1985 and the VOC Protocol of 1991 each contain the obligation to decrease emissions. This means, in the case of the 1985 Protocol, that the Parties have to reduce their emissions by around 30%, based on the value of a certain year of reference within a certain period.

   The Nitrogen Oxides Protocol of 1988, instead of setting a maximum level of emissions, which may not be exceeded, requires that emissions do not exceed the value of a determined reference year: it thereby creates a stabilization obligation. The VOC Protocol changes this stabilization obligation in accordance with Article 2 (3) lit. b) iii) VOC Protocol to a reduction obligation, whereby 15 Parties decided to reduce VOC emissions by around 30% and three Parties agreed to remain with the stabilization obligation.

3. Source-oriented (specific): the Nitrogen Oxides, the VOC and the second Sulphur Protocols are connected by their obligations, as already described in the text of the LRTAP Convention, to apply the best available technology using a detailed appendix on the status of what constitutes "best available technology" and its costs.

[23] This is based on Neumann (2000), 18 *et seq.*

4. Oriented to the effect of emissions: the second Sulphur Protocol classifies pollutants based on their critical polluting character and the capacity of special natural environments to neutralize emissions. A "multi-effects and multi-pollutant" Protocol is currently being prepared[24].

5. Pollutant-production and consumption-oriented: in the POPs Protocol of 1998, the Parties assume the obligation to stop the production and the consumption of certain substances, listed in Appendix I of the Protocol.

6. Use-oriented: the substances specified in Appendix II of the POPs Protocol can only be consumed and produced for specific uses. These uses are regulated in detail in the Appendix (e.g. medical use) to the Protocol.

7. Combination: the obligation to reduce the substances specified in Appendix III of the POPs Protocol is without a quantitative specification. However, it is a combination of the obligation to use the best available technology as of a fixed date (according to Annex V) and a maximum limit of pollutants. The Heavy Metal Protocol also follows this mixed form of obligations.

2.2 General Commitments under the Ozone Layer Protection Regime

In its structure the Vienna Convention on the Protection of the Ozone Layer is like the LRTAP Convention and the UNFCCC. All three are Framework Conventions, that is, they each have very general obligations. The Vienna Convention 1985 states that Parties shall take "appropriate measures" in order to "protect human health and the environment against adverse effects resulting or likely to result from human activities which modify or are likely to modify the ozone layer." (Article 2).

The term 'appropriate measures' in this environmental protection obligation is more indefinite and weaker than those of the LRTAP. Like the LRTAP (Article 2.6), the Vienna Convention 1985 is a compromise between the recognition of the necessity to protect the ozone layer on the one hand, and the fear of economic and social consequences of binding obligations on the other. The emphasis is, therefore, as in the LRTAP Convention, on procedural and co-operation obligations. Thus, the Vienna Convention 1985 follows in its regulatory system the model of the LRTAP.

Special substantive environmental protection obligations for complying with the framework have lead to the creation of the protocols. In fact, the Vienna Convention 1985, in contrast to the LRTAP Convention, contains an express reference to subsequent protocols. The one additional protocol, the Montreal Protocol, was even agreed upon before the Vienna Convention 1985 entered into

---

24 www.unece.org/unece/env/multipro.htm.

force, and subsequently, it has been amended several times. Thus, the general and vague substantive content of the Vienna Convention was very soon concretised by an implementing treaty.

In the Vienna Convention 1985 and Montreal Protocol, three types of environmental obligations may be identified[25]:

- A general requirement (not an obligation) of action under the Convention (Article 2)
- The quantified environmental protection obligations in the Montreal Protocol (Article 2) concerning the production and consumption of ODS
- Quantified trade restrictions for ODS in relation to both non-Party states and Parties in the Montreal Protocol (Articles 4 and 4A)

The Montreal Protocol entered into force on 1 January 1989. Since then, its substantive environmental protection obligations have been reinforced four times[26]:

- 29 June 1990: the London Amendments (entered into force on 10 August 1992)
- 25 November 1992: the Copenhagen Amendments (entered into force on 14 June 1994)
- 17 September 1997: the Montreal Amendments (entered into force on 10 November 1999)
- 3 December 1999: the Beijing Amendments (entered into force on 25 February 2002)

The Montreal Protocol contains many specific and complicated regulations for setting upper limits of production and consumption of ODS beyond those which are totally prohibited. Moreover, it differentiates between regulations based on groups of pollutants and those based on production and consumption. In addition, the ozone regime differentiates between groups of countries: industrial and developing countries (so-called Article 2 countries) on the one hand, and developing countries with low consumption (so-called Article 5 countries) on the other hand.

---

[25] This is based on Neumann (2000), 81.

[26] See the London Amendment and Adjustments, 1990 (ratified by 163 parties), Copenhagen Amendment and Adjustment, 1992 (ratified by 141 parties), Montreal Amendment, 1997 (ratified by 81 parties), Beijing Amendment and Adjustment, 1999 (ratified by 34 parties). For details on these amendments, see Earth Negotiations Bulletin (2002), 2.

The trade in ODS is, in accordance with Article 4, forbidden in principle. Furthermore, modifications of the Montreal Protocol in 1997 introduced a prohibition of the trade of ODS between Parties (Article 4 A).

2.3 General Commitments under the UNFCCC

The commitment under the UNFCCC is rather general and poorly defined. Based on Article 2, sentence 1 of the UNFCCC, the ultimate goal of the Convention is: "to achieve the stabilization of greenhouse gas concentrations in the atmosphere at a level that would prevent dangerous anthropogenic disturbance with the climatic system". The only significant substantive environmental protection obligation in the Convention is found in Article 4 (2) (a) under the heading "commitments" and applies only to Annex I Parties (developed country Parties and country Parties with economies in transition (CEITs).

The Kyoto Protocol to the UNFCCC, specifying the details of the climate change regime, was adopted in December 1997 in Kyoto, Japan (opened for signature on 16 March 1998). The Protocol aims to reduce the combined greenhouse gas emissions of industrialised countries by at least 5% compared to 1990 levels by end of the time period 2008–2012.

The Kyoto Protocol identifies policies and measures that can be taken by all country Parties (Article 2) and sets out specific quantified commitments for Annex B countries (certain industrialized countries) on six GHGs (Article 3). Parties must communicate their measures (Article 7), which are subject to review at a later point in time (Article 8). Other Articles in the Protocol concern definitions, methodologies, non-compliance, dispute resolution and the customary regulations on amendments, annexes, voting, depository, signature and ratification, entry into force, reservations and withdrawals[27].

II Common Guiding Principles

The three regimes are based on comparable basic principles. Among these principles, we will concentrate on duties to cooperate, information exchange, and data reports. Some are applied nearly the same way in each of the regimes, while others evolved with each new text. Cooperation, information exchange and data reports are the pillars of the three regimes.

[27] Gupta (2000), 10.

1 Cooperation, Exchange of Information and Good Neighbourliness

Cooperation, exchange of information and the principle of good neighbourliness[28] are the basic principles of every environmental agreement, including the three regimes dealt with in this article. All three of the Conventions discussed in this chapter are based on them.

These international principles have been systematically applied and repeated in legal texts and political declarations, ever since they were affirmed for the first time in the Stockholm Declaration of 1972: co-operation in Principles 22 and 24, exchange of information in Principle 20, and good neighbourliness in Principle 21[29].

These principles are now each considered as part of international customary law[30].

1.1 The LRTAP Convention

The LRTAP Convention specifically mentions the Stockholm Declaration, in particular Principle 21. In fact, the core of the Convention and the main focus of its text are the principles of cooperation, exchange of information and consultation.

The provisions on "Fundamental Principles" contains, in addition to the already described general commitments, three articles concerning information exchange and consultation. Article 3 states that the Parties "shall exchange information, consultation, research and monitoring; (...)". Article 4 requires the Parties to "exchange information on and review their policies, scientific activities and technical means (...)". Article 5 provides for consultations between Parties emitting air pollution and Parties suffering from the effects of air pollution.

The provision in the Convention on "Research and Development" contains the invitation to cooperate on research and development of techniques for reducing air pollution (Article 7) and the following article is entirely dedicated to

---

[28] The UN Charter mentions the principle as one of its ends. It is not clear whether or not 'good neighbourliness' is a international customary rule or a principle, and whether its foundation is found in Article 21 of the Stockholm Declaration; see Hach (1993), 46–47. Virally (1980), 26,170 *et seq.*, 208 *et seq.*: "Le droit international de voisinage doit être conçu comme *l'ensemble des règles conventionnelles ou coutumières internationales régissant les rapports mutuels entre Etats voisins dans la proportion limitrophes de leur territoires"*.

[29] This was already established in the *Trail Smelter* case (USA v. Canada), in which an international adjudicatory body asserted the principle that a State should not permit the use of its territory in such manner as to cause injury in or to the territory of another *(sic utere tuo ut alienum non laedas)*. Romano (2000), 261 *et seq.*; Hach (1993), 44–55.

[30] Gündling (1991), 93; Hach (1993), 37–82; Kiss and Beurier (2000), 110 *et seq.*

an "Exchange of Information". The following provision on "Implementation and Further Development of the Cooperative Programme" concerns exclusively the implementation of the "Cooperative Programme for the Monitoring and Education of the Long-range Transmission of Air Pollution in Europe" (EMEP), supplemented by the Protocol on Long-Term Financing of the EMEP. Further articles in the Convention address administrative matters, such as the Executive Body, the Secretariat, settlement of disputes, signatories etc.

The protocols, with the exception of the protocol concerning the EMEP, deal with the limitations of emissions of the various substances, national controls and cooperation between the Parties (for example, the Nitrogen Protocol, Article 2 (b), POPs Protocol, Article 5 on "Exchange of Information and Technology").

The LRTAP Convention is mainly based on cooperation and exchange of information, and underlines the themes relied on in the development of international legal frameworks in the 1970s. International environmental protection in global terms started in this decade at a time when Parties still thought it possible to protect the environment by cooperation and information exchange only and did not wish to restrict national sovereignty. States still thought it possible to achieve a satisfactory system of protection of the environment by simple cooperation and transparency.

1.2 The Ozone Protection Regime

The ozone protection regime specifically refers to the precautionary principle as well as Principle 21 of the Stockholm Declaration in the texts of both the Vienna Convention 1985 and the Montreal Protocol. The discussions during the elaboration of the Convention's text were also based on customary law[31], on the prohibition of the abuse of State sovereignty (Principle 6 of the Stockholm Declaration: the prohibition of the discharge of toxic substances), and on the cooperation principle (Principles 22 and 24 of the Stockholm Declaration).

The principle of cooperation is mentioned in the Vienna Convention in Article 2, which states that "Parties shall cooperate by means of systematic observations, research and information exchange in order to better understand and assess the effects of human activities on the ozone layer" and "adopt appropriate legislative or administrative measures and co-operate in harmonizing appropriate policies", and cooperate in order to implement the convention. Article 2 specifies, however, cooperation shall be done by Parties "in accordance with the means at their disposal and their capabilities". The possibilities of developing loopholes on the basis of this provision are large.

The importance of sovereignty is well protected in the Convention. In paragraph 3 of Article 2, it is specified that the provisions shall "in no way affect the right of Parties to adopt, in accordance with international law, domestic

---

31 Neumann (2000), 80 *et seq.*

measures additional to those referred to in paragraphs 1 and 2 above, nor shall they affect additional domestic measures already taken by a Party, provided that these measures are not incompatible with their obligations under this Convention". The cooperation principle concerns all fields: research and systematic observation (Article 3); legal, scientific and technical (Article 4); and the transmission of information (Article 5).

Thus, the ozone protection regime, in its initial shape, relies mainly on cooperation and information exchange, but not to the same extent as in the LRTAP system.

1.3 The Climate Change Regime

As to the emphasis on co-operation, the climate change regime does not differ from the other two regimes. Article 4 of the UNFCCC on "Commitments" is clear on this point. It states that Parties must "... promote and cooperate in the development, application and diffusion, including transfer of technologies, practices and processes (...)" and "promote and cooperate in the full, open and prompt exchange of relevant scientific, technological, technical, socio-economic and legal information related to the climate system and climate change, and to the economic and social consequences of various response strategies" (Article 4.1 (c) and (h)).

Furthermore, the climate change regime introduces a provision for the transfer of support to developing countries in order to permit them to have access to environmentally sound technologies and know-how (Article 4.5). The Kyoto Protocol confirms the co-operation provision in its Article 2.1(b).

The UNFCCC, too, contains general obligations as to co-operation (Art. 4 (1)(c)) and exchange of information (Art. 4 (1)(h), also Art. 5), but this regime is essentially buttressed by the fact that this is also part of the financial transfer mechanism in favour of the developing countries (Art. 4 (3)) and supported by scientific and technical bodies established under the Convention (subsidiary body for scientific and technological advice -SBSTA-; subsidiary body for implementation -SBI-). These obligations are further strengthened by the Kyoto Protocol (see in particular Arts. 10 and 11). This progress may be explained by the fact that the climate regime is the most recent. In the 1970s States tried to resolve environment problems mainly on the basis of voluntary engagements and "soft" cooperative obligations. In the 1990s it became clear that this was insufficient and that the "new" principle of using the best available technologies of the 1980s quickly became a demand for *transferring* this best available technology, above all, to developing countries.

A comparison of the three regimes demonstrates clearly the evolution of international environmental law into a system greatly dependent on the guiding principles of cooperation, information exchange and good neighbourliness.

2 Application of "Best Available Technologies" and its Transfer

The principle of using the "best available technologies" (BAT) can be applied on a national level in a Party's own country (1) or be subject to international transfer (2). The weakness of this principle lies in the question of finance (3).

2.1 Use of BAT at a National Level

In its Article 6, the LRTAP Convention contains *expressis verbis* the obligation to use BAT in order to fulfil the obligations set out in Article 2, which state that "each contracting Party undertakes to develop the best policies and strategies including air quality management systems, and, as part of them, control measures compatible with balanced development, in particular by using the best available technology which is economically feasible and low- and non-waste technology."

The obligation to use economically feasible BAT (that is: *best means*) at the national level, as in the LRTAP Convention, is typical of the conservative conventions of the 1970s, which respect state sovereignty. Without doubt, the obligation presents the possibility for flexibility in order to adapt commitments to new developments as well as to allow compliance with specifications as laid down in the conventions and protocols. However, the prominence of issues of availability and economic viability means that the BAT principle is little more than an "escape clause" providing for a large margin of political manoeuvring. This makes enforcement and control of commitments difficult.

The question arises as to whether or not any real obligation exists in practice. What defines "economically feasible"? Which data should be taken into account in order to assess it? Should the BAT be *feasible* on a national or international level? There is no known case in which these questions have been addressed[32]. Parties have limited themselves to informing each other during official workshops, seminars and meetings of executive organs and "Working Groups on Technologies" on new technologies, their costs and applications.

By comparison, the obligation to transfer BAT as contained in the ozone and climate change regimes is more far-reaching. It not only concerns the use of BAT at a national level in countries with the financial and technological capacity for its implementation, but is also meant to assist poorer countries to use BAT as well.

2.2 Transfer of BAT

For the purpose of this transfer, a financial mechanism is provided through the Multilateral Fund, which will be examined below.

32 Neumann (2000), 23.

The LRTAP regime does not provide for BAT transfer. The first Sulphur Protocol only refers to it in its preamble without mentioning it in its substantive text. More recent protocols[33] contain detailed regulations on technology transfer, but only as "soft rules" that are currently formulated in text such as: "The Parties shall facilitate the transfer ..." and "conform to their legislation". This non-specific formulation may be justified by the fact that technology transfer shall be based equally on public and private initiatives (see for example the POPs Protocol, Articles 5 and 6 on exchange of information and technology and public awareness). In regard to private initiative, the State usually has only a limited influence, thus a binding obligation for the transfer of technologies is not possible under these regulations.

Documents enabling a follow-up concerning the transfer practices include publications of the United Nations Economic Commission of Europe (ECE), reports of the executive organs of the regimes and statements of the organizations and persons involved in the transfer projects. Multilateral technology transfer issues are often discussed in seminars under the direction of the ECE and the effectiveness of the transfers is often influenced by bilateral "target-oriented" seminars.

It is the tendency of the LRTAP regime to be less involved in the dissemination of technological information than the other regimes. Yet, a greater emphasis is placed in the LRTAP regime on the organization of bilateral or small multilateral group meetings with a maximum of four Parties, and involvement by the private sector[34].

Although the ozone protection regime does not contain, *expressis verbis*, an obligation to use BAT, it goes a step further by expressing the general requirement of transferring BAT. Article 10A of the Montreal Protocol states that "each Party shall take every practicable step (...) to ensure that the best available, environmentally safe substitutes and related technologies are expeditiously transferred to Parties operating under paragraph 1 of Article 5". Such a transfer shall respect "fair and favourable conditions".

In the climate change regime the use and the transfer of BAT places a greater focus on developing countries. Article 4.5 of the UNFCCC states that "The developed country Parties and other developed Parties included in Annex II shall take all practicable steps to promote, facilitate and finance, as appropriate, the transfer of, or access to, environmentally sound technologies and know-how to other Parties, particularly developing country Parties, to enable them to implement the provisions of the Convention. In this process, the developed country Parties shall support the development and enhancement of endogenous capacities and technologies of developing country Parties. Other Parties and organiza-

[33] Hanf (1993) speaks of "second generations protocols", 36 *et seq.*
[34] Neumann (2000), 58.

tions in a position to do so may also assist in facilitating the transfer of technologies".

2.3 The Financing and Administration/Implementation of BAT Transfer

The ozone layer protection regime created a Multilateral Fund (FM) for the transfer of technologies (Article 10 Montreal Protocol) in order to enable Article 5 countries to fulfil their commitments. The Fund shall be financed by Article 2 countries. An executive committee was created (Article 10.5) and investment projects may be conceived and executed by implementing agencies, including the World Bank, UNDP, UNIDO and UNEP and others, which receive up to 13% of each project's revenues turnover as commission[35].

The climate change regime also confers the execution of programme financing to international institutions, including the Global Environment Facility, UNDP and the World Bank (Article 21.3 UNFCCC).

In both regimes, specific institutions are responsible at a multilateral level for technology and information transfer. In the ozone layer regime, the Technology and Economic Assessment Panel (TEAP) and the Technical Options Committee (TOC) have this responsibility. In the climate change regime these responsibilities are placed with the Subsidiary Body for Scientific and Technological Advice (SBSTA) and the Intergovernmental Technical Advisory Panels (ITAPs).

The TEAP in the ozone layer regime is responsible for technical and economic reports. It has only an advisory function, but is very important for the development of the regime. The TOCs, which meet regularly in small workshops, are not of great practical importance in spite of the amount of documentation that they produce. In fact, this informal transfer outside of the Multilateral Fund has practically no significance.

SBSTA and ITAPs in the climate change regime support and coordinate the transfer of technologies and know-how. The transfer under the Convention will be financed by a financial mechanism. It shall be non-profit, or shall be granted under preferential conditions, corresponding to the financial resources of the developing countries. This will operate under the direction of international institutions.

The effectiveness, in particular the control and evaluation of technology transfers differs considerably among the three regimes. Transfers in the LRTAP regime are mainly based on seminars. Although important as a form of capacity building, a lack of evaluation of their impact is regrettable[36]. Under the ozone protection regime, technology transfer is to a large extent achieved through the

[35] Neumann (2000), 104.
[36] Neumann (2000), 56.

projects financed by the Multilateral Fund[37]. In this respect, evaluation methods are being developed. The leverage of the Fund is considerable. As to technology transfer undertaken outside the Fund, the inadequacy of controls has given rise to considerable criticism[38].

Article 11 (3) of the UNFCCC provides a control for financial decisions. This "modern" characteristic responds to the lack of control over decisions which has been criticized under the ozone layer regime.

3 Data Reports to Ensure Compliance

Non-compliance with the provisions of a multilateral environmental agreement has special characteristics in comparison to other types of treaties. First, it could harm everyone – individuals, states (Party or not to the treaty) and economies in general. Thus, the principle of reciprocity does not fit well as a reaction to non-compliance: suspending one's own compliance with the treaty will only make the situation worse. In order to prevent or react to non-compliance, most multilateral environmental treaties provide for a variety of measures. A first level consists of promotional measures. Building up a domestic compliance system to implement an environmental treaty requires sufficient financial, administrative and technical resources. Thus, capacity building is the first step to help Parties to comply with their obligations[39].

Second, many environment treaties call for Parties to provide regular data reports. In this context, we may consider three steps: (self-)reporting, verification and assessment and response to non-compliance[40]. The differences are mainly found in the second step: verification of the reported information. All three regimes have each established a committee for this purpose, but with different competences. Concerning assessment and response, the three regimes, as is the case for most environmental treaties, were created as a reaction to the need to facilitate their compliance abilities and to make recommendations. Sanctions in the form of some kind of value deprivation have not played any practical role.

3.1 The LRTAP Convention

Article 9 of the LRTAP Convention provides for a "Co-operative Programme for Monitoring and Evaluation of the Long Term Transmission of Air Pollution in Europe" (EMEP). It has three main purposes: (1) collection of emission data; (2)

[37] Neumann (2000), 102 *et seq.*
[38] Neumann (2000), 111.
[39] Wang and Wiser (2002), 182.
[40] Wang and Wiser (2002), 183.

measurement of air and precipitation quality: and (3) modelling of atmospheric dispersion, using emission data, meteorological data, and information describing the transformation and removal processes[41]. The Convention created a separate Protocol for its financing with the basic objective to allocate the costs of the monitoring programmes.

The EMEP has been in operation since 1977, commencing prior to the adoption of the Convention in 1979. It was one of the first procedures for collecting information that does not depend on self-reporting by states only. Initially, the EMEP was created as an instrument for cooperation among the Member Countries of the UN Economic Commission for Europe and later was integrated as an instrument for verification in the LRTAP regime[42]. As the EMEP is primarily based on data transmitted by states, a cooperative procedure of quality control of these data has been developed and is an important precursor of the regime's compliance procedures.

The EMEP invites Parties to collect data on polluting substances' emissions on a national level. This collection of data is not designed to serve as a control mechanism, but exists in order to record scientific knowledge on prospective pollutant dispersion in the air and on their concentration in deposits and sedimentation. In fact, it is not possible to trace the source of a particular pollutant. Thus, the Convention mentions that the characteristics of the pollution it covers are "at such distance that it is generally not possible to distinguish the contribution of individual emission sources or groups of sources" (Article 1 lit. b).

Article 9 (1)(i) provides that it is desirable to extend the national EMEP networks "to make them operational for control and surveillance purposes".

At the beginning of the 1980s, there were 80 measuring stations in 23 European countries; in 1994 there were 96 stations in 28 countries. The geographical locations of the stations cover Europe with the exception of the Ukraine, Belarus, Romania and Belgium. As there are insufficient data collection stations, data results are simulated using computer models. These evaluations are not very reliable as the quality of measured data differs from one station to another and is based on that which is transmitted by each individual national Party. Thus, beyond discerning a rough divergence, an accurate assessment of the data is not possible[43]. In the case of rough divergences, there will be no possibility to prove which data are really inaccurate. As a result, the EMEP is often not regarded as an effective control programme. Delays in transmission of data by the Parties and a lack of funding constitute additional reasons which render the reports unreliable.

Another method to control the accuracy of the data was the introduction of "Quality Assurance Managers" (QA-Managers). Their mandate is to ensure that

---

41 Sand (1999), 190.

42 Bothe (2002), 85.

43 For details concerning this method, see Neumann (2000), 48 *et seq.*

the statistics transmitted consistently reflect reliable data. "Standard Operating Procedures" (SOPs) have been developed, which are equally valid for all Parties as well as for the EMEP.

3.2 Ozone Layer Protection Regime

The ozone layer protection regime has extensive and detailed obligations (see Article 7 Montreal Protocol) on data reports as many different substances are addressed. There are detailed guidelines and procedural rules concerning the treatment and the presentation of the data that are agreed upon by the Meetings of the Parties (Article 11 (4)(d) Montreal Protocol).

The ozone layer regime established a special type of sanction for developing country Parties in the case of non-compliance with the requirement to submit reports, which may deprive States of their status of being a "developing country" (Art. 5 country) under the regime. This means that they would no longer receive any financial support for preparing their reports[44]. Despite this, the vast majority of the missing reports are from developing countries.

This withdrawal of Article 5 country status is in every case only a last resort measure, and in practice only one country has been deprived of this status (and had it later returned when it subsequently transmitted the report in question)[45].

Developed country Parties are subject to the non-compliance procedure under Article 8 of the Montreal Protocol. The transmission of data reports may be considered as a secondary commitment, arising from the primary commitment of ODS reduction. It may, thus, also be the object of the non-compliance procedure described below.

The system may be considered as a model for cooperative control[46]. It requires, however, that the data arrive in time. In fact, frequently more than half of the reports have not been transmitted on time and often they contain many inconsistencies. For developing countries, an appropriate sanction may often only be possible where financial aid is given for drawing up their reports.

3.3 The Climate Change Regime

Articles 4 (2)(b) and 12 of the UNFCCC regulate the transmission of data reports. The purpose of the reports is, among other things, to assess whether or not Annex I Parties will be able to achieve the first steps set out in Article 4 (2)(a) and (b). The element of discipline in the reporting procedures of the Parties is stricter in the climate change regime than in the two other regimes[47], notwith-

---

[44] Instead of getting money for drawing their report, they should contribute to the financing mechanism previewed in Article 10 of the Montreal Protocol for all other Parties.

[45] Neumann (2000), 88.

[46] Bothe (2002), 86.

[47] Neumann (2000), 135.

standing the reports' complexity and the degree of ambiguity in the regime's rules. This sometimes leads to an unsatisfactory level of discipline in reporting in the upper categories of the reports. In relation to the punctuality of the submission of reports, there is a success rate similar to the other regimes.

The Kyoto Protocol provides for much more elaborate reporting duties and a very strong incentive to comply with them: Parties which have failed to honour their reporting duties are not entitled to participate in the so-called flexible mechanisms, an instrument which, for all practical purposes, makes compliance less costly.

A special characteristic of the climate change regime reports is the "in-depth-reviews" (IDRs) of national registers[48]. This formal and institutional control of data reports is singular in international environmental protection[49] and was decided on by the Parties at their first meeting. The legal foundation for these reports is found in Articles 4 (a) and (d), 7 (2)(a), (d) and (e), 9 (2)(b) and 10 (2) of the UNFCCC. Each national communication is to be evaluated and reported on, at the latest, one year after the report's transmission. The panel that reviews the reports is composed of members of the UNFCCC Secretariat, experts from the Parties, and representatives from non-governmental organizations. The Parties have stated that the in-depth reviews are more helpful than a control mechanism would be and provided that the original report is transmitted on time, it seems that these reviews will be useful[50,51].

3.4 Measures to Ensure Compliance

Progress control can take place in a centralized or decentralized fashion. In the latter case, one of the actors takes the role in the interest of all of them. A typical example is the dispute settlement system used by the World Trade Organization. In the air pollution system we find the centralized system, with an institutionalized organ[52]. Compliance control is undertaken in three steps: collection of information; evaluation of information; and subsequent measures. Most modern environmental treaty regimes provide for an implementation review report. Some of them contain specific individual case-related procedures for the verification of compliance or non-compliance by a Party. These procedures normally are initiated by one or more Parties, the concerned Party itself (in cases of inability to comply), or by the Secretariat of the regime in question[53].

48 Decision 2/CP.1 available at www.unfccc.int.
49 Neumann, (2000), 136.
50 Neumann, (2000), 137.
51 See for example UNEP, IEA, OECD (2001).
52 Bothe (2002), 90.
53 Bothe (2002), 90 *et seq.*

The LRTAP regime as well as the ozone layer protection regime provides for such specific procedures, but the Montreal Protocol, although the later regime, stands as the first example of a specific centralized compliance procedure.

Non-compliance Procedure in the Ozone Layer Protection Regime

The compliance regime of the Montreal Protocol is generally considered a major success[54]. In fact, the regime is very far-reaching concerning its compliance control tools. The compliance organ in the regime is the "Implementation Committee" composed of ten representatives from Parties (see Article 8 of the Montreal Protocol). The Committee can make recommendations to the Meeting of the Parties on the basis of which the Meeting may adopt decisions that concern critical questions[55]. The Committee generally meets twice each year. The compliance procedure can be initiated by the allegation of non-compliance of a Party by one or more other Parties, the Party concerned, or the Secretariat. The task of the Committee is "to identify the facts and possible causes relating to individual cases of non-compliance ..., as best it can, and make appropriate recommendations to the Meeting of the Parties". The objective of the procedure is an "amicable solution ... on the basis of the respect for provisions of the protocol" [56]. If the Parties do not find a solution, further measures can be taken which are enumerated in an indicative list: support for collecting data; technology transfer; financial aid; information transfer; training; the issuance of cautions to the concerned Party; and, suspension of rights and privileges under the Protocol, in accordance with the international rules on suspension of treaties (see Article 60 of the Vienna Convention on the Law of Treaties). Expressly included are rights and privileges that concern industrial rationalization, production, consumption, trade, technology transfer, financing mechanisms (Multilateral Fund) and institutional agreements[57].

Despite praise for the system, it is difficult to measure precisely the practical success of the Committee. The Committee checks the completeness of the reports and pushes the Parties to fulfil their reporting responsibilities; but this alone does not ensure success.

Compliance Procedure in the LRTAP Regime

In resolution 1997/2, the Executive Body established according to Article 10 of the Convention introduced a compliance control procedure for the whole LRTAP regime by creating a special committee. The Committee is composed of

---

[54] Széll (1997), 304.
[55] Széll (1997), 304.
[56] Neumann (2000), 93.
[57] Neumann, (2000), 93 *et seq.*

representatives from seven Parties, each of which must be a Party to at least one of the LRTAP Protocols. The task of the Committee is to supervise the fulfilment of reporting responsibilities, to verify assertions of non-fulfilment, find constructive solutions, obtain expert verification of data, and, if necessary, draft reports.

The Committee was established by a decision of the Executive Body[58]. Whether this is legally binding may be doubtful. But even a legally binding system would probably not contribute more to the implementation of the treaty. In any event, the system is mainly based on cooperation[59]. The more recent LRTAP Protocols (since the VOC Protocol) contain legal bases for the introduction of such a centralized compliance procedure.

The Committee is responsible "for the review of compliance by the Parties with their obligations under the protocols of the Convention" (see Preamble of Resolution 1997/2). Similar to the procedure used under the Montreal Protocol, there are several possibilities to initiate the procedure: the allegation of non-compliance by one Party; written and substantiated notification by one or more Parties; by the Party concerned itself (in case of inability to comply); and initiation by the Secretariat in cases where a Party does not respond after a three month period to a request of the Secretariat on details of a report.

The Secretariat can ask for supplementary information, investigate the situation on the ground in the country in question (with the authorisation of the Party) and verify the information on its own. At the end of the investigation, the Secretariat must report on its findings to the Meeting of the Parties and make recommendations. The decision on whether measures should be taken is left to the discretion of the Meeting of the Parties by consensus (apparently without the vote of the concerned Party)[60].

The measures decided by the Parties must be of a "non-discriminatory nature"[61]. This requirement is a unique feature of the LRTAP procedure. The Montreal Protocol system does not contain such a requirement. As to measures that can be taken (based on the text of Resolution 1997/2), they must respect the legal principle of proportionality in that the measures decided against the Party in question may only serve the purpose of helping that state to fulfil its responsibilities under the treaty. As under the Montreal Protocol, the enforcement measures in cases of non-compliance under the LRTAP regime include the suspension of rights and privileges; but in contrast to the ozone layer protection

[58] The Second Sulphur Protocol already provided a draft concerning a compliance procedure before the Resolution was made. The Executive Body suggested to the Parties of the Protocol to substitute the draft by the centralized procedure. Such a decision is not yet adopted, but the compliance procedure of the Sulphur Protocol was never applied in practice in any event. See Neumann (2000), 66..

[59] See Neumann (2000), 65 *et seq.*

[60] Neumann (2000), 67.

[61] Neumann, (2000), 67.

regime, there do not exist any institutional aides. Other enforcement measures include the suspension of bilateral technology transfers and economic sanctions. However, the suspension of technology transfers may be counter-productive for the purpose of environmental protection[62].

c) Climate Change Regime. The main focus of the UNFCCC is on "implementation" rather than on "compliance" and the main policy instrument to induce compliance in the UNFCCC is capacity building as we saw in the previous section on technology transfer[63]. Nevertheless, Article 13 of the UNFCCC addresses compliance in a larger sense. At first glimpse, Article 13 seems similar to Article 8 of the Montreal Protocol; but in fact, Article 13 is less strict. It provides that "[t]he Conference of the Parties shall, ..., consider the establishment of a multilateral consultative process ...". There are three main differences between the two regimes. Whereas the Montreal Protocol states that its Meeting of the Parties should "*consider and approve*", Article 13 of the UNFCCC only requires its Conference of the Parties to "consider" the establishment of a multilateral consultative process. The UNFCCC provides that the process should be available to Parties at their request, whereas the Montreal Protocol is not so limited. Furthermore, the Montreal Protocol sets out the unambiguous notion of "mechanisms for determining non-compliance ... and for treatment of Parties found to be in non-compliance". Article 13 is vague and its "multilateral consultative process" only is intended to resolve questions regarding the implementation of the Convention[64]. Finally, the multilateral consultative process of the climate change regime is more advisory than supervisory in nature.

The Kyoto Protocol provides for a more elaborate, far-reaching and complex system of substantive obligations. It would not be effective without a more stringent compliance mechanism. Yet, in the negotiating atmosphere reigning at Kyoto, time was not ripe for a consensus on this difficult matter. Article 18 KP on compliance only contains an enabling provision for the COP/MOP to adopt relevant secondary norms. The essential text has been agreed upon by the so-called Marrakech Accords of November 2001. They create a new institution, the Compliance Committee.

The UNFCCC Secretariat in Bonn will serve as the Secretariat for the Committee[65]. The Committee will be composed of two branches[66]: a facilitative branch and an enforcement branch. The facilitative branch will assist all Parties, while the enforcement branch will determine whether Parties have:

---

62 Neumann (2000), 68 *et seq.*
63 Wang and Wiser (2002), 184.
64 Széll (1997), 305.
65 Ibid., 77, section XVII.
66 Ibid., 67, section IV.

- met their emissions targets;
- complied with their monitoring and reporting obligations; and,
- met the eligibility tests for participating in the flexible mechanisms[67].

Compliance proceedings under the Committee may be triggered by:

- the Party concerned;
- any other Party; or,
- the expert review team charged to evaluate and assess data submitted by the Parties according to Article 8 KP.

Under the procedure used by the climate regime, the Secretariat cannot trigger compliance proceedings.

The system will be based on the principles of due process, predictability and impartiality, and response measures will be applied in a graduated manner based on the principle of proportionality[68]. A preliminary examination of cases before the Committee will be undertaken in order to decide if the issue has sufficient merit to be pursued.

In terms of participation in the process, interested third Parties may file with the Committee information that is relevant to the case. There is even the possibility of public participation. Intergovernmental and non-governmental organizations will be permitted to submit technical and factual information to the Committee. Compliance hearings will be, with possible exceptions, open to the public and information on the proceedings may be made publicly available by the UNFCCC Secretariat[69].

The Party concerned can be represented during the proceeding by a lawyer or other representative and in cases concerning compliance with emissions targets, an appeal procedure is available. Concerning decisions of the enforcement branch, relating to Article 3(1) of the Protocol, the COP/MOP will hear the appeal[70].

Thus, the climate change regime possesses a two-pronged system for ensuring compliance: a soft consultative system for all Parties to the UNFCCC and a strict one for the Parties to the Kyoto Protocol. As to developed countries which have quantified emission limitation or reduction obligation under the KP, there are still stricter sanctions available.

---

67 Ibid., 68, section V.
68 Wang and Wiser (2002), 192.
69 Wang and Wiser (2002), 190.
70 Marrakech Accords (see above note 84), 74, section XI, para 3.

4 Conclusion

The three regimes are linked in several aspects. The fields they cover overlap even if every system tries to be delimited from the others. It appears nevertheless that every area addressed by one Convention may also be a concern of one or both of the others. For instance, the problem of climate change is also caused and worsened by substances dealt with under the LRTAP Convention and the Vienna Convention of 1985. In fact, scientifically, it is difficult to distinguish among air pollution issues in the different conventions.

Less ozone in the stratosphere hampers the photosynthesis of plants[71] and so hinders the functioning of sinks in regard to CO_2 abatement. Conversely, global warming contributes to the impoverishment of stratospheric ozone through the modification of the atmospheric temperature structure caused by an accumulation of greenhouse gases. On the one hand, theses gases warm up the ground, on the other hand, they cool off the ground-near stratosphere. When methane reaches the stratosphere, its oxidation forms water vapour. This combination of water vapour and reduced temperature in the stratosphere can render the formation of clouds in the polar stratosphere more likely, which, in turn, increases the capacity for the destruction of chlor-ozone in the stratosphere[72].

While the LRTAP Convention and its protocols and the ozone protection legal framework are related, both have also important influences on the climate change regime. It would be impossible to have an efficient climate change regime without respecting the two others. Furthermore, by reducing the greenhouse gases responsible for global warming, the issues of acid rain and the ozone hole are being fought at the same time.

Most global damage is not traceable to a single cause, but rather to a host of interacting disturbances to the global equilibrium. However, it appears difficult to include all pollutants in order to be exhaustive in addressing these environmental problems. The issues of acid rain and ground-level ozone alone require the regulation of more than eighty chemical substances. It is also clear that some substances are easier to deal with than others[73]. In the climate change regime, the negotiations have concentrated above all on CO_2, leading to the public sometimes forgetting that other GHGs exist.

Furthermore, the three regimes are based on the same principles, where the use of BAT and the demand for transparency of information play crucial roles. All three regimes provide for the transfer of BAT and require regular data reports

---

[71] Frese (1987), 576.

[72] Sherwood Rowland (1997), 6.

[73] For instance, sulphur is easier to manage than nitrogen. Its effects are well understood and most emissions can be controlled by regulating the combustion of fuels in power generation and large industrial plants. Hanf (2000), 24.

as a control method. However, all the regimes suffer the same weaknesses: the transfer of technology and the preparation of data reports have to be financed, and data reports must be punctual and accurate. Also, if all three regimes provide financing mechanisms, the effective transfer of BAT will be deficient. There is generally no possibility of follow-up control regarding whether the BAT in the country receiving technology is applied or not. Short-term economic advantages are still the guiding reasons for preferring one technology to another. Data reports should get this weakness under control, but, in practice, they are often not significant or valid enough and are rarely transmitted in time.

Even if compliance with reporting obligations under the ozone layer protection regime is relatively high[74], good reporting is only the first step towards the goal of the treaty as the fulfilment of the reporting duty does not in and of itself ensure the protection of the ozone layer. This is surely a general problem of environmental treaties that require regular reports. Even strict compliance with these requirements does not guarantee the effectiveness of the treaty. Complete and regular reporting does not ensure the reduction of emissions.

A similar conclusion may also be reached regarding compliance with the general provisions of the three framework conventions. They are so vague that any small progress achieved as to the general regulations may be called "compliance"[75]. One thing is sure. Taking into account the existence of acid rain and depletion of the ozone layer, the vast scientific research and debate on climate change as well as the possible benefits of the various mechanisms for emission reductions (including the use of carbon sinks) are presented in another light. A reduction in the use of fossil fuel and natural resources is needed, if not to mitigate climate change, then to reduce acidification. The use of environmentally *best available technologies* is desirable, if not for the mitigation of climate change or acidification, than for a reduction in the depletion of the ozone layer. Policy makers seem to forget the interaction of the commitments of the different environmental treaties to which they belong. Thus, discussions on natural resources, such as fossil-fuel energy, and on emissions of pollutants should always be held with a comparative point of view.

An analysis of the regimes' historical developments demonstrates a need for greater regulation and incentives in environmental protection regimes. The goodwill and optimism of the 1970s has been replaced by the realisation that even the existence of sanctions is not sufficiently persuasive to ensure full participation and engagement in the treaties. It can be hoped that further analysis of the mechanisms for monitoring and evaluating compliance will lead to programmes that are more precise and binding, with the desired result of an improved rate of compliance with the legal rules protecting the atmosphere.

---

[74] Bothe (1996), 23.

[75] Bothe (1996), 15.

Probably, the future of strict compliance with environmental agreements will lie in flexible instruments and the inclusion of individual actors in international regimes. Whereas state responsibility only plays a marginal role in international environmental protection, civil liability for transboundary pollution may gain more importance[76]. Polluting activities are mainly not state enterprises, but are conducted by private actors and the capacity of states to control these private actors which are polluting the environment is limited. Too much state control may in any event not be desirable.

Compliance control on reporting duties will remain of immense importance. Flexible instruments, such as emissions trading, can only operate effectively if there is proper control of data. The climate change regime is the best example. The three Kyoto flexible mechanisms pose novel challenges to the Kyoto Protocol's compliance regime and vice versa. One of the key purposes of the compliance system is to instil confidence and integrity in the Kyoto flexible mechanisms[77].

References

Bothe, M. (1996). 'The Evaluation of Enforcement Mechanisms in International Environmental Law', in Wolfrum, R. (ed.), *Enforcing Environmental Standards: Economic Mechanisms as Viable Means?*. Berlin: Springer Verlag, 13–22.

Bothe, M. (1997). 'Compliance Control beyond Diplomacy – the Role of Non-Governmental Actors', *Environmental Policy and Law*, 27/4: 293–297.

Bothe, M. (2002).'Vollzugsdefizit im Völkerrecht – Überlegungen zu 30 Jahren Umweltrecht', *Tradition und Weltoffenheit des Rechts*, Berlin: Springer, 83–96.

Earth Negotiations Bulletin (2002). Volume 19, No. 18/29 July 2002. International Institute for Sustainable Development. http://www.iisd.ca

Frese, W. (1987). Menetekel nicht nur am südlichen Himmel: Ozonlöcher', *Selecta*, 10: 576.

Gündling, L. (1991). Protection of the Environment by International Law: Air Pollution', in Lang, W., Neuhold, H., Zemanek, K. (eds.), *Environmental Protection and International Law*, London: Graham & Trotman, 91–114.

Gupta, J. (2000). *On behalf of my delegation*. Center for Sustainable Development of the Americas and International Institute for Sustainable Development.

Hach, R. (1993). *Völkerrechtliche Pflichten zur Verminderung grenzüberschreitender Luftverschmutzung in Europa*. Göttingen: Heymann.

Hanf, K. (2000). 'The Problem of Long-Range Transport of Air Pollution and the Acidification Regime', in Underdal, A. and Hanf, K. (eds.), *International Environmental Agreements and Domestic Politics: The Case of Acid Rain*. Aldershot: Ashgate Publishing Ltd, 21–48.

Heß, R. (1878). 'Schäden durch Hütten- und Steinkohlerauch an Nadelhölzern' in Ministry of Agriculture, NRW, Germany, *Saurer Regen – Gefahr für unseren Wald*, 74 et seq.

---

[76] Bothe (2002), 96.
[77] Wang and Wiser (2002), 195.

Kiss, A. and Beurier, J.P. (2000). *Droit International de l'Environnement*. Paris: Pedone, 2nd ed.
Neumann, M. (2000). *Die Durchsetzung internationaler Umweltpflichten*. Baden-Baden: Nomos Verlagsgesellschaft.
Romano, C. P.R. (2000). *The Peaceful Settlement of International Disputes*. London, The Hague: Kluwer Law International.
Sand, P.H. (1999). 'Regional Approaches to Transboundary Air Pollution', in Sand, P.H. (ed.), *Transnational Environment Law – Lessons in Global Change*. Kluwer Law International, 189–213.
Sand, P.H. (1999). 'Protecting the Ozone Layer', in Sand, P.H. (ed.), *Transnational Environment Law – Lessons in Global Change*. Kluwer Law International, 207–213.
Sherwood Rowland, F. (1997). 'Evolution de l'Atmosphère'. *Notre Planète (UNEP)*, 9/2: 4–6.
Sonnemann, G. (1986). *Ist unsere Atmosphäre noch im Gleichgewicht?*. Köln: Aulis Verlag.
Széll, P. (1997). 'Compliance Regimes for Multilateral Environment Agreements – A Progress Report'. *Environmental Policy and Law*, 27/4: 304–307.
Trunko, L. (1987). Schadsymptome an Waldbäumen: Sind Aussagen über die unmittelbare Ursache möglich?'. *Natur und Museum*, 117/9: 278–288.
United Nations Environment Programme (UNEP) (1996). *Handbook for the International Treaties for the Protection of the Ozone Layer*. Ozone Secretariat.
UNEP, IEA, OECD (2001). *Technology without Borders, Case Studies of Successful Technology Transfer*.
Virally, M. (1980). *Voisinage et Bon Voisinage en Droit Internationl*. Paris: Pédone.
Wang, X. and Wiser, G. (2002). 'The Implementation and Compliance Regimes under the Climate Change Convention and its Kyoto Protocol'. *RECIEL* 11/2: 181–198.

All Treaties and Protocols under: http://www.fletcher.tufts.edu/multi/texts
A glossary on technical and scientific terms concerning acid rain and ozone can be found under: http://www.fluorocarbons.org/g-info/glossary/glossary.htm

Chapter 4

JÜRGEN LEFEVERE[1]

Greenhouse Gas Emissions Trading: A Background

Introduction

After President Bush's rejection of the Kyoto Protocol to the United Nations Framework Convention on Climate Change (Kyoto Protocol) in March 2001, many believed it to be the end of the most hotly debated international environmental agreement in recent history. Developments since March 2001 have, however, shown quite the contrary. On 23 July 2002 the entire international Community, without the United States, resolved key outstanding differences on the implementation of the Protocol with the conclusion of the 'Bonn Agreements'[2]. This agreement allowed the adoption of the 'Marrakech Accords' in the early morning of 10 November 2002[3], containing the detailed texts needed to implement the provisions of the Kyoto Protocol. With the resolution of the outstanding issues on the implementation of the Protocol, the path has been cleared for key developed countries to ratify. On 4 March 2002 the 15 Members of the European Union decided to ratify the Kyoto Protocol before 1 June 2002, in line with their previous commitment to allow its entry into force at the Word

---

[1] At the time of writing Jürgen Lefevere was Programme Director of the Climate Change Programme of the Foundation for International Environmental Law and Development (FIELD), London. He has been involved in a number of studies for the European Commission's Environment Directorate General that have laid the foundations for the Green Paper on Emissions Trading (March 2000), the proposal for a Directive establishing a scheme for greenhouse gas emission allowance trading within the Community (EATD Directive) (October 2001) and the proposal for a Directive to link the EATD directive with the project-based mechanisms (July 2003). Part of this chapter is an updated version of a contribution to the 2001 Yearbook of European Environmental Law.

[2] Decision 5/CP.6, The Bonn Agreements on the implementation of the Buenos Aires Plan of Action, FCCC/CP/2001/5, 36–49.

[3] Report of the Conference of the Parties on its Seventh Session, held at Marrakech from 29 October to 10 November 2001, FCCC/CP/2001/13 of 21 January 2002, and add. 1–4.

MICHAEL BOTHE AND ECKARD REHBINDER (ED.), Climate Change Policy, 103–129.

Summit on Sustainable Development in Johannesburg in August/September 2002. On 31 May the EU adopted its ratification decision and the European Community and its Member States jointly submitted the ratification instrument to the UN[4]. The US having rejected the Kyoto Protocol, and Japan having ratified the Protocol on 4 June 2002, ratification by Russia is now crucial for its entry into force[5].

The Kyoto Protocol, and in particular its introduction of the concept of 'flexible mechanisms' as a tool to tackle climate change, has provided an important impetus to the development of an innovative and, especially in the EU, largely untested instrument of environmental regulation: emissions trading. With the adoption of the EU's emission allowance trading Directive (EATD) in October 2003, discussed in Chapter 9, this instrument has, however, been given a central role in the EU's climate change policy[6].

This chapter discusses the background to the concept of emissions trading and how it was introduced in the Kyoto Protocol. It will first explain the concept of trading as it is set out in the Kyoto Protocol. It continues by giving a brief overview of the EU burden sharing agreement and its negotiating history, before providing an overview of the concept of emissions trading, as well as its ethical dimensions.

2 What is Emissions Trading?

The term 'emissions trading', although most commonly used, does in fact not correctly describe the instrument. It is not the emissions that are being traded, but the right to emit a particular quantity of greenhouse gases, laid down in a permit, credit or allowance. The title of the EATD, further discussed in Chapter 9, which refers to 'greenhouse gas emission *allowance* trading' is therefore more accurate.

All approaches to trading are based on the same concept: a target is given to each source. In most trading regimes the targets are set by the regulator, not the market. If a source does better than this target it can sell its surplus allowances. If it does worse, then it has to buy allowances on the market. The source will base its

---

[4] Council Decision 2002/358/EC concerning the conclusion, on behalf of the European Community, of the Kyoto Protocol to the United Nations Framework Convention on Climate Change and the joint fulfilment of commitments thereunder.

[5] Article 25(1) of the Kyoto Protocol states that it 'shall enter into force on the ninetieth day after the date on which not less than 55 Parties to the Convention, incorporating Parties included in Annex I (developed countries) which accounted in total for at least 55 per cent of the total carbon dioxide emissions for 1990 of the Parties included in Annex I, have deposited their instruments of ratification, acceptance, approval or accession'.

[6] Directive 2003/87/EC of the European Parliament and of the Council of 13 October 2003 establishing a scheme for greenhouse gas emission allowance trading within the Community and amending Council Directive 96/61/EC, OJ L275 of 25.10.2003, p. 32.

decisions on whether to buy or to sell allowances on the market price of the allowances and its marginal costs of abatement. If the market price is higher than the marginal costs to reduce emissions at the source, the source will choose to reduce its emissions further and sell the allowances that are freed up by doing so. If the market price of allowances is lower than the marginal costs to reduce emissions at the source, then the source will choose not to reduce its emissions, but maintain its emissions or even buy allowances on the market to increase its emissions. A well functioning trading regime will level the marginal reduction costs across all sectors of industry, by allowing sources with high marginal reduction costs to invest in reductions in sources with lower marginal reduction costs through buying allowances freed up by these sources. By allowing sources to optimally use all cheap abatement options it can significantly lower the compliance costs and ease the achievement of targets. The potential benefit depends upon differences in the marginal cost of reducing emissions among participating sources due to the ability to use different control options, remaining life of the facility, or other reasons. To realise the potential savings, the trading programme must include enough buyers and sellers to create a competitive market.

A well-designed emission allowance trading programme shifts the location and the timing of the emission reductions, but ensures that the target is achieved. The programme design must ensure that such shifts do not create environmental problems, such as local pollution 'hot spots'. Interest in emissions trading has grown in part because the cost of compliance with environmental regulations has grown, so the potential cost savings (often 50% or more) have become more significant. In addition environmental regulations have shifted from very local issues, such as water pollution, which often have only one or a few sources, to regional and global concerns where the number of sources is large and the local impacts of shifts in emission reductions are not significant. These environmental problems are better suited to emissions trading.

3 Market-based Instruments: A Complement Rather Than a Substitute

Emission allowance trading is often seen as one of the prime examples of a 'market-based mechanism'. The use of market-based mechanisms has for some time now been promoted by academics and critics of the current regulatory system as an alternative to the 'out-dated' 'command and control'-type legislation. Proponents of market-based mechanisms argue that the current approach, which brings emissions under government control through permitting and stringent monitoring requirements, is too rigid, fragmented, costly, bureaucratic, not transparent and fails to stimulate innovation[7]. Market-based mechanisms, they argue, would achieve the same level of pollution reduction as traditional instruments, but do so at lower costs, reduce government intervention, provide incentives for technology development and in some cases even generate additional revenue for the government[8].

Discussions on the introduction of market-based mechanisms in the EU have for a large part been fuelled by the experiences in the United States[9]. In Europe, both Member States and the EU have however, until recently, been reluctant to introduce these mechanisms. Attempts to do so in the last few years have largely focused on an increasing use of environmental taxation as an instrument to reduce fossil fuel and energy consumption and reduce greenhouse gas emissions[10]. Proposals to introduce environmental taxation at the Community level have, however, for a long time been thwarted by the EC Treaty's requirement that these measures must be adopted on the basis of unanimity among the Member States[11]. This does not mean that no regulatory innovation has taken place. More popular 'second generation' regulatory instruments have been the 'environmental', 'voluntary' or, better, 'negotiated agreements', which are increasingly used at both national and EU level[12]. Negotiated agreements can, however, not be classified as 'market-based instruments', but are rather attempts to build more flexibility into the current regulatory framework. They could thus be labelled as 'negotiated command and control instruments'[13].

---

[7] For an overview of the various new approaches to environmental regulation and how they fit into existing approaches, albeit more from a US approach, see Stewart (2001). See also the overview of literature pro and contra the use of new instruments in footnote 1 of that article.

[8] Johnson (2001).

[9] For an analysis of the two major trading schemes in the United States, see Schwarze and Zapfel (2000).

[10] Countries that have introduced different forms of taxation include the United Kingdom, Germany, Austria, the Netherlands and Belgium.

[11] Proposal for a Council Directive introducing a tax on carbon dioxide emissions and energy, COM(1992) 226, 30 June 1992; Amended proposal for a Council Directive introducing a tax on carbon dioxide emissions and energy, COM(1995) 172, 10 May 1995; and Proposal for a Council Directive restructuring the Community framework for the taxation of energy products, COM(1997) 30, 12 March 1997. After agreement in Council in March 2003 this Directive was finally adopted in October 2003. Council Directive 2003/96/EC of 27 October 2003 restructuring the Community framework for the taxation of energy products and electricity, OJ L 283 of 31.10.2003, p. 51.

[12] For an overview of recent agreements see Barth and Dette (2001).

[13] Stewart (2001), n. 7 above, 60.

Recent experience with the implementation of market-based mechanisms in both the US and in Europe has, however, shown that simply replacing existing 'old' regulatory instruments with new market-based instruments is often not an option, for a number of reasons. The investment in designing and implementing the 'old' instruments is often considerable; permitting regimes have been in place in most Member States since the early 1970s. These regimes have been considerably improved over the years. The introduction of integrated permitting in a number of Member States from the mid-1980s[14] and the Community-wide consolidation of this concept in the Directive on Integrated Pollution Prevention and Control (IPPC Directive)[15] has made the regime more sensitive to cross-sector pollution and introduced attention to a source's impact on the environment as a whole. The combination of permitting with general binding rules, allowed on the basis of Article 9(8) of the IPPC Directive, and the increased use of negotiated agreements[16] have introduced more flexibility and reduced the costs of achieving the targets. In many cases both the regulated sectors and the regulator are of the opinion that the system works and therefore do not see the benefit of a major overhaul of the rules.

It is thus not surprising that recently proposed and introduced market-based instruments can mostly be found in 'new' areas in which behaviour was previously either unregulated or in areas where the current set of regulatory tools has been found insufficient to reach the goals set by the regulator. Climate change is one of those 'new' areas. It is therefore in this area that, within the EU, the instrument of emissions trading has also found its broadest application. Before the adoption of the EATD it was already implemented in Denmark and the United Kingdom, and trading regimes were in various stages of development in other Member States[17].

But perhaps the most important argument that shows that simply replacing existing 'old' regulatory instruments with new market-based instruments is often not an option follows from recent experience with the design and implementation of emissions trading regimes. It is increasingly clear that market-based mechan-

---

[14] Such as the integrated environmental permit based on the Dutch Environmental Management Act and the Integrated Pollution Control permit based on the UK's Environment Act.

[15] Council Dir. 96/61/EC Concerning Integrated Pollution Prevention and Control, [1996] OJ L257/26.

[16] The use of negotiated agreements for the implementation of EC environmental law is, however, limited, and must usually be backed-up by a regulatory framework. See Communication from the Commission to the Council and the European Parliament on Environmental Agreements COM(96) 561, 27 Nov. 1996.

[17] See the various contributions in the Special Issue on Emissions Trading (2000). A regularly updated overview of the various trading initiatives can also be found on the UNCTAD homepage: http://www.unctad.org/ghg/etinfo/etinfo.htm.

isms are no substitute for legal controls on conduct, backed up by effective government enforcement and sanctions. Instead, they are designed to complement, rather than substitute, command and control measures. Most market-based mechanisms indeed rely for their success upon an underlying program of government regulatory control[18]. Emission allowance trading is a prime example of this, as will become clear from the rest of this chapter. To function well, emission allowance trading is in practice often built on top of existing permitting regimes. Emissions trading may thus even be described as a 'command and control *plus*' instrument, with often even stronger government oversight, in particular in relation to the monitoring and reporting of emissions, and high non-compliance sanctions. The true value of this market-based instrument therefore lies not in replacing the existing command and control regimes, but in building flexibility, cost effectiveness and incentives for technology development into those regimes.

4 International Emissions Trading: the Kyoto Protocol

Probably the most important reason for the current popularity of emission allowance trading as an instrument to reduce greenhouse gas emissions is its inclusion in the Kyoto Protocol. During the negotiations on the Protocol, international emissions trading was strongly promoted by a group of countries named the Umbrella Group[19], led by the United States. The European Union resisted the inclusion of international emissions trading in the Protocol, and instead favoured the inclusion of a binding list of different policies and measures[20]. The irony is that the United States has now rejected the Protocol, the design of which very much follows the US ideas, whereas the European Union and its Member States are furthest advanced in the development and implementation of emissions trading as an instrument to combat climate change.

Annex B to the Kyoto Protocol sets specific targets for the reduction or limitation of the emissions of greenhouse gases by developed country (Annex I) Parties. These targets apply to the period 2008–2012 (the first commitment period) and are based on a percentage of a country's greenhouse gas emissions in 1990. A pre-condition for the United States accepting this cap on developed country emissions was the inclusion of emissions trading, as part of a set of

---

[18] Johnson (2001), n. 8 above, 422 and Schwarze and Zapfel (2000), n. 9 above, 293.

[19] The Umbrella Group is a non-official group of countries that often have coordinated positions, but all speak for themselves. Its membership is not fixed. Members of the group include the United States, Russia, Japan, Canada, the Ukraine, Australia and also Iceland, Norway and New Zealand, although the latter three have since 2001 occasionally worked together as the informal Nizeland Group.

[20] Lefevere (2000a); Grubb, Vrolijk and Brack (1999).

'flexible mechanisms'. These flexible mechanisms include the Clean Development Mechanism (CDM) (Article 12), Joint Implementation (JI) (Article 6) and International Emissions Trading (IET) (Article 17)[21]. On the insistence of the European Union, the Kyoto Protocol also allows Parties to group together to form an emissions 'bubble' under a common target, re-dividing this target among its members (Article 4).

All three flexible mechanisms are forms of emissions trading. JI and CDM are 'project-based mechanisms', allowing for the generation of credits for reductions made by investments in specific projects. Rather than reducing emissions 'at home', these mechanisms allow investments in reductions or sequestration (storage of greenhouse gases in for instance trees) 'abroad'. Reduction credits are generated through comparing actual emissions of those projects with the 'baseline' emissions, the emissions that would have occurred in the absence of the project. The credits generated by these projects can be used by developed countries towards the fulfilment of their emission reduction or limitation commitment under the Kyoto Protocol. The difference in design between these two instruments is that JI projects take place within developed countries, that all have a target, whereas CDM projects take place in developing countries, that have no limit on their greenhouse gas emissions.

The Article 4 'bubble' allows a group of developed countries to agree on a common reduction target and subsequently redistribute this target among the different countries. In the negotiations a number of Parties argued that the Article 4 can also be seen as a form of emissions trading. The EU has, however, consistently (and correctly) argued that this is not the case. Rather, it concerns a redistribution of the targets in the Kyoto Protocol, which is done without direct financial compensation, but for development, equity or solidarity reasons. When the group of countries as a whole fails to meet the bubble arrangement, each of its members will be held to their individual target under the bubble. The EU has formed such a bubble.

The 'purest' form of international emissions trading is based on Article 17 of the Kyoto Protocol. Due to time pressure in the last night of the negotiations in Kyoto, this Article only lays the barest of foundations for the use of this mechanism, stating that:

---

[21] See for an in-depth discussion of the different flexibility mechanisms: Missfeldt (1998); Grubb (1998) and Werksman (1998).

> The Conference of the Parties shall define the relevant principles, modalities, rules and guidelines, in particular for verification, reporting and accountability for emissions trading. The Parties included in Annex B may participate in emissions trading for the purposes of fulfilling their commitments under Article 3. Any such trading shall be supplemental to domestic actions for the purpose of meeting quantified emission limitation and reduction commitments under that Article.

The elaboration of the rules for the implementation of this provision was one of the key topics in the negotiations after the adoption of the Kyoto Protocol. These rules were finalized with the adoption of the Marrakech Accords in November 2002[22]. Issues that needed to be resolved in these negotiations included 'supplementarity', 'hot air', eligibility, liability, legal entity participation and 'fungibility' – these issues and their resolution also play a role in the design of the EATD and are therefore worth a brief analysis.

The issue of 'supplementarity' and 'hot air' are closely related, and were among the most controversial topics during the negotiations. Supplementarity relates to the extent to which a country can use the flexible mechanisms, i.e. buy emission reductions that were achieved in a different country, to achieve its targets. Many countries, including the EU, argued that the primary goal of the Kyoto Protocol was to stimulate domestic reductions within the developed countries. Excessive use of the Kyoto mechanisms would allow these countries to buy their way out of their obligations, without actually making any reductions domestically. This could not only be a blow for technological innovation, but in combination with the 'hot air' issue could even allow an increase in global emissions.

'Hot air' relates to the large 'surplus' limitation targets given to Russia and some of the other Economies in Transition (EiTs). Due to the economic collapse in Russia and the large reduction in emissions caused by the transition to a market economy in EiTs, these countries are likely to have large excesses in emission credits to sell under the flexible mechanisms, in particular under emissions trading. Now the US has decided to stay out of the Kyoto Protocol, the balance between 'supply' of credits from the three mechanisms may even outweigh the 'demand' in developed countries. The transfer of this oversupply in credits, representing emissions which would otherwise not take place, to other developed countries could increase the global greenhouse gas emissions and lead to a collapse in the international price for these credits, thus discouraging real reductions[23].

---

[22] N. 3 above. The rules on emissions trading can be found in Decision 18/CP.7, Modalities, rules and guidelines for emissions trading under Article 17 of the Kyoto Protocol, FCCC/CP/2001/13/Add.2, 50–54.

[23] See Grubb, Hourcade and Oberthür (2001). The report quantifies the impact of US withdrawal on carbon prices and the Kyoto Protocol's environmental goals.

The supplementarity issue was only partly resolved in the Bonn Agreements, reached in July 2001[24]. Countries agreed that 'domestic action shall thus constitute a significant element of the effort made by each Party included in Annex I (developed countries) to meet its quantified emission limitation and reduction commitments under Article 3, paragraph 1' and Parties are required to report on their implementation of the supplementarity principle[25]. The hot air issue was not officially resolved. Behind the scenes there are, however, discussions on the idea that Russia and other EiTs 'recycle' revenues received from emissions trading, using these for renewable energy and energy efficiency projects. Such a 'Green Investment Scheme' could provide *de facto* restrictions on the amount of surplus emission rights that could be sold at any one time and ensure that real reductions are achieved as a result of the purchase of those rights[26].

The discussion on the eligibility requirements for participating in emissions trading focussed on the need to ensure certainty about a country's ability to account for its emissions. Without information on what a country's emissions are, the risk of overselling credits that do not represent genuine reductions or surplus credits can be too high. To allow a trading system to function, countries must have in place the necessary accounting systems and registries to account for these trades. The eligibility issue was finally settled in Marrakech, where Parties agreed on a list of requirements, including the requirement to have in place a national system to estimate its emissions, a national registry, and requiring Parties to submit specific information in their annual inventories[27]. Although participation in the Kyoto Protocol's compliance regime is no longer an explicit requirement, this requirement follows implicitly from the procedure for the suspension of a Party's eligibility to participate in emissions trading[28].

The liability issue was resolved in the Bonn Agreements. The EU and the developing countries initially advocated a system of 'buyer liability'. The fear was that countries could recklessly oversell their credits and find themselves in non-compliance at the end of the first commitment period. By using a system of 'buyer liability', the risk of overselling would be borne by the buyer of the credits. The buyer would thus have an incentive to check whether credits bought represented genuine reductions, or accept the risk, which would likely be translated in a lower price for 'risky' credits. The Umbrella Group on the other hand advocated a system of 'seller liability'. Following this approach, the credits bought on the

---

24 N. 2 above.

25 See also FCCC/CP/2001/13/Add.2, 2 and FCCC/CP/2001/13/Add.3, 27, para. 21.

26 See Tangen (2002).

27 FCCC/CP/2001/13/Add.2, 52–53, para. 2.

28 This requirement was deleted in the last night in Marrakech on the insistence of Japan, who did not want to pre-empt the discussion on the legally binding nature of the compliance decision. The compromise text can be found in FCCC/CP/2001/13/Add.2, 53, para. 3 and at page 4, para. 5.

market would be valid, even if its seller was in non-compliance. The consequences of overselling would be carried by the seller. This approach is also followed in virtually all domestic trading regimes. The difference between domestic trading regimes and international emissions trading is, however, that domestic regimes usually operate under strict monitoring, reporting and verification requirements and have strong and enforceable penalties for those who are in non-compliance. Within such a regime the risk of overselling is minimal, and if it occurs it is unlikely to have a large impact on the market in view of the large number of buyers and sellers. This is, however, different in a global regime. The number of countries that are allowed to trade is more limited, and there is a risk that one country can oversell large quantities. Under the Kyoto Protocol not only the monitoring, reporting and verification requirements are more cumbersome and include more uncertainty and margins for interpretation, but also the compliance regime is less effective, lacks strong penalties and it takes a long time before the non-compliance is actually established[29]. The compromise that was finally found is based on a seller-liability system, but contains a 'safety-net' in the form of a 'commitment period reserve'. This commitment period reserve limits the net amount of credits that a developed country can sell to 10% of its 'assigned amount' (the total amount of greenhouse gas emissions in CO_2 equivalent that a country is allowed to emit during the first commitment period) or to any credits that it holds in surplus of five times its most recently reviewed annual greenhouse gas emissions inventory[30].

The issues of legal entity participation and 'fungibility' related to the different perceptions of the character of the flexibility mechanisms. Key developing countries, with India and China at the forefront, saw the use of those mechanisms restricted to trades between sovereign states, and only on an occasional basis. In their view, once a trade took place, a country could use the credits it purchased for compliance, but not sell these credits on to other countries. A too liquid market could also contribute to violations of the supplementarity principle and lead to

[29] The Kyoto Protocol's compliance regime requires countries that are in non-compliance to draw up a compliance action plan and requires them to reduce their surplus emissions in the subsequent commitment period at a rate of 1.3 times their compliance gap. See Decision 24/CP.7, Procedures and mechanisms relating to compliance under the Kyoto Protocol, FCCC/CP/2001/13/Add.3, 64–77. For an elaborate overview of the Kyoto compliance regime and its history see Lefeber (2001).

[30] See for a more detailed analysis of the background and functioning of the commitment period reserve Haites and Missfeldt (2001), Missfeldt and Haites (2002).

larger developed country claims to emission rights in future commitment periods[31]. In particular the Umbrella Group believed the opposite. These countries saw an active participation of companies and traders in a large and liquid international market in credits. On this market, credits from CDM and JI projects could be sold on to other companies and traders or to countries. While both Article 6 (JI) and Article 12 (CDM) of the Kyoto Protocol refer to the possibility for countries to authorize legal entities to participate in these mechanisms under the responsibility of that country, Article 17 (IET) is silent on the issue of legal entity participation. In July 2001 Parties agreed in Bonn to allow legal entities to participate, under the responsibility of the authorizing country[32]. The fungibility question was more difficult to resolve. In Marrakech a solution was found that addressed the fear of developing countries that the excessive use of trading would increase a country's claim to emission rights in future commitment periods, but at the same time creates a large degree of fungibility between the credits generated under the various flexible mechanisms[33].

The Kyoto Protocol in combination with the Marrakech accords thus lays the foundation for international greenhouse gas emissions trading. The international rules allow legal entities to participate in this trading, but this is done under the responsibility of the countries that authorize them to do so. The international rules do not elaborate how this authorization takes place, and how countries control the participation of legal entities under their responsibility. This requires the further elaboration by those countries wishing to do so, either individually at the domestic level or collectively at a regional level. The various proposals for domestic emissions trading that are currently being elaborated and implemented can be used for this purpose.

5 The EU Burden Sharing Agreement

The previous paragraph referred to the Article 4 'bubble' provision in the Kyoto Protocol. Before starting a more in-depth discussion of the concept of emissions trading, it is useful to give a brief overview of the implementation of this Article through the EU Burden Sharing Agreement or 'EU Bubble'. The EU Burden

---

31 To address this fear, the text of Decision 15/CP.7, Principles, nature and scope of the mechanisms pursuant to Articles 6, 12 and 17 of the Kyoto Protocol, explicitly states that 'the Kyoto Protocol has not created or bestowed any right, title or entitlement to emissions of any kind on Parties included in Annex I (developed countries)'. FCCC/CP/2001/13/Add.2, 2.

32 FCCC/CP/2001/13/Add.2, 53–54, para. 5.

33 This system is too complex to describe here, but can be found in UNFCCC Decision 19/CP.7, Modalities for the accounting of assigned amounts under Article 7, paragraph 4, of the Kyoto Protocol, FCCC/CP/2001/13/Add.2, 55–72.

Sharing Agreement plays an important role not only in the Community's ratification of Kyoto, but also in the implementation of the EATD.

The origins of the idea of burden sharing can be traced back to the elaboration

Table 1: Member State targets under the EU burden sharing agreement

Austria	13%
Belgium	–7.5%
Denmark	–21%
Finland	0%
France	0%
Germany	–21%
Greece	+25%
Ireland	+13%
Luxembourg	–28%
The Netherlands	–6%
Portugal	+27%
Spain	+15%
Sweden	+ 4%
United Kingdom	–12.5%

of the EU negotiating position in preparation for the third Conference of the Parties (COP3) in December 1997 in Kyoto, Japan, at which the Kyoto Protocol was adopted[34]. The EU Bubble was found necessary to allow the Community to adopt a common negotiating position for a challenging target under the Kyoto Protocol. Identical targets for each Member State were, in view of the widely different energy generation infrastructure in each Member State, not seen as feasible. Coming up with a differentiated set of targets at the international level would, on the other hand, significantly complicate and even jeopardize the success of the international negotiations. Under Dutch leadership the Environment Council, against expectations of many observers, managed to reach agreement on an internal EU burden sharing agreement at its meeting on 2 March 1997[35]. The environment ministers agreed to propose a 15% cut in emissions of a basket of three greenhouse gases (carbon dioxide (CO_2), methane (CH_4) and nitrous oxide (N_2O)), by 2010, as the EC negotiation position in the talks under the UNFCCC. Three more gases (the hydrofluorocarbons (HFCs), perfluorocarbons (PFCs) and sulphur hexafluoride (SF_6)), that were originally

[34] For an overview of the background to the EU burden sharing agreement see Ringius (1997). An overview of the negotiations and results of COP3 can be found in Yamin (1998).
[35] Ibid., 6.

included in the proposals, were not included in the final reduction goal. The burden sharing agreement, the adoption of which initially seemed impossible, set specific post-2000 emission targets for each Member State[36]. The success of this victory was overshadowed by the fact that the total emissions on the basis of the agreed burden sharing amounted to only two-thirds of the 15% reduction agreed for the Community as a whole. Also, the agreement did not entail a unilateral commitment, but was dependent on what countries would agree upon at the meeting in Kyoto in December 1997.

During the negotiations in Kyoto, the EC did succeed in including its bubble concept into Article 4 of the Protocol, but the scope of gases was expanded from the three gases originally proposed by the Community to six gases and the reduction target that was finally agreed upon was only 8% over the period 2008-2012 rather than by 2010.

To allow the Community to use the 'bubble' possibility, it had to renegotiate its initial burden sharing agreement of 23 March 1997, adapting it to requirements of the Kyoto Protocol. On the Environment Council meeting on 16 and 17 June 1998, the Member States agreed to divide the 8% emission reduction for the European Community as a whole over the Member States as set out in Table 1.

The EU's Burden sharing agreement was made legally binding under Community law, through its inclusion in the ratification Decision, adopted by the Council on 4 March 2002[37].

6 The Concept of Emissions Trading

This section gives an overview of the various approaches to the concept of trading that are relevant in the area of climate change. A clear view of these approaches is essential to understand the background to and implications of the approach chosen by the EATD, which will be further discussed below. Although the issues involved in the design of a trading regime are numerous, this section focuses on the three most important: the choice between an absolute or a relative trading regime, the issue of allocation and the question of coverage.

[36] See the conclusions of the 1990th Council meeting, Brussels, 3 March 1997; as well as the conclusions of the 2017th Council meeting, Luxembourg, 19–20 June 1997 and the Conclusions of the 2033rd Council meeting, Luxembourg, 16 October 1997.

[37] N. 4 above.

6.1 Absolute v. Relative Regimes

Two distinct approaches to emissions allowance trading can be distinguished: trading regimes with absolute targets and trading regimes with relative targets.

Trading regimes with absolute targets are usually 'cap and trade regimes'. These regimes set a total cap, an absolute quantity of emissions measured in weight over a specific period of time, on all emissions from the sources covered by the regime. This total is subsequently allocated free or by auction, in the form of individual allowances, each usually representing an amount of CO_2 equivalent, to the various sources under the regime. After the allocation sources can choose to reduce their emissions and sell their allowances, maintain their emissions, or increase their emissions and buy allowances. Choices to buy or sell are made on the basis of the market price of the allowances and the marginal costs of emission reductions at the source. At the end of the trading or compliance period, sources have to match up their actual emissions with sufficient allowances. Sources that are unable to do so must buy allowances, whereas sources that hold an excess of allowances can sell these. The Kyoto Protocol is a clear example of a trading regime with absolute targets. It lays down the amount of emissions during the first commitment period for each of the developed countries, and allows them to trade surplus amounts. The EATD is also an absolute regime – it requires Member States to set absolute targets for the trading sectors and subsequently allocate the allowances to the sources.

Trading regimes with relative targets are usually 'baseline and credit' regimes. Unlike absolute regimes, relative trading regimes do not set a fixed absolute cap on the emissions from the sectors covered by the regime, although the regulator usually has an absolute target in mind when setting the relative target. The relative target is usually set through defining a baseline, which is expressed in the emissions efficiency in relation to the activity of the source, measured in weight per unit of input, output or activity. The same baseline can often be set across a wide range of similar installations. Installations that can reduce their emissions cheaper than the market price of allowances, will reduce their emissions and obtain allowances which they can sell. Installations for which the reduction of emissions is more expensive than the market price of allowances will maintain or increase their emissions and buy additional allowances on the market. A key difference with an absolute trading regime is that allowances are not allocated up-front, but when a source demonstrates that it performs better than its baseline. It is possible to use relative targets for the free of charge allocation under a regime with absolute targets. The relative targets are used to calculate the share of the total quantity of allowances allocated to each participant under the cap.

Absolute trading regimes are attractive to both policy makers and environmental groups since they give certainty on the environmental outcome of the trading system. The regime's cap determines the total amount of emissions from the sectors covered by the regime, the market determines where the necessary reductions in emissions will take place. The fact that the Kyoto Protocol itself sets an absolute cap, and the international emission trading under the Protocol is based on absolute units, are reasons for policy makers to choose for an absolute system in the design of their domestic or regional trading regime.

Relative regimes are usually more attractive for the sources. The absolute effects of most of the abatement techniques are usually dependent on the activity of the installation. Any reductions in emissions caused by the installation of new technology can often easily be outdone by an increase in activity. Industry usually argues that production growth itself should not be punished, but that the efficiency of production should be increased. Relative targets do exactly this. An additional reason for industry to support trading with relative targets is that a number of Member States, including Germany, the Netherlands and the United Kingdom, have already negotiated agreements on CO_2 emissions with industry sectors, most of which contain relative targets[38]. Both industry and regulator are reluctant to reopen these agreements and replace them with a trading regime with absolute targets. The downside of relative targets is, however, that the environmental outcome cannot be predicted with full certainty. This risk, usually determined by economic circumstances, is borne by the regulator.

Trading regimes with absolute and relative targets are not necessarily incompatible, and relative trading regimes could be used to attain an absolute target such as the one set under the Kyoto Protocol. Predictions can be made of the emissions in absolute numbers from a relative trading regime, the accuracy of which is dependent on the predictability of the behaviour of the sources. It is also possible to periodically adjust relative targets to increase the chance that an absolute objective is met. The domestic Nitrous Oxides (NO_x) trading regime that is currently being developed in The Netherlands for instance, will set a relative target, in amounts of NO_x emissions per unit of energy consumed. If it becomes clear that through the application of this relative target the absolute target that the Dutch government has set itself for NO_x emissions in 2010 will not be met, the relative target can be adjusted downwards in 2006 by a maximum of 20%[39].

Relative and absolute regimes can also be linked, although these links are in practice often complex and difficult to operate. At the end of the compliance period of a relative regime, when the volumes of emissions or the activity rate of the sources is known, the results of the relative regime can be translated into absolute numbers and a link can be created with an absolute trading regime. The

---

[38] See Barth and Dette (2001), n. 12 above.

[39] For a brief overview of the Dutch NO_x trading proposal, see FIELD and IEEP (2002). An up-to-date account of the development of the Dutch NO_x trading regime (in Dutch) can be found on: http://www.emissierechten.nl.

greenhouse gas allowance trading regime in the United Kingdom for instance allows trading with sectors that have closed negotiated agreements with the government, many of which are based on relative targets[40]. The negotiated agreement sector is linked with the absolute trading sector through a 'gateway'. Sources with absolute targets can sell their allowances to sources with relative targets, but only when there has been a net flow into the relative sector will any relative sector participant be permitted to transfer allowances out to the absolute sector. The gateway thus allows the relative sector to use allowances generated by the absolute sector to facilitate its compliance, but it cannot increase the total amount of allowances available to the absolute sector.

6.2 Allocation of Allowances

Through allocation the total amount of emissions allowed under a trading regime is distributed in the form of emission 'rights'. In most trading regime these emission rights are allocated free or auctioned to the sources covered by the regime, although this is not necessarily always the case. These emission rights could for instance also be allocated free to a state's citizens or other legal entities. The method of allocation is not important for the operation of the emissions trading regime itself. The initial allocation only decides the allocation of the economic value of the emission rights. The market's 'invisible hand' will redistribute those rights during the trading period. The method of allocation is, however, often the most controversial part of a trading regime – it distributes valuable assets among economic operators, each of whom will seek to obtain the maximum amount for the lowest cost before the start of the regime. It is because of this that discussions on allocation often tend to take up the largest part of the negotiating time needed to establish a trading regime, and can often be the reason for significant delays in the entry into force of even failure of the adoption of a trading regime[41].

The allocation issue is usually a key argument for regulators to opt for a baseline and credit regime. In such a regime the allocation is not done through transferring allowances to sources up-front, but through defining the baseline. Using a baseline and credit approach usually eliminates auctioned allowances as an option, as well as free allocations to entities that are not sources of the regulated emissions, and an absolute cap on total emissions. Since the baseline

[40] Relevant documents for the UK trading regime can be found at the DEFRA UK Emissions Trading Scheme Website: http://www.defra.gov.uk/environment/climatechange/trading/index.htm.

[41] Presentation by McLean (2000). See also the summary record of this meeting, available on the Internet at: http://europa.eu.int/comm/environment/climat/wg1_minutes.pdf.

for a source determines the free allocation it receives, industry in general also prefers a baseline and credit approach. The baseline can often be determined for a broad range of sources within a regime and is independent of a source's activity, which makes these baselines significantly easier to negotiate. This in turn is likely to increase the political acceptability of a regime and significantly reduce the time needed for its adoption. Following the same reasoning used by supporters of a baseline and credit regime to argue that it is not incompatible with an absolute target, the setting of baselines in combination with predictions of the sector's activity can, however, also be used for allocation under a regime with absolute targets.

Two main allocation methods are usually distinguished: free allocation and auctioning. A free allocation may be based on historic emissions (or other historic variables) – grandfathering – or on a baseline related to current activity as in a baseline and credit system. Auctioning usually stands for the auctioning of the allowances to the sources covered under the regime, where the amount of allowances allocated to a source depends on the price it is willing to pay for those allowances.

Grandfathering is usually industry's preferred allocation method. Under grandfathering sources are unlikely to incur much extra costs from the allocation, but will hold a valuable asset, which can be used by the source or sold on. Grandfathering is, however, also the most complicated allocation method. Many different interests will need to be accommodated, including the need not to disadvantage sources that significantly reduced their emissions in the past ('early movers') or benefit those that have not done so. Because of the value of the allowances, each of the sources will seek to maximize the amount of allowances allocated to it through the allocation process, irrelevant of whether it actually expects to need these allowances to cover its emissions. If very different types of sources are covered by the regime the selection of the baseline or benchmark used to help set more objective grandfathering criteria may be even more complicated. Grandfathering may also have important implications under the Community's state aid rules: valuable assets are given free of charge to industry. These state aid aspects of allocation will be further discussed below.

Auctioning is often favoured by economists as the most efficient and easiest allocation method, but in practice often not applied due to the opposition by industry. This opposition can be explained by the fact that use of a resource which was previously free of charge must now be paid for through an auction, often at prices unknown at the time of the design of the regime. During the stakeholder discussions on the ET Proposal, a number of industry representatives even stated that auctioning, by requiring industry to pay for allowances, would be taking away industry's 'right to pollute' and therefore require financial compensation to industry. This argument is unfounded. Industry, certainly under Community legislation, does not have a 'right to pollute', similar to the fact that countries under the Kyoto Protocol do not have a right to pollute[42]. Although in many cases these emissions have not been regulated under either Community or national law, this cannot mean that industry's emissions cannot be further reduced without compensation[43]. To compensate for industry's objection against the large cost burden that auctioning may impose on industry; suggestions have been made to re-distribute the revenue of the auction to the sources, after the auction[44]. This would, however, bring a difficult distribution debate into the allocation discussion, which would take away one of the key advantages of the use of auctioning in the first place.

In practice sometimes combinations of these allocation methods are used – government may decide to grandfather part of the allowances and hold regular auctions to sell off the remaining part[45]. A variety on the auctioning approach was used by the UK government in the establishment of the targets and coverage of its trading regime. The UK government made available 215 million UK pounds in incentive money to the sources willing to participate in the trading regime. This incentive money was paid to sources in relation to their committed reductions per tonne of CO_2 equivalent. The allocation took place through a 'descending clock auction'. In this auction the government announced an opening price per tonne, to which firms could offer commitment to reduce a specific amount of tonnes by 2006. As long as the total tonnage offered times the price was larger than the amount of the incentive money, the government revised its price downwards, with bidding rounds continuing until the clearing price was reached. The firms that remained in the auction are those included in the trading regime[46].

6.3 Coverage

The question of coverage relates to the gases and sources that are covered by the emission limitations of the trading regime. Although the Kyoto Protocol includes a range of six greenhouse gases from a wide range of sources, as listed in Annex A

42 See below the section on arguments against emissions trading.

43 See Lefevere (2000b). See also Krämer (2001).

44 These 'Hahn/Noll auctions', named after the authors of the paper that first introduced this idea, are, however, rarely used in practice. See also Schwarze and Zapfel (2000), n. 9 above, 289, in particular footnote 21.

45 This approach has been used in a number of US trading programmes. See Schwarze and Zapfel (2000), n. 9 above, 289.

46 Information on the UK auction and its results can be found on DEFRA's emissions trading website, n. 40 above.

to the Protocol, it is by no means obvious that all these gases and sources are covered by a trading regime, in particular in domestic regimes.

The choice on the coverage of gases by a trading regime usually depends on the coverage of the sources and the measurability of the emissions of gases by those sources, which is determined by the diffuse nature of the source of the emissions and the uncertainty related to the estimation or measurability of the quantities of those emissions. Some greenhouse gases are only covered in small quantities by the sources covered by a regime – inclusion of these gases would therefore not make much sense. Some greenhouse gas emissions from specific sources also have a considerable degree of uncertainty in relation to the measurement or estimation of their emissions. While CO_2 emissions from energy can be measured with an uncertainty of around 10%, the uncertainty in measuring CH_4 emissions from the combustion of biomass have been estimated at around 100% and for N_2O emissions from agricultural soils even around 200%[47]. Bringing emissions that have a high degree of measurement uncertainty attached to them within a trading system could complicate the trading system and undermine its practical feasibility and political acceptability. It is for those reasons that the EATD, discussed below, initially limits the coverage of gases to CO_2 only. The Danish trading regime is similarly limited to CO_2 only[48]. The UK trading regime on the other hand leaves the choice to sources to either include only CO_2 emissions, or to include their emissions from all six greenhouse gases. The inclusion of all six gases is, however, subject to the source being able to demonstrate that it can actually monitor the emissions in a sufficiently accurate manner[49].

Discussions on the coverage of sources are usually characterized by defining a trading regime as either an upstream or a downstream regime[50]. This characterization can be unclear as different authors use different interpretations of these terms. The two terms can, however, be used to indicate the extremes of a spectrum of choices related to coverage, rather than single options, and it does provide a useful aid to illustrate the various options. A typical upstream regime for fossil-fuel CO_2 emissions would require fuel producers or importers to hold allowances equal to the carbon content of the fossil fuel sold or produced. Rather than limiting the amount of emissions, the regime would limit the carbon content of fossil fuels used in a country. The advantage of this regime is that virtually all fossil fuel combustion CO_2 emissions would be covered, no matter how diffuse the actual sources of emissions actually are. The regime would be administratively easy to manage, in view of the limited amount of producers or importers of fossil

---

[47] Denne (1999).

[48] Information on the Danish trading regime can be found at the website of the Danish Energy Agency: http://www.ens.dk/uk/energy_reform/emissions_trading/index.htm.

[49] N. 41 above.

[50] Hargrave (1999).

fuel. Although the allowance requirement would not be linked with the actual CO_2 emissions, those emissions can be easily determined on the basis of the carbon content of each fuel and the allowance holders would translate their requirement to purchase permits for each unit of fossil fuels sold into a price signal towards the final consumers. A disadvantage is that the permit requirement may be too distanced from the actual emissions, and the price signal passed through by the producers or importers may be seen as a fuel tax. In some countries the number of fossil fuel producers and importers may be too small to create a competitive market. The upstream design also presumes an auction of allowances, since the participants have virtually no options for reducing the carbon content of the fuels, so the result is higher prices for fossil fuel products. A free allocation would provide the fossil fuel producers and importers with a windfall profit. An auction captures this windfall and allows the government to use the revenue to mitigate the impact of the higher prices.

A purely downstream regime includes the actual sources of the emissions. A downstream regime with full coverage would not only include household emissions from natural gas boilers, emissions from cars, but also methane emissions caused by the digestive processes of ruminant animals. It is obvious that a purely downstream system with extensive coverage would be administratively too difficult to manage, because of the enormous amount of sources and the often tiny amount of emissions from each source. It is for this reason that most of the trading regimes that are currently being designed are actually hybrids of these two extreme approaches. These regimes cover only larger industrial sources, mostly only their direct emissions, although the peculiarities of the UK trading regime include indirect emissions caused by electricity consumption as well[51].

The upstream-downstream choice as a point of regulation exists for many, but not all sources of GHG emissions. It exists for CO_2 emissions fossil-fuel combustion. In principle it exists for manufactured gases such as HFCs, but in practice the number of downstream sources is too large, so the only practical option is an upstream design. Process emissions such as CO_2 emissions from cement, N_2O emissions from adipic acid, and PFCs from aluminium production can only be regulated at the point of production.

7 The Ethical Dimension of Emissions Trading

Although the EATD received a cautious welcome from European environmental NGOs, the concept of emissions trading is by no means widely accepted[52].

---

[51] N. 40 above.

[52] See Climate Action Network Europe (2001).

Environmental NGOs have repeatedly expressed severe doubts on the implementation of the flexible mechanisms under the Kyoto Protocol[53]. Former French minister for the Environment Voynet, is even rumoured to have called the concept of emissions trading 'diabolique'. This section explores some of the arguments pro and contra emissions trading that have been raised in the wider debate on the acceptability of emissions trading as an instrument for environmental protection[54].

7.1 Arguments in Favour of Emissions Trading

The most often heard argument in favour of emissions trading is that it can significantly reduce compliance costs. Through allowing reductions to take place where they are most effective and cheapest, optimal use is made of reduction opportunities, which in turn reduces the burden on industry. This in turn makes the instrument attractive to industry and thus an interesting option for the regulator.

Related to this cost argument is the fact that emissions trading, more than other instruments, turns pollution reduction from a 'burden' into a 'business'. Sources, investors and brokers become actively interested in making money out of the flexibility offered to them. This can in turn give an important boost for the development and distribution of new technologies. Sources with the largest and cheapest reduction potential are often located in less developed countries and regions. Emissions trading allows for the transfer of technology and capital into those countries and regions and can provide for an important stimulus for the reduction of emissions from those sources. The technology that is transferred is often not only aimed at reducing the emissions of the particular substance, but may lead to improvements in the entire production process and thus reduce the total environmental impact of the source. More active industry participation in seeking out reduction opportunities may also stimulate the development of new technology and new reduction options.

Emissions trading regimes with sufficiently strict caps and a good compliance regime are also likely to gain support from environmental NGOs. This support is not because of the trading regime itself, but because of the certainty of the environmental impact of these regimes – they ensure that the emissions from the sector as a whole do not exceed the amount set out in the cap. Emissions trading can thus provide an important benefit over environmental taxation, which does

---

53 CAN-E in its reaction to the Commission proposal, n. 52 above, describes the Kyoto trading system as 'riddled with flaws'.

54 See for a discussion of the ethical aspects of emissions trading in a broader context, including JI and CDM, also Ott and Sachs (2000).

usually not provide this certainty of the environmental outcome. While environmental taxation allows the polluters to pay the tax and continue to pollute, the finite availability of emission allowances under a cap and trade regime, combined with a strong enforcement regime ensures that it becomes unattractive for sources to pollute more than the total cap.

7.2 Arguments Against Emissions Trading

A number of the key arguments against emissions trading have a moral or ethical basis. It can be argued that emissions trading turns the polluter pays principle on its head. A trading regime no longer requires the polluter to pay, but gives whoever pays the right to pollute. Emissions trading thus 'turns pollution into a commodity to be bought and sold' and by doing so it 'removes the moral stigma that is properly associated with it', which 'makes pollution just another cost of doing business, like wages, benefits, and rent'[55]. Following this reasoning, an analogy to emissions trading would be to sell permits to inflict injury on other humans or living beings. As long as the permit price is paid, the behaviour is allowed. The argument that emissions trading is morally wrong can be illustrated by explaining the moral difference between fee and a fine[56]. Payment of a fee exempts the behaviour for which the fee was paid from being classified as morally wrong. Payment of a fine on the other hand confirms the morally wrong character of the behaviour by confirming it as a violation of a society's rules. Emissions trading allows the polluter to pay a fee, rather than a fine, exempting him from the guilt associated with the behaviour.

The 'morally wrong' character of emissions trading thus hinges on the perception whether all emissions of greenhouse gases should be stigmatized as wrong behaviour, or whether society should accept some level of pollution[57].

It is widely accepted that greenhouse gas emissions lead to climate change[58]. The Kyoto Protocol is only a very first step in achieving the significant reductions necessary to avoid dangerous climate change impacts. Both the UNFCCC and the Kyoto Protocol recognize the need to revise their targets in the light of new information on the causes and impact of climate change[59]. The UNFCCC sets as its objective: 'to achieve, in accordance with the relevant provisions of the Convention, stabilization of greenhouse gas concentrations in the atmosphere at a level that would prevent dangerous anthropogenic interference with the climate

---

55 Sandel (1997).
56 Sagoff (1999).
57 Scott (1998).
58 See IPCC (2001).
59 See Article 4.2 (d) of the UNFCCC and Article 9 of the Kyoto Protocol.

system. Such a level should be achieved within a time-frame sufficient to allow ecosystems to adapt naturally to climate change, to ensure that food production is not threatened and to enable economic development to proceed in a sustainable manner'. While it is unclear what constitutes 'dangerous anthropogenic interference with the climate system', it is clear that the current global emission patterns fall within this category. Does the fact that the Kyoto Protocol requires only a limited reduction in the greenhouse gas emissions from developed countries give these countries the right to emit up to that level and exempt them from classifying their behaviour as morally wrong? The Decision on the principles, nature and scope of the Kyoto Protocol mechanisms in the Marrakech Accords contains an interesting phrase, inserted at the insistence of India, which states that the UNFCCC's supreme body, the Conference of the Parties, recognizes that 'the Kyoto Protocol has not created or bestowed any right, title or entitlement to emissions of any kind' on developed countries[60]. Both the Convention and the Protocol also recognize the responsibility of emitters of greenhouse gases for the damage caused by their behaviour, by requiring them to 'assist the developing country Parties that are particularly vulnerable to the adverse effects of climate change in meeting costs of adaptation to those adverse effects'[61]. It can thus be argued that while the Kyoto Protocol only requires a limited reduction in greenhouse gas emissions by developed countries, this does not exempt their behaviour from being morally wrong – it merely recognizes the need to reduce those emissions and lays down a first step for doing so.

Greenhouse gas emissions are, however, to some extent, 'inevitable and unavoidable consequence of economic activity'[62]. As with many other substances it is not the mere emission of greenhouse gases, but the quantity of those emissions that produces the negative impact on the environment. To allow economic activity, society can decide to accept a specific level of greenhouse gas emissions, and exempt emissions to that level from being 'immoral'. This is also the underlying rationale of the classic and widely accepted non-tradable permit as an instrument to regulate environmental pollution. Pollution up to the level set out in the permit is accepted, to allow for economic activity. Any pollution beyond the permit's limits is 'wrong' and is penalized with a penalty. Emissions trading works within this system, but allows emissions to be traded within that limit (the cap). Following this argumentation, emissions trading is not about penalizing inherently wrongful acts, but a method for allocating a scarce resource, within the boundaries set by society[63]. Emissions trading moves from a situation where emissions of GHGs are largely unregulated and free to one where they have a cost to the regulated sources.

---

[60] N. 31 above.
[61] See Article 4.4, 4.8 and 4.9 of the UNFCCC and Article 3.14 of the Kyoto Protocol.
[62] Sagoff (1999), n. 56 above.
[63] Ibid.

Another argument against emissions trading that is linked to the previous relates to possible equity concerns that emissions trading may cause. Especially when applied on a global scale, emissions trading could allow the rich to buy their way out of their obligations and continue their wasteful lifestyles, while the poor, in need of money for their development, are forced to sell their emission rights. Emissions trading would thus consolidate the economic power of the rich, by allowing them to buy emission rights from the poor, without which their development will be even more difficult. Part of this argument can be countered by the argument that the emission rights from the poor are not simply taken away, but are bought at a price, which will allow them to develop and implement new technology[64]. The other part of the argument, the need for the rich to change their lifestyle and reduce their emissions at home is precisely the reason why the European Union and others insisted on the elaboration of the supplementarity principle during the negotiations on the Marrakech Accords, as discussed above.

An argument related to the need to require 'domestic action' is that emissions trading, rather than promoting technology development, may actually provide a disincentive to technological innovation. Polluters with relatively advanced technology will no longer have an incentive to explore further reduction opportunities 'at home'. Instead they will invest 'abroad', using existing technology to clean up polluting sources, and by doing so merely 'pick the low-hanging fruits'. This in turn has an equity implication in case of emissions trading on a global scale. Developing countries are currently exempted from limitations on their greenhouse gas emissions. When they are subject to target in the future, they may be confronted with the fact that all their cheap reduction options have already been claimed by developed countries, which may increase their costs of complying with those targets. This argument, however, assumes a fixed stock of 'low hanging fruit'. In practice the stock of reduction opportunities is likely to change continuously over time, due to the replacement of equipment, the construction of new facilities, and the development of new technology. In addition, as with the equity argument, the validity of this argument is linked to the incentives or requirements given to polluters to reduce their emissions 'at home', as well as to the stringency of the 'cap'. If the cap is set sufficiently stringent, there may not be sufficient reduction options available to buy, which will increase the price to a level where it becomes attractive to reduce emissions domestically.

The stringency of the cap is also vital for the argument that emissions trading may actually increase emissions rather than reduce them. In the discussion of the 'hot air' debate above, it was shown how the economic collapse and transfer to the market economy in many EiTs has created a vast surplus in credits, which can now be traded under the Kyoto Protocol. Without trading, the EiTs would be

[64] Ibid. and Sandel (1997), n. 55 above.

unable to fill up their Kyoto 'assigned amounts'. With trading, they can sell these off to countries that do need these credits and thus increase the global amount of emissions. This shows that the setting of the cap, also in domestic trading regimes, is crucial for the credibility of a trading regime. Without a sufficiently strict cap, a trading regime may lead to adverse environmental effects[65].

A final argument against emissions trading relates to the issue of 'hot spots'. Classical use of the permitting instrument evaluates the permissible pollution on an installation by installation basis, and sets emission standards on the basis of the costs and technological feasibility to limit emissions in combination with the local environmental conditions[66]. Emissions trading in its purest form does not include such an assessment of the local impact of the emissions, but is only concerned with the aggregate 'cap' on the emissions from all sources. If, as a result of trading, one source significantly increases its emissions, this may have severe local impacts and lead to the creation of pollution 'hot spots'. Research in the United States has shown that these pollution 'hot spots' tend to be located in low income or minority areas, which in turn raises serious environmental justice concerns[67]. The question is, however, whether these hotspots were actually created by the trading programmes. The concentration of emissions in hot spots can also be tackled in the design of the trading regime, through limitations on the use of allowances by sources located in specific areas or through local environmental quality standards[68]. More importantly, the 'hot spots' argument does not apply to the emission of substances which do not have a local effect, such as most greenhouse gas emissions, in particular CO_2. The polluting effect of greenhouse gases consists of their build-up in the atmosphere, which causes global warming. The impact of global warming is independent from the location where these greenhouse gases were emitted. In many instances, the emission of greenhouse gases is, however, closely linked to the emission of other substances. An increase in the emission of CO_2 may, for instance, lead to an increase in the emission of NO_x. To avoid the indirect creation of hot spots, it is important that other emissions that do have a local impact are covered by sufficiently strict emission requirements.

On the whole, it thus seems that strong arguments can be made in favour of emissions trading as an instrument to combat greenhouse gas emissions. Care must, however, be taken with the design of the instrument to ensure that it

---

65 This is also the reason why CAN-E's comments on the Commission's ET proposal focus on the need to set clear and ambitious targets, n. 52 above.

66 See, for instance, the IPPC Directive, which incorporates the requirements permitting authorities to set emission limit values on the basis of the Best Available Techniques in combination with the local environmental quality. N. 15 above.

67 Stewart (2001), n. 7 above, 101.

68 This is done in some of the US trading programmes. See Schwarze and Zapfel (2000), n. 9 above, p. 292.

actually delivers a positive environmental result and avoids equity and environmental justice implications.

8 Conclusion

The Kyoto Protocol has given emissions trading a firm place as an innovative policy instrument in the area of climate change. This chapter has given a background to the concept of emissions trading, as it was introduced in the Kyoto Protocol, setting out the various core elements of and choices to be made when developing an emissions trading regime. It has also given an overview of the arguments in favour and against emissions trading. The translation of this concept into EU environmental policy, in particular the EATD, will be discussed in more depth in Chapter 9.

References

Barth, R. and Dette, B. (2001). 'The Integration of Voluntary Agreements into Existing Legal Systems'. *Environmental Law Network International Review*, 1:20–29

Climate Action Network Europe (CAN-E) (2001). 'Reaction to the Commission proposal: 'Emission trading in the EU: let's see some targets!'. http://www.climnet.org/EUenergy/ET.html.

Denne, T. (1999). *Scoping Paper 6: Aggregate Versus Gas by Gas Models of Greenhouse Gas Emissions Trading*. Washington: Center for Clean Air Policy (CCAP). http://www.field.org.uk/papers/papers.htm.

FIELD and IEEP (2002). *Assessment of the relation between Emissions Trading and EU Legislation, in particular the IPPC Directive*. http://www.field.org.uk/PDF/FINALReport31Oct.pdf.

Grubb, M. (1998). 'International Emissions Trading under the Kyoto Protocol: Core Issues in Implementation'. *Review of European Community and International Environmental Law*, 2:140–146.

Grubb, M., Vrolijk, C. and Brack, D. (1999). *The Kyoto Protocol: A Guide and Assessment*. London: Earthscan.

Grubb, M., Hourcade, J.-C. and Oberthür, S. (2001). *Keeping Kyoto: A Study of Approaches to Maintaining the Kyoto Protocol on Climate Change*. http://www.climate-strategies.org.

Haites, E. and Missfeldt, F. (2001). 'Liability Rules for International Greenhouse Gas Trading'. *Climate Policy*, 1/1:85–108.

Hargrave, T. (1999). *Identifying the Proper Incidence of Regulation in a European Union Greenhouse Gas Emissions Allowance Trading System*. http://www.field.org.uk/papers/papers.htm.

Intergovernmental Panel on Climate Change (IPCC) (2001). *Third Assessment Report – Climate Change 2001*. http://www.ipcc.ch.

Krämer, L. (2001). Grundlagen aus europäischer sicht, Rechtsfragen betreffend den Emissionshandel mit Treibhausgasen der Europäischen Gemeinschaft', in Rengeling, H.-W (ed.), *Klimaschutz durch Emissionshandel*. Köln: Carl Heymanns Verlag, 30.

Johnson, S.M. (2001). 'Economics v. Equity II: the European Experience'. *Washington and Lee Law Review*, 58:421.

Lefeber, R. (2001). 'From the Hague to Bonn to Marrakesh and Beyond: A Negotiating History of the Compliance Regime Under the Kyoto Protocol'. *Hague Yearbook of International Law*, 14:25–54.

Lefevere, J.G.J. (2000a). 'Atmospheric Pollution'. *Yearbook of European Environmental Law*, 1: 361–382.

Lefevere, J.G.J. (2000b). *Comment during the 2nd meeting of ECCP Working Group 1, 19 July 2000, reflected in the summary record of that meeting*. http://europa.eu.int/comm/environment/climat/eccp.htm.

McLean, B. (2000). Presentation at the first meeting of the European Climate Change Programme (ECCP)/Working Group 1.

Missfeldt, F. (1998). 'Flexibility Mechanisms: Which Path to Take after Kyoto?'. *Review of European Community and International Environmental Law*, 7/2:128–139.

Missfeldt, F. and Haites, E. (2002). 'Analysis of a Commitment Period Reserve at National and Global Levels'. *Climate Policy*, 2/1:51–70.

Ott, H. and Sachs, W. (2000). 'Equity and Emissions Trading – Ethical Aspects of Emissions Trading'. *Wuppertal Papers*, 110.

Ringius, L. (1997). 'Differentiation, leaders and fairness: Negotiating climate commitments in the European Community'. *CICERO Report*, 8. http://www.cicero.uio.no/media/99.pdf.

Sagoff, M. (1999). *Controlling Global Climate: The Debate over Pollution Trading*. Report from the Institute for Philosophy and Public Policy 19:1. http://www.puaf.umd.edu/IPPP/winter99/controlling.global.climate.htm.

Sandel, M.J. (1997). 'It's Immoral to Buy the Right to Pollute'. *New York Times*, 15 December 1997.

Scott, J. (1998). *EC Environmental Law*. Harlow, UK: Longmans, European Law Series.

Schwarze, R. and Zapfel, P. (2000). 'Sulfur Allowance Trading and the Regional Clean Air Incentives Market: A Comparative Design Analysis of Two Major Cap-and-Trade Permit Programs?'. *Environmental and Resource Economics*, 17:279–298.

Stewart, R.B. (2001). 'A New Generation of Environmental Regulation?'. *Capital University Law Review*, 29:21–182.

Special Issue on Emissions Trading (2000). *Review of European Community and International Environmental Law* 9:3.

Tangen, K., a.o. (2002). 'A Russian Green Investment Scheme: Securing Environmental Benefits from International Emissions Trading'. Climate Strategies Report, October 2002, available at: http://www.climate-strategies.org/gisfinalreport.pdf.

Werksman, J. (1998). 'The Clean Development Mechanism: Unwrapping the 'Kyoto Surprise'. *Review of European Community and International Environmental Law* 7/2:147–158.

Yamin, F. (1998). 'The Kyoto Protocol: Origins, Assessment and Future Challenges'. *Review of European Community and International Environmental Law* 7/2:113–127.

Chapter 5

BARBARA BUCHNER, ALEJANDRO CAPARRÓS GASS
AND TARIK TAZDAÏT

The Clean Development Mechanism and Ancillary Benefits

1 Introduction

Climate change is a global environmental problem (Nordhaus 1999), and in order to find an effective long-term approach to reduce the risk of climate change, as many countries as possible need to be involved (including, most importantly, the large emitters of greenhouse gases (GHGs)). Notwithstanding progress in international climate policy, some of the largest key players are currently not participating in the international efforts to control climate change. A number of studies have analyzed ways of re-involving the US by highlighting incentives for participation, which are provided by the present constellation of countries that have ratified the Kyoto Protocol (UNFCCC, 1997) and by testing various incentive strategies and climate regimes which could induce the US to join the international efforts to reduce the risk of climate change[1]. This paper focuses on another group of key players: the developing countries. These countries will probably be the largest GHG emitters in the near future and their participation in global climate change control is therefore crucial in the longer term. However, although developing countries face a drastic increase in their GHG emissions, mitigation actions against climate change do not rank high among their priorities. The obvious reason lies in the necessity for them to continue the development process, which is characterized by pressing needs

[1] See, for example, Buchner *et al.* (2002), Buchner, Carraro and Cersosimo (2003a), Buchner, Carraro and Cersosimo (2003b) and the papers presented at the RFF/IFRI Conference held in Paris on 19 March 2003 (http://www.rff.org/Post_Kyoto_Conference.htm).

MICHAEL BOTHE AND ECKARD REHBINDER (ED.), Climate Change Policy, 131–148.

other than emissions control. As a consequence, one of the major problems of climate change negotiations resides in the capacity of industrialized countries to induce developing countries to join the process of emission reductions. The Kyoto Protocol provides the prospect for integration based on the Protocol's so-called flexibility mechanisms, which have the effect of decentralizing multilateral negotiations into bilateral negotiations. The most relevant flexibility mechanism[2] in this regard is specified in article 12 of the Kyoto Protocol and is known as the Clean Development Mechanism[3] (CDM). The CDM enables developing countries to benefit from technology transfers in exchange for their emission reductions (an independent institution must validate this transfer). More precisely, the CDM stipulates that countries included in Annex I of the United Nations Framework Convention on Climate Change (essentially OECD and countries with economies in transition) can respect their commitments on GHG limitation (they are established in Annex B of the Kyoto Protocol for each country and account for an overall drop of emissions of 5% by 2012 compared to 1990 levels) by obtaining emission reduction certificates as counterparts for their investments in emission reduction projects in developing countries. It is, however, specified that these projects must contribute to the sustainable development of the host countries. Therefore, the CDM implies bilateral negotiations between a developing country and an Annex I Party with a requirement of 'double additionality': (i) environmental additionality (emission reductions must be supplemental to the 'no-action' alternative, which presents difficulties due to the unobservability of the basic scenario) and (ii) contribution to development.

In parallel to negotiations, theoretical literature[4] on the topic of co-operation within the climate change framework was developed (Barrett 1990 and 1994, Carraro and Siniscalco 1993, Chander and Tulkens 1995 and 1997, Péreau and Tazdaït 2001). One of the basic features of this literature is that it concentrates on multilateral negotiations. Nevertheless, and as the Kyoto Protocol supports bilateral negotiations between heterogeneous countries, a thorough study of this

---

[2] In fact, most of the issues discussed below and the results obtained are also applicable to the Joint Implementation (JI) mechanism, defined in article 6 of the Kyoto Protocol (UNFCCC 1997). This mechanism enables a similar schema to the CDM but between Annex I countries (it is expected to enhance cooperation between OECD countries and economies in transition). Nevertheless, to keep the discussion simple we will refer exclusively to the CDM.

[3] The Rio Earth Summit (1992) was followed by the Berlin Conference (1995) and only in Kyoto (1997) a first agreement on the Clean Development Mechanism was reached (although the Kyoto Protocol is not yet enforced since ratification is pending). This agreement was thereafter elaborated and adopted in what should be its final version in Marrakech (2001).

[4] For a recent survey see Finus (2001 and 2003) or Caparrós, Pereau and Tazdaït (2003).

type of negotiation is essential. In addition, previous work on the convenience of technology transfers between countries of the north and countries of the south within the climate change framework (Yang 1999, Kohn 2001) has focused on the conditions that make the transfers attractive for northern countries. The acceptance from southern countries has been presumed almost without question, or based on the benefits obtained due to increased emissions abatement. Nevertheless, if the increased abatement effort in the south goes together with a reduced abatement effort in the north (as is the case under the CDM), one might wonder if the southern countries would continue to be interested in accepting the offer.

It is in this context that we propose to highlight the strategic dimensions arising in bilateral negotiations[5] regarding technology transfers between countries with different characteristics. Our objective is to analyze the specificities of technology transfers as an instrument of co-operation, in order to deduce some lessons for the CDM. In order to provide a comprehensive evaluation of the meaningfulness of technology transfers in general and the CDM in particular, we will also include the local co-benefits arising from these instruments. Recently, the issue of ancillary costs and benefits, which are said to arise as 'by-products' of GHG mitigation policies, is gaining more and more attention. This is illustrated by the special role that the Intergovernmental Panel on Climate Change (IPCC) Third Assessment Report (2001) assigns to this topic. Indeed, GHG mitigation policies seem to create a number of non-climate side benefits, which have become increasingly important in the design of climate change policies.

For this reason, we will in the next section introduce the subject of ancillary benefits and briefly discuss the main issues that have been identified in this field. In Section 3, we present a model describing the bilateral bargaining process for adoption of green technologies by developing countries. In Section 4 we compare our model with the CDM. Finally, in Section 5, we develop our previous model incorporating local ancillary benefits for host countries into the analysis. Section 6 draws conclusions and indicates future possible research steps.

2 Growing Interest in the 'Side Effects' of Abatement Measures

For a long time, debates on environmental policy have focused on the cost-effectiveness of potential abatement measures and policies with a focus on direct abatement costs. This is in particular pertinent to the issue of global warming

---

[5] There is literature that analyzes the CDM based on agency theory (Millock 2000, Millock and Hourcade 2001). However, this literature is focused on the practical implementation of the CDM, and not on the interaction between negotiation and implementation.

since countries are interested in the present impacts of climate change controls on their economies and hence consider the long-term benefits arising from reduced global warming risk as less important. As a consequence, research in the context of climate change control started to look for additional reasons that could set incentives for GHG abatement and thus could make it more profitable to engage in mitigation activities 'here and now'.

A recent strand of this literature puts emphasis on the potential side effects that are induced by abatement measures[6] (for example: enhanced local air quality, reduced traffic congestion or restoration of degraded lands). While the main focus surrounding the climate change debate is on the costs of these measures, they can at the same time be accompanied by additional positive and negative externalities, which might change the evaluation of climate policies. Within the last few years, interest in these so-called 'ancillary' or co-benefits and costs has continually been growing, and recently received special attention through the IPCC Third Assessment Report (IPCC 2001). The study points to 'extensive benefits' that mitigation actions may yield in areas outside of climate change and sets a milestone by explicitly acknowledging the existence of these ancillary benefits. Furthermore, it emphasizes that the inclusion of ancillary benefits could create 'no regret opportunities' which might allow the reduction of GHG emissions 'at no or negative net social cost'. After the IPCC's first, more cautious,assessment of ancillary benefits in the Second Assessment Report (IPCC 1996), this recent and important assessment report proved to be helpful in increasing awareness of additional benefits from GHG mitigation and probably induced the beginning of intensified research efforts on this topic.

A particularly challenging question in the context of ancillary effects concerns their policy implications. If it is true that environmental policy has additional benefits, their inclusion could change the way in which this policy is viewed. In particular, the consideration of ancillary effects could change the evaluation of single policies and measures. Since environmental policy is mostly characterized by specific targets (e.g. reduced GHG emissions in climate policy), without specifying the choice of instruments, combinations of policies are often used to cope with commitments. Any policy instrument or policy mix should be evaluated including the additional impacts that it could have in order to verify whether the chosen policy approach really remains appropriate in light of the new understanding of abatement activities.

---

[6] For an overview on research studies see e.g. OECD (2000) and for a discussion of the main issues arising in this research field see e.g. Buchner (2001). For other examples of ancillary benefits/costs associated to climate change policies see chapter 9 of IPCC (2001).

As has been pointed out among others by Davis, Krupnick, and McGlynn (2000) and Pearce (2000), the concept of additional policy effects is not unique to climate policy. Nevertheless, it requires special attention in this context because of the underlying complexities. Due to the global nature of climate change, the uncertainties surrounding the issue itself and its long-term impacts, the estimation of ancillary benefits becomes particularly important. This importance is emphasized by the heterogeneity of GHG sources throughout the economy and their complex economic effects. Ancillary benefits can play a special role in climate policy since they generally embody two characteristics, which are rarely noted in global warming debates: they are local and occur in a short-term dimension – instead of global and in the long term[7].

Let us now briefly discuss some of the methodological and conceptual issues arising in the context of ancillary benefits. Most importantly, as a precondition for the incorporation of ancillary effects into the broader policy context, a problem of the initial research studies has to be overcome: the plurality of terminology. The term 'ancillary benefits' has been simultaneously used with 'multiple', 'joint', 'co-benefits or 'secondary benefits'. In order to enable ample awareness of the issue in consideration, an accurate definition is needed.

Based on the terminology used by the IPCC, this paper will refer to 'ancillary benefits' as the 'incidental side effects of policies aimed exclusively at GHG mitigation'[8]. Abatement policies can have additional impacts which do not impact GHG emissions, but arise as a consequence of their reductions. Ancillary benefits can thus be seen as changes in social welfare other than those caused by the emission reduction itself. This definition implies that side effects do not have to be positive; in fact, abatement measures can cause also negative, harmful effects, to which we refer to as 'ancillary costs'[9] (see the chapter by Caparrós and Jacquemont for a discussion of the potential negative impacts on biodiversity of forestry alternatives to fight climate change).

---

[7] Nevertheless, since pollution crosses over boundaries, ancillary effects do not necessarily have to be local. Moreover, impacts as acidification of ecosystems have lasting, long term implications.

[8] In contrast, again in the sense of IPCC, the term co-benefits embodies '(monetised) effects that are taken into account as an explicit (or intentional) part of the development of GHG mitigation policies' (Davis, Krupnick, and McGlynn 2000: 2). For a detailed discussion of issues in the context of terminology see chapter 2 of Davis, Krupnick, and McGlynn (2000).

[9] Apart from the division into positive and negative ancillary effects, we can group them also into effects causing direct and indirect changes in outcome. The former refers to the 'usual' perception of ancillary effects whereas the latter represents the possibility that ancillary effects also can take the form of avoided costs. Finally, we have to take into account both the ancillary effects arising from mitigation and from adaptation policies.

The complexities occurring within the context of global warming do not only make an ancillary benefits analysis appropriate, they also indicate that special caution has to be taken when applying this approach. Each 'potential' ancillary impact requires a thorough assessment of whether it qualifies as an ancillary effect. In fact, various difficulties surround estimations of ancillary effects, for instance, valuation issues, location issues, uncertainty considerations, the need to achieve the broadest-possible comprehensiveness of effects, the dependence on the choice of model, differentiation between developed and developing countries[10] and equity considerations[11]. As a consequence of these difficulties, it becomes obvious why the estimates of ancillary impacts still suffer from credibility problems. In addition, the true impact of climate policies can only be revealed if the 'counterfactual' situation is appropriately specified, namely what would happen in the absence of any explicit policies. These underlying assumptions are usually referred to as 'baselines' and determine the scale, scope and quality of ancillary effects estimates. In particular, these baseline issues are responsible for the environment in which climate change policies will have their effects. Thus, baseline issues are a very sensitive and vital element in the analysis and need to be carefully specified in a credible and consistent way since they have far-reaching implications[12]. In order to overcome the various problems in the context of ancillary impacts, it is of crucial importance to address all the issues considered in this section, and to include them in an overarching conceptual framework, which could then serve as the basis of an improved approach to ancillary effects estimations[13].

---

[10] For a comprehensive discussion of developing issues in the context of the analysis of ancillary effects see O'Connor (2000) and Krupnick, Burtraw, and Markandya (2000).

[11] Among other factors, the presence of ancillary effects causes the special joint production character of GHG mitigation policies. Schleicher and Buchner (2002) use this characteristic of GHG mitigation policies as an argument to distinguish between the investments and the costs related to abatement measures. The reason for this distinction is that it is difficult to allocate the overall abatement as soon as abatement investments generate as joint products additional effects besides the reduction of GHGs. In particular, there arises the problem of splitting the investment costs between the abatement effect and the additional impacts, as ancillary benefits for consumers, induced technological change and increased economic activity. As in all joint production situations this allocation is to a large extent arbitrary and thus questions according to Schleicher and Buchner (2002) the 'usefulness of the concept of an abatement cost function'.

[12] For a detailed discussion of the baseline issues see Morgenstern (2000), Krupnick, Burtraw and Markandya (2000) and Davis, Krupnick, and McGlynn (2000).

[13] In particular, the framing of ancillary effects analysis should not only include a proper definition and illustration of effects, but above all an economic and institutional system that is crucial to the identification and determination of these effects (Krupnick, Burtraw and Markandya 2000).

Notwithstanding the large uncertainties and other difficulties surrounding the assessment of ancillary effects, there is compelling evidence that mitigation policies are indeed accompanied by side effects and can even serve as a means to support them. In particular, the inclusion of ancillary effects can affect the geographic, sectoral and even technological focus of policy-making. The relevant magnitude that has been specified in various case studies differs substantially due to different methodologies, data restrictions and country-specific characteristics. Nevertheless, the early findings of Ekins (1996) are likely to be confirmed by recent studies: ancillary benefits appear to be ‘a significant fraction of or even larger than the mitigation costs’, whereas the ancillary costs are likely to be on a smaller scale (Davis, Krupnick and McGlynn 2000).

Despite this implication of a potential change in the ranking of policy options and the increased importance of integrated approaches with multiple targets, the process of policy-making has not yet reacted in an appropriate way. In particular, relatively little work has been done on how this issue affects the choice of specific policy tools or sectors for GHG abatement[14]. Two factors are responsible for this paradox: on the one hand, the lack of adequate techniques aimed at incorporating the ancillary effects; and on the other hand, missing institutional structures which support this inclusion. As a precondition to deal seriously with the policy implications of ancillary effects analysis, an accurate evaluation of various abatement measures with respect to ancillary effects is necessary. Only after such an assessment is it legitimate to require changes in policy design. In this context, a new approach to the estimation of ancillary effects is required since the specific characteristics of ancillary effects add new complexity (above all through the new importance of the spatial, geographic dimension). However, in the meantime, it is important to continue research on ancillary impacts in order to strengthen their role in environmental policy. For this reason, after having introduced the basic underlying model on bilateral negotiations about technology transfers, the next sections will analytically investigate the implications of including local ancillary benefits into the CDM. In particular, we would like to verify whether our analysis supports the above findings and confirms that taking account of ancillary effects can change the performance of the policy instruments under consideration.

[14] Only recently, Pearce (2000), Krupnick Burtraw and Markandya (2000) and Davis, Krupnick, and McGlynn (2000) started to give ideas for a policy framework by pointing at the main issues that have to be considered within the interactions between ancillary effects and the policy context.

3 The Model

This section will introduce the basic game-theoretic framework used throughout our analysis[15]. Let us consider two countries (countries 1 and 2) sharing an international environmental resource. Call q_i the emission reductions agreed by country *i*, defined in relation to the situation where it does not undertake any emission abatement measures at all. The welfare function for country *i* can be written:

$$\pi_i(q_1 + q_2) = B_i(q_1 + q_2) - D_i(q_i), \quad i = 1, 2 \tag{1}$$

where $B_i(q_1 + q_2)$ represents the benefits for country *i* due to the emission reductions and $D_i(q_i)$ the costs resulting from the emissions abatement measures.

Let us suppose that the benefit function[16] of country 1 ($B_1(q_1 + q_2)$) grows at a decreasing rate, i.e. $B_1'(\cdot) > 0 \geq B_1''(\cdot)$. To simplify, we suppose that for country 2 the benefits follow[17]:

$$B_2(q_1 + q_2) = b_2 \cdot (q_1 + q_2) \tag{2}$$

In the subsequent analysis, we overlook the possibilities of reducing pollution at a negative cost ('no-regrets' options and potential double dividends). By doing this, any emission reduction measure has – by definition – a depressive effect on the aggregate output of the applying country. This economic cost is represented by $D_i(q_i) = \alpha_i C_i(q_i)$, where α_i is a technological parameter. In addition, and following an assumption largely accepted in economic literature on transfrontier pollution, we suppose that each additional abatement unit is more difficult to obtain than the preceding one (i.e. that the marginal abatement cost increases). Several computable general equilibrium models have corroborated this assumption (Manne 1993, Oliveira-Martin, Burniaux and Nicoletti 1993). We thus suppose: $C_i'(\cdot) > 0$ and$C_o''(\cdot) > 0$.

In our model, if country 1 has a technological advantage, that will result in the relation $\alpha_1 < \alpha_2$. Computable general equilibrium models applied to the estimation of emission reduction costs confirm this assumption (Manne 1993, Oliveira-Martin, Burniaux and Nicoletti 1993). These models show that the

---

[15] See Hoel (1991) for a precedent to our model.

[16] This benefit arises from the reduction of the undergone environmental damage.

[17] This assumption is based on the opinion of Cararro and Siniscalco (1993, 1994) for whom the reaction-functions of the players should be almost orthogonal in most cases of cross-border pollution. That is, the abatement efforts of one country should not be compensated by the increase in emissions in another country.

evolution in time of abatement costs and marginal abatement cost functions implies a progressive downwards displacement. This is related to the introduction of new lower-carbon-intensive technologies, as a logical response of the productive systems to fossil energy price increases (in proportion to their carbon contents). The argument can be summarized as follows: if two initially identical countries have implemented different environmental policies in the past, technological innovations will occur at different moments in time and each country will be located, in the long term, on a different cost curve. However, our subject is not limited to this simple statement. If such differences exist, the advanced countries have the possibility to help the others, thus benefiting from their advantage while transferring green technologies. We can refer to this issue as 'technological jumps'. We are now able to specify the central assumption of this study: green technology transfers reduce the host country's marginal abatement cost function. This enables abatement at a lower cost, which produces a return on investment for the donor country.

To simplify the subsequent discussion, while preserving the spirit of what has just been said, we adopt the following assumptions. Country 1 uses the best available technology ($\alpha_1 = \underline{\alpha}$). It transforms the final commodity that it produces in abatement capital, incorporating the green technology at a constant cost per unit equal to 1. Thus, if it grants a transfer of t_1 to country 2, this will cost it t_1 final good units. The coefficient α_2 depends on the received technological assistance. Without external intervention ($t_1 = 0$), country 2 uses ineffective technology. Then, as the external assistance intensifies, its abatement technology improves. Hence, we suppose that it follows a function $\alpha_2 = \alpha(t_1)$, with $\alpha(0) = \bar{\alpha}, \alpha'(\cdot) < 0$ and $\alpha''(\cdot) > 0$.

The game proceeds according to a two-stage structure. Country 1 plays first, proposing to country 2 a certain amount of technology transfers. The alternative answers are: (i) the proposal is accepted and an agreement is signed, or (ii) it is refused and the negotiations fail. In the case of a failure, the game ends with the simultaneous choice of the abatement efforts that each country wishes to undertake (country 2 can in this case only use its original abatement technology). On the contrary, if an agreement is signed, countries determine simultaneously, during the second period, their environmental policies taking into account the technology transfers set up in the first stage. Given the dynamic aspect of the game, the equilibrium concept selected is the 'subgame-perfect equilibrium' (Selten 1975). Thus, the resolution of the game is carried out analyzing firstly the second stage of the game and then moving on to the first stage. However, we will not fully analyze and solve the model here. For a full development of the model see Caparrós and Tazdaït (2003).

4 The CDM

The game described above gives a highly theoretical vision of the interaction between environmental policy and green technology transfers. It is only of interest if it is possible to provide concrete contents (i.e. if it possible to define an institutional framework able to guide it). Let us summarize the situation. Without transfers, countries 1 and 2 use their own abatement technology ($\underline{\alpha}$ and $\bar{\alpha}$ respectively). They internalize their national environmental damage via an unspecified economic instrument (e.g. a pigouvian tax or an emission trading system) so that their marginal abatement costs are respectively b_1 and b_2 (see conditions (3) and (4)). Based on this reference situation, both countries agree on the need for a green technology transfer of $\bar{t}_1$ units from country 1 towards country 2 (the solution of the game is $(\bar{q}_1, \xi_2(\bar{t}_1), \bar{t}_1)$). But, how can it be done? This question engages us on the topic of economic instruments.

A parallel must be made between the mechanism described above and the CDM, proposed in the Kyoto Protocol. This comparison raises some interesting questions about its practical definition and application. Under the CDM, Annex I countries have the possibility of financing emission reduction projects abroad. This agreement indicates that signatory countries recognize the advantage that may arise by sharing know-how on emissions control for homogenizing abatement technologies and reducing GHG abatement costs. This precisely coincides with our problem. However, the element, which makes immediate comparison with our model difficult, is that under the CDM the obligations of the investor in terms of national abatement are reduced in exchange for the amount of emission reductions financed abroad. Nothing similar exists in our model.

Green technology transfers can be used in two ways. The position adopted in the Kyoto Protocol implies choosing a fixed ceiling for global emissions and then, by means of the CDM, distributing this quota in order to limit overall abatement costs. According to this point of view, the CDM constitutes only an efficiency economic instrument. By adopting the opposite position (no compensation for the investor apart from 'the pleasure' provided by the improvement of the environment), the technological assistance releases additional resources for further abatement.

The Nash equilibrium in the second stage of the game (i.e. known t_1) is:

$$\underline{\alpha} C_1'(q_1) = B_1'(q_1 + q_2), \tag{3}$$

$$\alpha(t_1) C_2'(q_2) = b_2. \tag{4}$$

The strategy of country 2 does not depend on the effort of its neighbor. On the contrary, the strategy of country 1 from now on depends on the effort of its

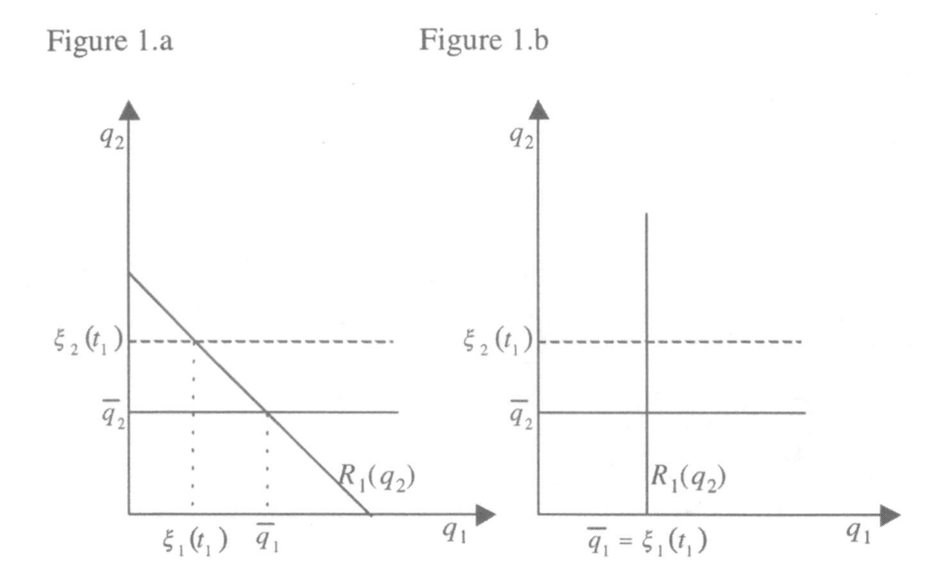

neighbour. Its reaction functions are defined by: $q_1 = R_1(q_2)$, with $R_1'(q_2) \in [-1, 0]$, and $q_2 = \xi_2(t_1)$, with $\xi_2'(t_1) > 0$. Hence: $q_1 = \xi_1(t_1) = R_1(\xi_2(t_1))$.

Under the condition that its marginal abatement benefit is decreasing, the optimal emissions of country 1 are a diminishing function of the transfers that it grants. From the point of view of the country that grants the technological assistance, the issuance of additional emission reduction credits is thus justified.

To reconsider the use of technology transfers, let us arbitrarily suppose in figures (1.a) and (1.b) respectively that $R_1'(q_2) = -1$ and $R_1'(q_2) = 0$. In both cases, and in the absence of green technology transfers, the Nash equilibrium would correspond to the solution $(\bar{q}_1, \bar{q}_2)$. When $R_1'(q_2) = -1$, country 1 authorizes aid of a given amount t_1. Because of the displacement of its reaction function, the abatement of country 2 increases from $\bar{q}_2$ to $\xi_2(t_1)$. On the other hand, country 1 reduces its abatement from $\bar{q}_1$ to $\xi_1(t_1)$. Globally, abatement does not change, since by assumption $\bar{q}_1 - \xi_1(t_1) = \xi_2(t_1) - \bar{q}_2$. It follows that transfers are in this case only useful to move the abatement efforts from country 1 towards 2. This corresponds with the definition of the CDM in the Kyoto Protocol. The profit for country 1 comes from the abatement cost reduction which it undergoes, net from the cost of the technology transfer itself. It is necessarily positive since otherwise it would set $t_1 = 0$. For country 2, the profit is positive if the benefits associated with the external aid compensate the additional costs related to the increase in its abatement efforts.

When $R_1'(q_2) = 0$, country 1 grants aid of a given amount t_1(different from the preceding case). Country 2 increases its abatement level from $\bar{q}_2$ to $\xi_2(t_1)$.

Country 1 does not modify its abatement effort (its reaction function is vertical). Globally, abatement increases. The transfers granted are used in this case to increase global emission reductions. The advantage for country 1 results from the improvement of environmental quality, net of the technology transfers costs. For country 2, the profit results from the aggregation of three effects: (i) improvement of environmental quality; (ii) reduction in its abatement costs ascribable to the technological aid; and (iii) the additional abatement costs.

The question arises whether the host country will be favorable to technology transfers. To answer, let us rewrite the welfare function of country 2 (taking into account the fact that the abatement of country 1 depends on the transfers which it grants):

$$\pi_2(t_1) = b_2 \cdot (\xi_1(t_1) + \xi_2(t_1)) - \alpha(t_1)C_2(\xi_2(t_1)). \tag{5}$$

Differentiating and simplifying using (4), we obtain:

$$\pi_2'(t_1) = b_2\, \xi_1'(t_1) - \alpha'(t_1)C_2(\xi_2(t_1)). \tag{6}$$

The sign of this expression is ambiguous. When technological aid goes together with a lower abatement effort on behalf of country 1 (i.e. when $\xi_1'(\cdot) < 0$, the welfare of country 2 can decrease with t_1. Consequently, country 2 may refuse the technology transfers proposed in order to preserve its reservation payoff $\bar{\pi}_2$.

5 Local Ancillary Benefits

As described in section 2, climate change policies can have a positive influence on the economy in addition to the benefits raised by the mitigation of greenhouse effect damages. In order to provide a more comprehensive overview of mitigation policies, a recent research strand – supported also by the IPCC (2001) – argues that these effects need to be taken into account.

In our previous model, local ancillary benefits were neglected. Nevertheless, they are a fundamental part of the CDM defined by the Kyoto Protocol (UNFCCC 1997) and given the increasing interests in these effects we shall thus incorporate them in the analysis.

According to the Kyoto Protocol (article 12): 'The purpose of the CDM *shall be to assist Parties not included in Annex I in achieving sustainable development* and in contributing to the ultimate objective of the Convention, and to assist Parties included in Annex I in achieving compliance with their quantified emission limitation and reduction commitments under Article 3.' (emphasis added).

As shown, the CDM is expected to assist host countries in achieving sustainable development. This can be seen as an ancillary benefit from the point of view of climate policy (however, the use of the term *ancillary* is debatable according to the wording of the Protocol, since its importance is set equal to the main objective of the Protocol). An additional feature of this ancillary benefit is that it is – almost by definition – *local* in its nature (since only the host country will benefit directly from its own development).

As stated above, several examples of ancillary benefits associated with climate change policies, and therefore potentially to the CDM, can be found (e.g. enhanced local air quality, reduced traffic congestion, restoration of degraded lands, etc.) in chapter 9 of IPCC (2001). In almost all the cases described, the ancillary benefits achieved are local.

To incorporate local ancillary benefits in our previous model, let us define the benefit function of country 2 as (the benefit function of country 1 remains unchanged):

$$B_2(q_1 + q_2) = b_2 \cdot (q_1 + q_2) + a_2 \cdot (q_2) \tag{2'}$$

where a_2 stands for the local ancillary benefits in country 2 associated with the abatement effort undertaken in this country.

The optimal conditions in the second stage of the game are now:

$$\underline{\alpha} C_1'(q_1) = B_1'(q_1 + q_2), \tag{3'}$$

$$\alpha(t_1) C_2'(q_2) = b_2 + a_2. \tag{4'}$$

The decision of player 1 is unchanged. The decision process of player 2 still implies[18]:

$$q_2 = \xi_2(t_1), \text{with } \xi_2'(t_1) > 0.$$

In the first stage of the game, the decisions of country 1 remain unchanged, so that the results shown in the previous section are applicable. In the case of country 2, the welfare function is now written:

$$\pi_2(t_1) = b_2 \cdot (\xi_1(t_1) + \xi_2(t_1)) - \alpha(t_1)\ C_2(\xi_2(t_1)) + a_2 \cdot (\xi_2(t_1)). \tag{5'}$$

[18] See Caparrós and Tazdaït (2003) for more details.

Differentiating and simplifying using (4′) this time, we obtain equation (6) again:

$$\pi_2'(t_1) = b_2\, \xi_1'(t_1) - \alpha'(t_1)\, C_2(\xi_2(t_1)). \tag{6'}$$

Of course, the discussion of the sign of equation (6) is still valid. That is, even including explicitly local ancillary benefits, such as the contribution to local sustainable development, the host country might refuse technological transfers.

Nevertheless, a particular case is now relevant. If for country 2 benefits associated with greenhouse effect mitigation are irrelevant[19] ($b_2 = 0$), equation (6′) is strictly positive and the offer of country 1 will be accepted (since the welfare of country 2 rises together with the technological assistance received). The reason why this situation was not relevant in the previous section, with equation (6), is that in the preceding game, country 2 decided its optimal abatement effort in the second stage of the game based on equation (4). Therefore, if b_2 is set equal to zero, country 2 would undertake no emission reductions at all (since the marginal cost of its abatement effort has to be null and this is only possible with no abatement at all, given the assumption made that no-regret options do not exist). However, in the game with ancillary benefits, country 2 decides its optimal abatement effort based on equation (4′), which will not always remain at zero (since a_2 is supposed as being positive).

Hence, if local ancillary benefits in country 2 are positive and country 2 does not care about climate change, the offer of country 1 will be accepted regardless of the reaction function of country 1. This implies that industrialized countries should maximize local ancillary benefits (including local sustainable development in this concept) if they want to see their offers of green technology transfers accepted. It also implies that countries less concerned about climate change are more likely to accept the offers made by industrialized countries under the CDM. This might imply a race to the bottom in terms of the climate impact of the CDM if the rules on baselines and monitoring are not well established, since host countries are, in the configuration just analyzed, *only* concerned about local benefits. On the contrary, an alternative schema for the CDM, where the host country would also be interested in the climate impacts of the project (i.e. if emission abatements in the south would not imply an equal reduction in the

---

[19] We may wonder if a situation where country 2 does not care about climate change is possible at all. Two explanations can be found. Climate change damages are far in time and uncertainties about their extent remain. Thus, countries with high discount rates (as poor or extremely poor countries will probably apply) will almost neglect these damages and consequently not take into account benefits associated with mitigation strategies. An additional explanation could be found if countries of the South, or some of them, consider that climate change is not their business, since industrialized countries mainly caused the problem, and they simply do not incorporate this term in their objective function.

emission abatement effort in the north), could induce a stronger interest for the host country in self-monitoring the progress of the emission reductions achieved.

6 Conclusion

This paper analyses the conditions under which bilateral negotiations between northern and southern countries regarding green technology transfers in exchange for GHG abatement can be successful, deducing thereby some lessons for the CDM. In particular, due to the growing interest in the side effects of mitigation policies in order to obtain a more comprehensive view of different policies, one of the main objectives of this study has been to discuss the role of local ancillary benefits in the context of the CDM.

Our basic model shows that bilateral negotiations between northern and southern countries on the subject of technological transfers will not necessarily lead to an agreement. This is especially true if the abatement effort in the south goes together with a reduction in the abatement effort from the northern country. Nevertheless, after several rounds of multilateral negotiations, the international community has agreed on the creation of the CDM, which has precisely this latter characteristic. However, we have shown that in order to enable the CDM to be meaningful, it is necessary to include local ancillary benefits. The situation can be described as follows: to combat global warming effectively, an agreement based on the CDM has to generate local ancillary benefits for the developing country concerned. It is only under this condition that it is in the interest of the developing country to accept the technological transfer and to perform the corresponding abatement efforts. That is, the southern country will fight against the 'global' environmental problem, not because it is concerned by the problem, but because it obtains 'local' benefits from which other countries cannot take advantage. Hence, industrialized countries have to maximize local ancillary benefits (including local sustainable development) if they want to see their offers of green technology transfers accepted. An additional implication is that southern countries less concerned about climate change are more likely to accept the offers made by industrialized countries under the CDM. As shown above, this implies that countries more interested in CDM projects are likely to be those less interested in ensuring positive impacts on climate change, implying a kind of race to the bottom in the climate change mitigation quality of CDM projects (as long as they have strong local benefits, the *only* benefit considered by the southern countries interested in CDMs). To avoid this threat based on the current shape of the CDM, monitoring rules have to be strict, since , southern countries will have a reduced interest in self-monitoring the projects (this would be different if the emission reductions achieved in the south did not imply an equal reduction in the abatement effort of the north, since southern countries *mainly* interested in climate change mitigation would also be interested in developing CDM projects).

Thus, our results confirm recent research findings and highlight that ancillary effects may affect the allocation of environmental policy measures. As a consequence, this paper emphasizes that more attention should be paid to ancillary effects within the process of policy-making. Nonetheless it is true that there are still a lot of open questions in the context of ancillary effects analysis and consequently more research is needed to specify its exact influence on the ranking of policy options. Until this point, though, the available evidence of ancillary effects can already serve as an additional 'no regrets' argument to pursue integrated policies in favor of climate change control. Since the ultimate target in policy-making should consist of the achievement of sustainability, such an integrated strategy based on ancillary effects appears to be promising. In addition, ancillary effects play an important role because they can alter the relationship of benefits and costs of climate policies in the short term. Due to their special attributes they involve the 'here' and 'now' which is important for policy-making. As a consequence, including these effects in the evaluation of policy instruments can become a 'win-win' strategy: the transparency of impacts can be tremendously increased and a more comprehensive assessment of policies can become possible while they attract attention to the policy context.

References

Barrett, S. (1990). 'The Problem of Global Environmental Protection'. *Oxford Review of Economic Policy,* 6: 68–79.

Barrett, S. (1994). 'Self Enforcing International Environmental Agreements'. *Oxford Economic Papers,* 46: 878–894.

Buchner, B. (2001). 'Ancillary Benefits and Costs of Climate Change Policies'. *FEEM Working Paper*. Milan: FEEM.

Buchner, B., Carraro, C. and Cersosimo, I. (2003a). 'China and the Evolution of the Present Climate Regime'. *FEEM Working Paper*. Milan: FEEM.

Buchner, B., Carraro, C. and Cersosimo, I. (2003b). 'Regional and Sub-Global Climate Regimes'. *FEEM Working Paper*. Milan: FEEM.

Buchner, B., Carraro, C., Cersosimo I. and Marchiori C. (2002). 'Back to Kyoto? US Participation and the Linkage between R&D and Climate Cooperation'. *CESifo Working Paper*, 688:8 and *FEEM Working Paper*, 15:02. Milan: FEEM.

Caparrós, A. and Tazdaït, T. (2003). 'Green Technology Transfers and Climate Change'. Presented at *EAERE2003*, Bilbao (Spain), June 28–30.

Caparrós, A., Pereau, J.-C. and Tazdaït, T. (2003). 'Coalition et Accords Environnementaux Internationaux'. *Revue Française d'Economie*, 18: 199–232.

Carraro, C. and Siniscalco, D. (1993). Strategies for the International Protection of the Environment. *Journal of Public Economics,* 52: 309–328.

Carraro, C. and Siniscalco, D. (1994). 'Environmental Policy Reconsidered: The Role of Technological Innovation'. *European Economic Review,* 38: 545–554.

Chander, P. and Tulkens, H. (1995). 'A Core-Theoretic for the Design of Cooperative Agreements on Transfrontier Pollution'. *International Tax and Public Finance,* 2: 279–294.

Chander, P. and Tulkens, H. (1997). 'The Core of an Economy With Multilateral Environmental Externalities'. *International Journal of Game Theory,* 26: 379–401.

Davis, D., Krupnick, A. and McGlynn, G. (2000). 'Ancillary Benefits and Costs of Greenhouse Gas Mitigation: An Overview'. Presented at the *Expert Workshop on Ancillary Benefits and Costs of Greenhouse Gas Mitigation Strategies*, Washington (USA), March 27–29.

Ekins, P. (1996). 'The Secondary Benefits of the CO2 Abatement: How Much Emission Reduction Do They Justify?'. *Ecological Economics,*16: 13–24.

Finus, M. (2001). *Game Theory and International Environmental Cooperation.* Cheltenham: Edward Elgar.

Finus, M. (2003). 'Stability and Design of International Environmental Agreements: The Case of Transboundary Pollution', in H. Folmer and T. Tietenberg (eds) *International Yearbook of Environmental and Resource Economics, 2003/4.* Cheltenham: Edward Elgar, 82–158.

Hoel, M. (1991). 'Global Environmental Problems: The Effects of Unilateral Actions Taken by One Country'. *Journal of Environmental Economics and Management*, 20: 55–70.

IPCC (Intergovernmental Panel on Climate Change) (1996). *Climate Change 1995.* Cambridge: Cambridge University Press.

IPCC (Intergovernmental Panel on Climate Change) (2001). *Climate Change 2001: Mitigation.* Cambridge: Cambridge University Press.

Kohn, R.E. (2001). 'Unilateral Transfer of Abatement Capital'. *Resource and Energy Economics*, 23: 85–95.

Krupnick, A., Burtraw, D. and Markandya, A. (2000). 'The Ancillary Benefits and Costs of Climate Change Mitigation: A Conceptual Framework'. Presented at the *Expert Workshop on Ancillary Benefits and Costs of Greenhouse Gas Mitigation Strategies,* Washington (USA), March 27–29.

Manne, A.S. (1993). 'Global 2100: Alternative Scenarios for Reducing Carbon Emissions', in: OCDE (ed.), *Les coûts de la réduction des émissions de carbone: resultats tirés des modéles mondiaux.* Paris: OCDE, 55–66.

Millock, K. (2000). *Contracts for Clean Development – The Role of Technology Transfers.* Working Paper FEEM 70.2000. Milan: FEEM.

Millock, K. and Hourcade J.C. (2001). *Bargaining Institutions Under the Clean Development Mechanism.* Working Paper CIRED. Nogent-sur-Marne: CIRED.

Morgenstern, R.- D. (2000). 'Baseline Issues in the Estimation of the Ancillary Benefits of Greenhouse Gas Mitigation Policies'. Presented at the *Expert Workshop on Ancillary Benefits and Costs of Greenhouse Gas Mitigation Strategies,* Washington (USA), March 27–29.

Nordhaus, W. D. (1999). 'Biens Publics Globaux et Changement Climatique'. *Revue Française d'Economie*, 14: 11–32.

O'Connor, D. (2000). 'Ancillary Benefits Estimation in Developing Countries. A Comparative Assessment'. Presented at the *Expert Workshop on Ancillary Benefits and Costs of Greenhouse Gas Mitigation Strategies,* Washington (USA), March 27–29.

OECD (2000). *Ancillary Benefits and Costs of Greenhouse Gas Mitigation*. Proceedings of the Expert Workshop on Ancillary Benefits and Costs of Greenhouse Gas Mitigation Strategies, Washington (USA), March 27–29.

Oliveira-Martins, J., Burniaux J.-M. and Nicoletti, G. (1993). 'The Cost of Reducing CO2 Emissions: A Comparison of Carbon Tax Curves with Green', in: OCDE (ed.), *Les coûts de la réduction des émissions de carbone: resultats tirés des modéles mondiaux*. Paris: OCDE, 67–94.

Pearce, D. (2000). 'Policy Frameworks for the Ancillary Benefits of Climate Change Policy'. Presented at the *Expert Workshop on Ancillary Benefits and Costs of Greenhouse Gas Mitigation Strategies*, Washington (USA), March 27–29.

Péreau J.-C. and Tazdaït, T. (2001). 'Cooperation and Unilateral Commitment in the Presence of Global Environmental Problems'. *Environmental and Resource Economics*, 20: 225–239.

Schleicher, S.P. and Buchner, B. (2002). *Efficient Strategies for the Co-Benefits of the Kyoto Mechanims*. University of Graz.

Selten, R. (1975). Reexamination of the Perfectness Concept for Equilibrium Points in Extensive Games. *International Journal of Game Theory* 4, 25–55.

UNFCCC (United Nations Framework Convention on Climate Change) (1997). *The Kyoto Protocol to the UNFCCC*. http: / /www.unfccc.int" www.unfccc.int.

Yang, Z. (1999). 'Should the North Make Unilateral Technology Transfers to the South? North-South Cooperation and Conflicts in Responses to Global Climate Change'. *Resource and Energy Economics*, 21: 67–87.

Chapter 6

ALEJANDRO CAPARRÓS GASS AND
FRÉDÉRIC JACQUEMONT

Biodiversity and Carbon Sequestration in Forests: Economic and Legal Issues

1 Introduction

World leaders agreed at the 1992 Earth Summit in Rio de Janeiro on a comprehensive strategy for 'sustainable development' with the aim to meet our needs without compromising the ability of future generations to meet their own needs. In order to do so, at the same time as the United Nation Framework Convention on Climate Change (UNFCCC) was adopted, a second key agreement was reached: the Convention on Biological Diversity (CBD). Both instruments are global and comprehensive conventions that address the full range of human activities affecting particular natural systems. Moreover, they serve as frameworks for specialized subsidiary instruments called Protocols[1]. The ultimate objective of the UNFCCC is to achieve the 'stabilization of greenhouse gas concentration in the atmosphere ... within a time-frame sufficient to allow ecosystems to adapt naturally to climate change'[2]. The first goal of the CBD is the conservation of biological diversity on earth, understood as the variety of plants, animals, micro-organisms, their habitats, and ecosystem levels[3]. Both conventions are thus concerned with the conservation of the existing variety of ecosystems. Additional relationships between the conventions arise due to the fact that climate change can be seen as one of the major threats to biodiversity and that some of the actions proposed to mitigate climate change potentially create dangers for biodiversity (ESCBD 2000, FOEI *et al.* 2000). These relationships

---

[1] Cartagena Protocol on Biosafety and Kyoto Protocol on Greenhouse gas reductions.
[2] UNFCCC, Art. 2. (United Nations 1992a).
[3] CBD, Art. 1. (United Nations 1992b).

MICHAEL BOTHE AND ECKARD REHBINDER (ED.), Climate Change Policy, 149–182.

between the two conventions imply a high level of cooperation between them. Unfortunately, this has not always been the case[4], mainly because each convention focuses on a specific environmental issue. They adopt single-issue approaches, without taking into account the significant ecological interdependence existing between global environmental problems.

Forests contribute to climate change mitigation as terrestrial sinks and assist conservation efforts by acting as pools of biological diversity[5]. Indeed, carbon is cycled between the atmosphere, oceans, and terrestrial biosphere. Significant reservoirs of carbon are found in oceans, vegetation, and soil[6]. Terrestrial ecosystems, such as forests, capture significant amounts of carbon from the atmosphere through photosynthesis[7] and retain part of this carbon in soil and vegetation. Thus, reducing deforestation, enhancing afforestation and reforestation, and improving forest management activities, may increase the capacities of terrestrial sinks to absorb carbon.

Co-operative efforts between the UNFCCC and the CBD have only recently begun[8]. In terms of forestry, the UNFCCC and the Kyoto Protocol promote the use of forests as sinks, and aim at protecting forests[9]. These efforts appear to complement the CBD. However, the primary goal of the Kyoto Protocol is to reduce greenhouse gas (GHG) emissions to 5 percent below 1990 levels. By taking the view of solely promoting forests as natural sinks, the initiatives taken

---

[4] Nevertheless, these relations have appeared in official documents (e.g. FCCC/SBSTA/2001/INF.3 or FCCC/SBSTA/2001/L.14)

[5] Oceans are also significant carbon and biodiversity reservoirs. However, no activities for enhancing oceans as carbon stock are foreseen yet by the UNFCCC or the Protocol. Forest biological diversity means the variability among forest living organisms and ecological processes of which they are part; this includes diversity in forests within species, between species and of ecosystems and landscapes. (proposed definition by the ad hoc Technical Expert Group on Forest Biological Diversity under CBD (UNEP/CBD/SBSTTA 2001).

[6] Carbon is stored both above and below ground. Below-ground stocks are greater than above-ground, particularly in non-wet areas such as grasslands, savannahs, tundra, and croplands (CBD 2001).

[7] Plants absorb CO_2 from the atmosphere through photosynthesis. A portion of this CO_2 is released again but a part is used by plants to build up their biomass, liberating oxygen. This implies a reduction in atmospheric CO_2, but its effect is constrained by the life-cycle of the biomass, since the carbon stored is finally released through oxidation.

[8] These relations have in fact shown up in the official documents (for instance: FCCC/SBSTA/2001/INF.3 and FCCC/SBSTA/2001/L.14).

[9] Other LULUCF eligible activities are additional human-induced 'cropland management', 'grazing land management', 'revegetation' (see, Marrakech Accords, Annex, Draft Decision-/CMP.1 (LULUCF), FCCC/CP/2001/13/Add.1, Paragraph 6 (hereinafter 'Annex DD-CMP.1')). These activities in farming include using minimum tillage, mulching, improving efficiency of fertilizer use, restoring degraded agricultural lands and rangelands, recovering methane from stored manure, and improving the quality of the diet of ruminants.

under the Kyoto Protocol may neglect the potential impacts of these initiatives on other related forest instruments whose aim is the preservation of forests. Converting an old natural forest into a single species forest, or using unsustainable forest management practices, only with the view to capture more carbon, may have negative impacts on biological diversity (CBD 2001). Thus, problems of compatibility with the goals of other international agreements related to forests[10], such as the CBD, may arise from the implementation of the Kyoto Protocol.

In this chapter, we first present the treatment of carbon offsets and biodiversity criteria in the Marrakech Accords, and in the following decisions adopted at COP9, held in Milan (November 2003), regarding sinks projects under the Clean Development Mechanism (CDM). Afterwards, we study the implications of carbon incentives for existing forests as well as the expected influence of this policy on decisions relating to the type of forest for use of afforestation and reforestation. Finally, we investigate the relationship between the UNFCCC and the CBD to assess any possible influence on the results obtained in the first two sections.

2 The Marrakech Accords

Land use, land use change, and forestry (LULUCF) activities are actions that alter land use to reduce emissions of greenhouse gases (sources) and remove gases from the atmosphere through carbon sequestration (sinks). The carbon is stored in forests and other biomass and soil. Under the circumstances described in Articles 3.3 and 3.4 of the Kyoto Protocol (UNFCCC 1997), developed State Parties may undertake forestry management, cropland management, and other resource-centered activities that remove and store carbon as a means to help meet their greenhouse gas emissions reductions commitments. Sinks were one of the crunch issues that resulted in the failure to reach an agreement at COP 6 in The Hague in November 2000. The Parties were not able to decide if sinks activities should be limited in quantity, and if so, what this limit should be. At the resumed session of COP 6 (COP 6 bis) held at Bonn in July 2001, an agreement was

[10] Other related instruments are (non-exhaustive list): the Forests Principles (Non-Legally Binding Authoritative Statement of Principles for a Global Consensus on the Management Conservation and Sustainable Development of all Types of Forests), A/CONF.151/26 (Vol. III), (Rio de Janeiro) 14 August 1992) in 31 ILM (1992), 881; Agenda 21 (section II, chapter 11, 'Combating Deforestation'), A/CONF.151/26 (Vol. I), ibid.; the United Nations Convention to Combat Desertification in Countries Experiencing Serious Drought and/or Desertification Particularly in Africa (CCD), (Paris) 17 June 1994; Convention on International Trade of Endangered Species of Wild Flora and Fauna (CITES), (Washington) 3 March 1973; and Ramsar Convention on Wetlands of International Importance Especially as Waterfowl Habitat, (Ramsar) 2 February 1971.

reached, which was similar in terms of sinks to the one rejected at COP 6. This agreement was included, with some slight modifications, in the Marrakech Accords[11], which were adopted at COP 7 in Marrakech in November 2001 (UNFCCC 2001). Rules regarding sinks under the CDM were later adopted at COP 9, in Milan, in November 2003 (UNFCCC 2003), since Parties did not reach any agreement on this issue at Marrakech. In this article, we review the limitations on the use of sinks activities to obtain carbon credits established in the Marrakech Accords and the links between sinks and biodiversity included in the Accords.

Limits on the amount of credits under the Kyoto Protocol (Articles 3.3 and 3.4) that may be generated by using LULUCF activities during the Protocol's first commitment period from 2008 to 2012 are set out[12] in the Marrakech Accords[13]. In the Accords at Paragraph 10 of the Annex to Draft Decision CMP.1 on LULUCF (Annex DD-/CMP.1), credits generated from forest management activities[14] can be accounted up to the amount of carbon debits resulting from LULUCF activities permitted under Article 3.3 of the Protocol (such as afforestation, reforestation and deforestation), to a maximum amount of 9.0 megatons of carbon times five[15] (this enables Parties to compensate deforestation debits by forest management). Additional credits for forest management can be

[11] FCCC/CP/2001/13 and Addenda 1 to 4 (Marrakech Accords). If not specified otherwise, all the documents in these notes will refer to the Marrakech Accords.

[12] Caps were set in Bonn (COP 6bis) in Decision 5/CP6, FCCC/CP/2001/L.7 and incorporated, with some modifications, in the Marrakech Accords (see Annex (Definitions, modalities, rules and guidelines relating to LULUCF activities under the Kyoto Protocol) DD-CMP.1(LULUCF)). Nevertheless, in most cases, the formal decision remains FCCC/CP/2001/L.7 (Bonn Agreement), since several paragraphs have been included in the Marrakech Accords in the Annex, which is not yet formally adopted (this is a draft proposed for adoption in the first Meeting of the Parties to the Kyoto Protocol). However, according to Paragraph 12 of the Annex, the limits can be revised in the light of new country specific data (this provision was included in Marrakech). In Paragraph 4 of the Annex an additional limit for carbon debits is established, not allowing them to surpass credits in the case of afforestation and reforestation.

[13] Annex (Definitions, modalities, rules and guidelines relating to LULUCF activities under the Kyoto Protocol) DD-CMP.1 (LULUCF), Paragraphs 10, 11, and 14.

[14] 'Credits' is used where removals are larger than emissions on a unit of land and 'debits' when the opposite is true.

[15] The reason for multiplying by five is the five years duration of the commitment period. In the Bonn Agreement the limit was set at 8.2 megatons of carbon. (See Annex (Definitions, modalities, rules and guidelines relating to LULUCF activities under the Kyoto Protocol) DD-CMP.1(LULUCF), Paragraph 10.)

accumulated under Paragraph 11 of the Annex DD-CMP.1 up to the maximum amount established for each country in the agreement reached at COP 6 bis (Bonn Agreement)[16]. These credits may be generated either through domestic measures or by means of investments in emissions reduction measures via the Protocol's Joint Implementation (JI) mechanism (Article 6), which permits industrialized Parties listed in Annex I of the Convention (Annex I Parties) to accumulate credits by making investments to reduce emissions in other Annex I Parties. Under the Kyoto Protocol's Clean Development Mechanism (CDM) (Article 12), Annex I Parties may generate credits by making investments in Developing State Parties. In terms of LULUCF-related activities, only those concerning afforestation and reforestation are permitted under the rules of the CDM[17]. The applicable cap for LULUCF activities in the CDM is established in Paragraph 14 of the Annex DD-/CMP.1, requiring that the credits obtained by a Party through eligible LULUCF activities under the CDM do not exceed one percent of the base year emissions of that Party, times five.

All caps can be seen as absolute limits (in the case of Paragraph 14, the limit is set when the emissions base year is fixed, which in most cases is 1990). This will be a relevant feature in the analysis below. Only forest management is included in the caps set out in Paragraphs 10 and 11 of the Annex DD-/CMP.1 for domestic and JI activities, while afforestation and reforestation are only limited when undertaken as CDM projects. The other LULUCF activities[18] permitted under Paragraph 6 of the Annex DD-/CMP.1 are free of these limitations. These activities are 'cropland management', 'grazing land management' and 'revegetation'. No general cap on 'sinks' was adopted at Bonn and at Marrakech, but specific caps on domestic forest management activities, JI projects and on CDM afforestation and reforestation activities were adopted.

Regarding the references to biodiversity included in the Marrakech Accords, the Parties affirm 'that the implementation of LULUCF activities contributes to the conservation of biodiversity and sustainable use of natural resources'[19]. The Parties also 'request the [UNFCCC] Subsidiary Body for Scientific and Technological Advice (SBSTA) to develop definitions and modalities for including afforestation and reforestation projects under the CDM in the first commitment

---

16 The maximum amount for each country was set in the Appendix Z of the Bonn Agreement and appears now in the Appendix to the Annex. Under the Marrakech Accords, the limit for Russia was raised from 17.63 to 33 MtC.

17 Annex (Definitions, modalities, rules and guidelines relating to LULUCF activities under the Kyoto Protocol) to Draft Decision-/CMP.1(LULUCF), Paragraph 13.

18 Parties themselves decide which of these activities are applied during the first commitment period and the selection is fixed for all the first commitment period (Annex, Paragraph 7).

19 Draft Decision -/CMP.1 (LULUCF), FCCC/2001/13/Add.1, Paragraph 1(e).

period, taking into account ... environmental impacts on biodiversity and natural ecosystems'[20]. The terms 'contributes' and 'taking into account' cannot be understood as concrete limitations, especially not the latter. Hence, biodiversity conservation has not been included as a clear constraint to LULUCF activities in the Marrakech Accords.

Another reference to biodiversity can be found in the definition of forest management in the Annex DD-/CMP.1[21]. However, including biodiversity only in the definition of forest management, and not in the definitions of other activities such as afforestation or reforestation, risks the dangerous interpretation that biodiversity conservation is only necessary in the case of forest management. Nevertheless, Article 2 of the Kyoto Protocol, which requires Parties to implement policies and measures in accordance with their national circumstances, balances this assertion by referring to sustainable management practices in relation to afforestation and reforestation as a policy of protection and enhancement of sinks[22] (the interpretation of what sustainable management covers is left to the Parties).

The Executive Secretary of the CBD submitted to the UNFCCC, during the preparations for COP 6, a note outlining several issues concerning the impacts on biodiversity of climate change mitigation measures[23]. One of the main threats to biodiversity described in this note is the conversion of natural forests into plantations. Nevertheless, in theory, this latter threat should be avoided under the rules governing afforestation and reforestation. Under these rules, afforestation[24] can only take place in areas that have not been forested for 50 years and reforestation[25] is limited to lands that were not forests on 31 December 1989. Hence, it impedes plantations to be set up in areas covered by natural forests at the present time. Moreover, the option of clearing a natural forest and waiting until the first limit is no longer applicable is not attractive, since the land owner would need to wait for 50 years (this time span is enough to make future profits irrelevant for any meaningful discount rate). Nevertheless, the conversion of land into forests could also be done as forest management. This possibility has not been explicitly ruled out by the definition of 'forest management' in the

---

20 Decision 11/CP7 (LULUCF), FCCC/CP/2001/13/Add.1, Paragraph 2(e).

21 Annex (Definitions, modalities, rules and guidelines relating to LULUCF activities under the Kyoto Protocol) to Draft Decision-/CMP1(LULUCF), Paragraph 1(f).

22 Protocol, Article 2.1(a)(iii).

23 Executive Secretariat to the CBD, 'Climate Change and Biological Diversity: cooperation between the Convention on Biological Diversity and the United Nations Framework Convention on Climate Change' (CBD 2000), UNEP/CBD/SBSTTA/6/11 (download full text at www.biodiv.org).

24 Annex (Definitions, modalities, rules and guidelines relating to LULUCF activities under the Kyoto Protocol) to Draft Decision-/CMP.1 (LULUCF), Paragraph 1(b).

25 Ibid., Paragraph 1(c).

Marrakech Accords (nevertheless, as stated above, the definition of forest management includes a direct reference to biodiversity, so that aggressive strategies should be prohibited). In addition, Parties called for the elaboration of biome-specific forest definitions, with the aim of reflecting the ecological variability of these ecosystems, for the second and subsequent commitment periods. COP 10 should recommend their adoption on applying such biome-specific forest definitions by the COP serving as a meeting of the Parties (MOP) to the Kyoto Protocol at its first session[26]. However, the meaning of biome in this context is less than clear as its interpretation varies, which may entail the adoption of a new specific biome classification for the purpose of the Kyoto Protocol[27]. Furthermore, a change in forest definitions that include biome specificities may affect the amount of land detected as afforested/reforested or deforested, and thus entail new difficulties for reporting. In particular, errors may occur during the conversions from one definition to another, leading to unreliable results on changes in carbon stocks. Furthermore, these changes in definitions may lead to different methods of carbon accounting and of inventories for forests of different biomes, as carbon sequestration varies among different forest types[28].

Other specific definitions of afforestation and reforestation projects under the CDM shall be elaborated by SBSTA and should take into account the issues of, non-permanence, additionality, leakage, uncertainties and socio-economic and environmental impacts, such as impacts on biodiversity and natural ecosystems[29]. These definitions were discussed at UNFCCC COP 8 in New Delhi, with no real changes in the original position of the different Parties. The definitions were forwarded to UNFCCC COP 9, which finally took the decision to maintain the same definitions for afforestation and reforestation as those adopted at Marrakech[30].

The Marrakech Accords require Annex I Parties to report on their legislative and administrative procedures for ensuring that LULUCF activities under

---

26 Decision 11/CP.7, Paragraph 2(b).

27 See Technical paper on Biome-Specific Forest Definitions (FCCC/TP/2002/1) (download at http://unfccc.int/resource/doc/tp/tp0201.pdf) and, IPCC Proceedings, Expert Meeting on Harmonizing forest-related definitions for use by various stakeholders, Rome 22–25 January 2002, FAO (download at http://www.fao.org/forestry/fop/fopw/climate/doc/Y3431E.pdf). Moreover, forest biome reflects the ecological and physiognomic characteristic of the vegetation and broadly corresponds to climatic region of the earth. See some definitions and example at National Council for Science and the Environment (download at http://www.NCSEonline.org/NCE/CRReports/biodiversity/biodiv.6.cfm).

28 Ibid.

29 Decision 11/CP7 (LULUCF), FCCC/CP/2001/13/Add.1, Paragraph 2(e).

30 Annex, Decision 19/CP.9 FCCC/CP/2003/6/Add.2, Paragraph 1.

Articles 3.3 and 3.4 of the Kyoto Protocol contribute to the conservation of biodiversity and to the sustainable use of natural resources[31]. However, the scope of Paragraph 26 is limited since Parties only need to list their national laws and are not required to provide information on tangible results for the conservation of biodiversity in LULUCF projects. Furthermore, during the first commitment period, the reporting on national communication under Article 7.2 will not be an eligibility criterion for the use of the flexible mechanisms[32]. Therefore, with no means to ensure the accuracy of reporting on biodiversity preservation in LULUCF activities, the reliability of reporting requirements on sinks activities as set out in Article 7.2 of the Kyoto Protocol is questionable. As a result, biodiversity concerns in the Marrakech Accords do not constitute a strong barrier to counterbalance the economic incentives established under the Protocol.

3 Specific Rules for Afforestation and Reforestation Projects under CDM as Adopted at COP 9

As decided in Marrakech, sinks activities under the CDM projects, would be limited to afforestation and reforestation projects, and the credits resulting from such activities would not exceed one percent of the base year emissions of that Annex I Party, times five, and would not be banked in the second commitment period. Beside the adoption of specific definitions for afforestation and reforestation under the CDM, Parties had to agree on modalities and procedures in the first commitment period (2008 to 2012) that should take into account the issues of non-permanence, additionality, leakage, uncertainties and socio-economic and environmental impacts, including impacts on biodiversity and natural ecosystems. Such modalities would thus secure, as stated in Article 12 KP, that those emissions reductions resulting from CDM projects are 'real, measurable, and long-term' as well as 'additional'. Rules on these issues will be explained, before addressing the treatment of biodiversity concerns under CDM projects.

For CDM projects undertaken under the KP to be certified, the emission reductions from each project activity must be 'additional to any that would otherwise occur in the absence of the certified project activity' (Article 12.5(c)). The objective of the additionality criterion is to ensure that credits are generated only for projects that are truly additional and that would not occur otherwise,

---

[31] Annex (Guidelines for the preparation of the information required under Article 7 of the Kyoto Protocol/Reporting of supplementary information under Article 7, paragraph 2), Draft Decision-/CMP.1 (Article 7), FCCC/CP/2001/13/Add.3, Paragraph 26.

[32] Only reporting on inventory and registry under Article 7, paragraph 1 and paragraph 4 respectively is required as an eligible criterion.

since the certification of business-as-usual carbon sequestration under the CDM would undermine the global GHG emissions reduction target of the KP. Thus, the additionality criterion avoids that resources are inefficiently allocated to projects that would be implemented anyway (Sussman and Leining 2002). Furthermore, the baseline project can be deduced as to what would have occurred in the absence of the project. The two concepts of additionality and baseline, although different strictly speaking, are closely related in practice. The determination of the extent of project additionality requires a baseline, because, in order to calculate the net change of GHG emissions of an additional project, one has to compare the business-as-usual baseline scenario. This approach was adopted in Marrakech for other projects under the CDM 'a CDM project activity is additional if [...] emissions by sources are reduced below those that would have occurred in the absence of the registered CDM project activity'[33], and was finally retained in the definition for afforestation and reforestation projects under the CDM adopted in Milan: 'An afforestation or reforestation project activity under CDM is additional if the actual net greenhouse gas removals by sinks are increased above the sum of the changes in carbon stocks in the carbon pool within the boundary that would have occurred in the absence of the registered CDM afforestation or reforestation project activity'[34]. By nature baseline methodologies are uncertain, as they have to apprehend a counterfactual situation: what would have happened in the absence of the specific project? Thus, Paragraph 22 of the Milan Decision[35] lays down different methodologies on a project specific basis for calculating baseline, which are open to the choice of the project participant (for different baseline methodologies and their assessment see Sussman and Leining (2002)).

Another issue to be addressed, in order to secure that afforestation and reforestation project activity is environmentally beneficial, is leakage. Every project brings changes in activities that take place outside of the project boundary. The effects of these activity changes, outside the project boundary, on GHG emissions are called leakage, if they result in an emissions increase, or spill-over, if emissions decrease. Thus, sink projects have the potential of both, to offset any sequestration within the project boundaries by increasing emissions, or to increase carbon sequestration outside the project boundaries (Sussman and Leining 2002). Despite some support from Latin American countries to credit spill-over, only leakage was considered and defined: 'Leakage is the increase in GHG emissions by sources outside the boundary of an afforestation or reforestation project activity under the CDM which is measurable and attributable to the

[33] Decision 17/CP.7; Draft Decision-/CMP.1, FCCC/CP/2001/13/Add.2, Paragraph 43.
[34] Decision 19/CP.9, Draft Decision-/CMP.1, Annex, FCCC/CP/2003/Add.2, Paragraph 19.
[35] Ibid., Paragraph 22.

afforestation and reforestation project activity'[36]. There are two approaches to address leakage risks. Firstly, in designing the project in such way that leakage effects can be reduced. Secondly, one may estimate leakage effects in the baseline, or develop correction coefficients for leakage risk that adjust the GHG reductions achieved by the project. The first method of proper project design has been chosen in Milan[37].

One crucial issue regarding sinks projects is the non-permanence of sequestered carbon. The climate benefit depends heavily on the durability of the carbon fixation. Thus, it is also important that project lifetimes are long enough to safeguard continuity in carbon sequestration activity. Carbon stored in forest can be released due to natural causes, such as natural decay, fire, diseases or pest, or because of human activities, such as converting the forested land into agriculture. Two main proposals were proposed and discussed to address this risk, an insurance approach supported by Canada and a temporary credit proposal sponsored by the EU. The last proposal is a variation of a Colombian proposal to give each Certified Emission Reduction (CER) that results from a forestry project an expiration date that corresponds to the end of the project. The expiration of the temporary CER (t-CER) would be considered as an offset, and the buyer Annex I country would then retire an Assigned Amount Unit (AAU) or other credit units to compensate. Thus, the validity period would correspond to the lifetime project and carbon sequestered guarantee by the project participant. Monitoring the projects at regular intervals is required. In the event of a prematurely loss of the CO_2 sequestration, the project participant would have to compensate by retiring an alternative t-CER. At Milan, the Parties adopted broadly this approach by creating two optional schemes with t-CER units and long-term CER (l-CER) units to the choice of the project participants. The main difference between the two credits as their names show is the lifetime of the credit. The t-CER scheme allows as many t-CER as the number of tons of carbon that can be verified to have been additionally sequestered by the project since the project start date. At the end of the commitment period following the commitment period for which they were issued, they expire and have to be replaced prior to their expiry date by other credits in a special t-CERs replacement account of the national registry[38]. In sum, after 5 years, the t-CER expires and has to be replaced by another credit. If a new verification is done, new t-CER can be issued from the sink project every 5 years. On the contrary, the l-CER scheme issues l-CERs for every verified ton of carbon of the project and expires at the end of the crediting period of the project, with the limit of either (i) 20 years maximum, which may be renewed two times, or (ii) a maximum of 30 years[39] (however, these

---

[36] Ibid., Paragraph 1(e).
[37] Ibid., Paragraph 1(e).
[38] Ibid., Paragraphs 42, 43 and 44.
[39] Ibid., Paragraph 23.

limits also apply to projects under the t-CERs crediting method). Every l-CER has to be replaced by other credits as soon as the verification shows that the carbon stocks has decreased or if no verification report is provided within 5 years. A special l-CERs replacement account should be set up for this purpose in the national registry of the buyer[40]. Both schemes require verification every five years, and the liability rests on the buyer side, even when there is a reversal of net GHG removals for the l-CER scheme, and in both schemes the credits issued cannot be banked for the following commitment period[41]. To sum up, these rules fail to fix a minimum project lifetime with the consequence of potentially favoring short-term projects. Furthermore, they do not fully solve the non-permanence problem of sinks, which is natural to sinks component, in contrast to energy projects whose emission reductions are permanent. Small-scale forestry projects with simplified modalities can be undertaken under the CDM, the modalities rules shall be adopted at COP 10.

On the side of biodiversity preservation, the Milan decisions raise some alarm. Decision 11/CP.7 requested that the modalities developed for forestry projects under the CDM must take into account issues of socioeconomic and environmental impacts, including impacts on biodiversity and natural ecosystems. The sovereignty argument advanced by the potential host countries during the negotiations have watered down this concern, as there are no legal constraints, or standards, in the Milan decisions that secure internationally this preoccupation. Thus, it leaves open the door for large-scale monoculture plantations, which may be detrimental to local biodiversity. Regarding the definitions, by taking the definitions for afforestation and reforestation as agreed in Marrakech, and the requirements that land eligible for afforestation cannot have been forested for a period of at least fifty years, and that reforested lands cannot have contained forest on 31 December 1989, Parties intended to prevent a perverse incentive for non-Annex I Parties to clear forest in order to expend lands eligible for sink projects under Article 12. However, as recalled in the introductory part of decision 19/CP.9, the risks from the use of potentially invasive alien species and genetically modified organisms in forestry projects have to be evaluated by the host countries. The term 'host parties evaluate [...] potential risks' is not a real constraint for the host party to exclude or even to limit the use of such varieties in sinks projects. Nor is dissuasive the possibility open to the buyer countries to limit or exclude the use of credits generated by such projects. In order to restrict the use of such credits, the buyer countries need a means to know whether the credits issued are from sinks projects that employ invasive alien species or/and genetically modified organisms. Such information could be found in the project design document to be submitted to the designated operational

---

[40] Ibid., Paragraphs 47 and 48.
[41] Ibid., Paragraphs 41 and 45.

entity for validation, where the project participant is required to describe the techniques of the project activity 'including species and varieties [...]' [42]. Such information do not necessary follow the credit purchased by the buyer countries. Nevertheless, the decision allows in theory an Annex I Party, such as the EU, to exclude such credits from its Emission Trading Scheme.

Furthermore, although the project design document has to include a description of socio-economic and environmental impacts, "including impacts on biodiversity and natural ecosystems, and impacts outside the project boundary", undertaking a socio-economic or an environmental impact assessment is only mandatory if any 'negative impact is considered significant by the project participants or the host countries' [43]. Thus, the interpretation of what is 'significant' is left to the discretion of either the project participants themselves, or the host country, with no guarantee that future negative impacts would be considered before the sinks projects start.

Also, no international criteria or indicators to assess and compare sustainable development impact of sinks projects under CDM have been adopted. This concern is left to the host country, which determines itself whether afforestation or reforestation project activity under the CDM assists it in achieving sustainable development[44]. Finally, in the introductory part of Decision 19/CP.9 the reference to other multilateral environmental agreements, such as to create a synergy with the Biodiversity Convention, has been deliberately deleted to turn into 'Cognizant of relevant provisions international agreements that may apply to afforestation and reforestation project activities under the clean development mechanism', which bring a confusion on what should be the relevant convention. As a result, biodiversity concerns either in the Marrakech Accords or the Milan decisions do not constitute a strong barrier to counterbalance the economic incentives established under the Protocol, as will be seen in the next section.

4 Economic Incentives in the Marrakech Accords

Once the Kyoto Protocol enters into force, Parties will earn credits for LULUCF activities. Hence, Parties will probably establish incentive schemes to increase the amount of carbon units issued by means of these activities. We analyze in this section the expected outcomes of these incentives, focusing on forest manage-

---

[42] Decision 19/CP.9, Draft Decision-/CMP.1, Annex, Appendix B FCCC/CP/2003/Add.2, Paragraph 2(a).

[43] Decision 19/CP.9, Draft Decision-/CMP.1, Annex, FCCC/CP/2003/Add.2, Paragraph 12(c) and Appendix B Paragraph 2(j) and 2(k).

[44] Decision 19/CP.9, Draft Decision-/CMP.1, Annex, FCCC/CP/2003/Add.2, Paragraph 15(a).

ment and on afforestation and reforestation (since these are the activities expected to account for the lion's share).

4.1 Forest Management

During the discussion in this subsection we will assume the existence of a forest, that the type of forest is not going to be changed and that no real risk of disappearance exists. These are realistic assumptions in contexts where deforestation is no longer relevant and where forests are protected by effective laws. In this scenario the main[45] decision of the agent is the harvesting age (the rotation). Bringing this scenario to the Kyoto framework, we examine a forest management[46] alternative that could be incorporated by means of Article 3(4) of the Kyoto Protocol.

We discuss the different objective functions of the agent if he: (i) takes into account timber values exclusively, (ii) incorporates carbon sequestration values and (iii) also integrates biodiversity values. For a more complete presentation of the mathematical exposition see Caparrós and Jacquemont (2003).

The theoretically sound objective function to maximize timber values is the Faustmann formula, which maximizes the infinite flow of net revenues (Samuelson 1976). The Faustmann objective function can be written in a simple form as (assuming clear-cutting):

$$PV_w = G(T)e^{-rT} \cdot [1 + e^{-rT} + e^{-2rT} = \ldots] = \frac{G(T)e^{-rT}}{1 - e^{-rT}} \quad (1)$$

where, PV_w is the present value of timber; G is the timber net value function[47]; r is the discount rate; and T is the rotation period. The typical form of the timber value function increases with age up to a point where timber is too old for most

---

[45] An alternative forest management strategy to increase sequestration is to use fertilisation products. Fertilisation can have negative impacts on biodiversity (ESCBD 2000), but we do not consider it in this study (it supposes a change in growth functions).

[46] If the forest type is changed (i.e. changing the species) the analysis is closer to the situation described in the next section. As already discussed, this possibility has not been explicitly ruled out by the definition of forest management proposed in the Marrakech Accords. In addition, credits for Forest Management can only be earned inside Annex-I countries (OCDE and economies in transition).

[47] G(T) is used, instead of $P_wW(T)$ with P_w constant, as in the following heading, to allow for the influence of age on timber price.

commercial uses and its value decreases with age, mainly due to the influence of diseases and malformations[48].

Englin and Callaway (1993) and Van Kooten, Binkley and Delcourt (1995) proposed, independently, very similar approaches to modify the Faustmann formula in order to incorporate carbon sequestration values (see also Caparrós, Campos and Martín 2003). Basically, their proposal consists of valuating each ton of carbon sequestered with a given price (P_c), which is paid to the forest owner when carbon is sequestered and is paid by the owner when carbon is released. The question that arises is whether this approach is adequate to model the expected outcomes of the Marrakech Accords. According to the Marrakech Accords, and for units issued by means of Article 3.4 of the Kyoto Protocol, carbon sequestered in a given year is added to total allowances (added to the assigned amount of the Party, neglecting any temporality issue) and carbon liberated is considered as an emission[49] (subtracted from the assigned amount). Hence, if incentives are established to maximize national sequestration through forest management, the schema described above is appropriate[50].

Concentrating only on carbon related terms, the objective function can be written as follows (the overall objective function for the agent would be formed summing up (1) and (2)):

$$PV_c = \frac{\int_0^T P_c \cdot g'(t) \cdot e^{-rt} dt + \int_T^\infty P_c \cdot h'(t) \cdot e^{-rt} dt}{1 - e^{-rT}} \tag{2}$$

PV_c is the present value of carbon sequestration; P_c is the carbon price; g (t) is the carbon sequestered at each moment during the growth of the trees (incorporating above-ground biomass and below-ground biomass in the terminology of the Marrakech Accords)[51]; h (t) is the carbon sequestered in deposits after harvesting (including litter, dead wood and soil organic carbon[52]); and t is time.

As it occurs for timber in the first stages, the rate of carbon sequestration is positive and decreases with the age. However, the total amount of carbon in the

48 We assume $G'(T)>0$ in the first stages and $G''(T)<0$, assuming that $G'(T)\leqslant 0$ is possible for long rotations.

49 Annex (Definitions, modalities, rules and guidelines relating to LULUCF activities under the Kyoto Protocol) DD-/CMP.1 (LULUCF), Paragraph 17.

50 However, the cap imposed on Forest Management (see previous section) implies that incentives are only necessary if the cap is not surpassed without additional measures. If no incentive measures are established no negative impacts for biodiversity will occur by definition (since we only consider distortions not present in the current situation).

51 Annex (Definitions, modalities, rules and guidelines relating to LULUCF activities under the Kyoto Protocol) DD-/CMP.1 (LULUCF), Paragraph 21.

52 Idem.

forest does not necessarily decrease, as it occurs for the value of timber in old forests. Diseases and malformations influence the commercial value of timber but they do not significantly reduce the carbon content of the biomass[53]. Since the liberation of carbon is gradual, through oxidation, h (t) is a decreasing function[54]. This last function is not applicable if harvested wood products[55] (HWP) are not taken into account, since this option is equivalent to assuming instantaneous liberation of the sequestered carbon (the decision on HWP has been left for further negotiation rounds)[56].

However, a standing forest can have other values apart from timber and carbon sequestration. The theoretical incorporation of these values was proposed by Hartman (1976), such as using recreational values as an example (biodiversity is another example of a value of a standing forest). The Hartman objective function can be written (showing again only the non-timber part of the equation):

$$PV_B = \frac{\int_0^T P_b B(t) \cdot e^{-rt} dt}{1 - e^{-rT}} \tag{3}$$

PV_B is the present value for biodiversity values, B is the biodiversity function, and P_b is the biodiversity shadow price. The biodiversity value function typically increases monotonically with age (in general no reasons exist to expect it ever to decrease)[57].

The aggregate objective function (PV_S) is constructed summing up the different objective functions ((1), (2) and (3)). We assume that PV_S is society's objective function. First order conditions can be shown to be (Caparrós and Jacquemont 2003):

$$[G'(T) + P_c \cdot (g'(T) - h'(T)) + P_b B(T)] = r(1 + e^{-rT} + e^{-2rT} + \cdots)$$

$$\left[rG(T) + rP_c \left(\int_0^T g'(t) \cdot e^{-rt} dt + \int_T^{\infty} h'(T) \cdot e^{-rt} dt\right) + rP_b \int_0^T B(t) e^{-rt} dt\right] \tag{4}$$

Interpretation of this formula follows conventional lines, weighting the value of waiting an additional year (left hand side) against the interest forgone by not

[53] Hence, we assume $g'(t) \geqslant 0$ and $g''(t) < 0$. $g''(t) < 0$ might not be true for the first years, but it is an adequate assumption for the ages where the rotation is decided.

[54] With $h' \leqslant 0$ and $h'' < 0$, indicating that carbon liberation occurs mainly in the first moments after cutting.

[55] It refers to the accounting of the carbon stored in wood products after harvesting.

[56] FCCC/SBSTA/2001/L.12.

[57] Thus, we assume $B'(t) \geqslant 0$.

investing the future monetary stream associated to felling at T (right hand side). The first term of the left hand side accounts for the increase in the value of timber, the second (in brackets) for the increase in carbon sequestered and the third for the value of biodiversity. The first term of the right hand side accounts for the multiple rotation aspect of any version of the Faustman formula. The second term incorporates the monetary stream associated with future rotations for each of the three benefits considered.

However, that is the outcome only if the agent internalizes carbon sequestration as well as biodiversity. If this is the case, its decision will lead to the social optimum. The problem is that currently only $G(T)$ is internalized by private agents and that the implementation of the Kyoto Protocol will only internalize carbon values, leaving biodiversity out of the market and therefore out of the decision process. To correct this deviation, the conventional approach is to establish an incentive to ensure that the agent's strategy coincides with the social optimum. The problem arises from the difficulty to estimate the biodiversity function (B) and especially the price for biodiversity (P_b) that will probably imply that biodiversity will remain outside of the markets.

The influence of partial internalization of external costs and benefits on optimal management is not easy to establish regardless of the particular form of the different valuation functions involved and of the discount rate applied by the agent. However, a more detailed discussion of equation (4) enables one to determine which are the overall tendencies.

By setting $P_c = P_b = 0$ in equation (4), the original Faustman formula is recovered. The left hand side of equation (4) is simplified to G'(T), which declines with age. This is weighted up, on the right hand side of equation (4), with a non-discounted value ($rG(T)$), and a set of discounted values (accounting for the future rotations). Contrary to the rest of the benefits analyzed below, the timber part of the function offers a positive undiscounted value at the moment of felling. This usually implies that the timber part of equation (4) holds for relatively short rotations, especially with high discount rates as used by private agents (since $rG(T)$ is not discounted for the current rotation, it increases together with the discount rate r).

By setting $G(T) = P_b = 0$, the carbon part of equation (4) is isolated. The left hand side, waiting one additional year, has two positive terms, which only slightly decline with age (recall $h' < 0$). On the right hand side of equation (4), all future benefits are discounted since they are associated with the growth of future generations, while costs also appear[58]. This enables us to expect that, at least for high discount rates, the optimal rotation focusing on carbon sequestration will be reached at an older age than for the commercial benefits (if it is at all optimal to

[58] The integral of $h'(t)$ is negative since $h'(t) < 0$, in addition, since $h''(t) < 0$ the costs are close to the time of felling.

cut at any moment in time). This is especially true if gradual payment associated with carbon liberation is substituted by an instant payment for all the carbon contained in the timber felled, as is the case if harvested wood products are not considered (this would imply an undiscounted cost associated with felling).

Finally, by setting $G(t) = P_c = 0$, equation (4) focuses on biodiversity. The comparison is now between the full value for biodiversity in the year (left hand side) – not only the increase in this value – versus a discounted value representing the future benefits of biodiversity associated with the growth of the next generation of trees after felling (right hand side). In addition, since the value of biodiversity does not generally decrease with age ($B'(t) \geqslant 0$) the left hand side of equation (4) increases with time (contrary to the case described for timber above). The right hand side also increases, but at a lower rate (due to the influence of the discount rate), so that the two sides will in general never equalize. Hence, the expected outcome is that the optimal strategy, from a biodiversity point of view, would be to never cut (once more, this is especially true for high discount rates).

To sum up, the expected outcome if only carbon is internalized – in addition to the commercial values already provided – is that the optimal rotation age would increase, since the new terms incorporated in equation (4) tend to raise this rotation. But this increase is expected to be lower than the one that would occur if biodiversity were also internalized. Thus, providing incentives to private agents to take into account carbon sequestration when setting the rotation period of a managed forest should have positive effects for climate change and positive impacts on biodiversity (or at least no major negative impacts). Empirical studies confirm these expectations. Englin and Callaway (1993), Van Kooten, Binkley and Delcourt (1995) and Campos and Caparrós (1999) have shown that incorporating carbon sequestration in the objective function of the agent implies longer rotation periods, at least for high discount rates. Englin and Callaway (1995) analyzed numerically the environmental impacts of carbon sequestration maximization regimes, finding that, for high discount rates, the externalities associated with old forests are enhanced in regard to timber maximization strategies.

4.2 Afforestation and Reforestation

Under this heading, the choice between two types of forests when reforesting[59] agricultural land, will be analyzed. Based on the Marrakech Accords, Parties can issue credits through afforestation and reforestation by means of Article 3.3 of the Kyoto Protocol if the land is located in an Annex I country that ratifies the

[59] Afforestation and reforestation will be treated as synonyms, official definitions can be found in Paragraph 1(f) Annex DD-/CMP1.

Protocol (or eventually via Article 6 and JI), and by means of Article 12 (CDM) if the land is located in any Non-Annex I Party. Thus, as discussed in the previous section, incentives will probably be created to get forest managers to take carbon sequestration into account. For afforestation and reforestation undertaken inside an Annex-I country, the incentive schema will be probably similar to the one described in the previous section (associating payments with the actual carbon budget). For credits earned by CDM projects, and as discussed above, two methods have finally been accepted: the t-CERs and the l-CERs. Although the model discussed in this section follows Caparrós and Jacquement (2003), which was published before COP9, the general results obtained apply to both methods, as discussed below.

It is usually accepted that biodiversity increases when degraded and agricultural lands are converted into forests (IPCC 2000). However, this is only true in regard to indigenous forests and not when the 'reforestation' is actually the setting up of rapidly growing alien species plantations. It is also not true where pre-existing land uses have high biodiversity values (IPCC 2000). Matthews, O'Connor and Plantinga (2002) have quantified bird biodiversity associated with reforestation projects in the United States and have found further evidence of the potential negative impacts of reforestation regimes. See FOEI *et al.* (2000) for an overview of these potential negative impacts.

As stated, Caparrós and Jacquemont (2003) have formalized this decision process, and we will base our analysis on their results (see also Van Kooten (2000)). We will assume that the agent can choose between two types of forest: one with greater carbon sequestration potential (type 1) and another with greater biodiversity values (type 2). A typical example of this situation is when reforestation with a fast growing alien species[60] (forest, or plantation, type 1) is compared with a natural indigenous species alternative (forest type 2).

In this context, the following definitions are used: L = total land available for reforestation; $a(t)$ = agricultural land at t (state variable); $f_1(t)$ = reforested land of forest type 1 (state variable); $f_2(t)$ = reforested land of forest type 2 ($f_2(t)$) can be eliminated from the model as state variable, since $f_2(t) = L - a(t) f_1(t)$); u_i (t) = total area reforested at time t of forest type $i (i = 1, 2)$ (control variables); $k_i(u_i)$ = reforestation cost of type i (function of the amount of land reforested in a given year); r = discount rate; $A(x)$: a space-related function describing annual marginal net revenues associated to present agricultural uses (this variable includes commercial as well as non-commercial values). Also, $F_i(x)$ is defined as a space-related function for the annual marginal net revenues of forested land type $i (i = 1, 2)$. These functions are supposed to have three terms: $F_i(x) = P_w W_i(x) + P_c C_i(x) + P_b B_i(x)$. This applies where: $W_i(x)$, $C_i(x)$ and

[60] Fast growing species do not always yield higher carbon sequestration per hectare when mature, but since sequestration occurs faster the present value for theses species of the sequestration is generally higher, due to the effect of the discount rate.

$B_i(x)$ represent physical quantities associated with timber, carbon sequestration and biodiversity respectively, and P_w, P_c and P_b the prices, real or shadow ones, associated with these three components. For simplicity, these values are annualized, ensuring that the investment incentives are not changed, i.e. we include in our model the constant annual income that would equalize the actual future stream of incomes generated for each value over the entire reforestation cycle (Caparrós and Jacquemont, 2003). Thus, we abstract from the precise crediting method (t-CER or l-CER), since the relevant question for us is the additional income generated by the reforestation (which we assume irreversible) with one or another species if carbon sequestration and/or biodiversity values are internalized.

Nevertheless, the t-CER method and its five years period may imply an additional incentive for short-term projects (which would need fast growing species to get meaningful carbon sequestration yields). This is especially true if agents expect the Kyoto process to fail in the objective of starting a second commitment period. Under this assumption, the t-CERs could be a short-term strategy to escape the obligation to undertake any emission reduction. In addition, remark that rules and modalities have just been decided for the first commitment period and that only a maximum number of years for the crediting period has been decided (see above). That is, no minimum length has been decided for a forestry project to be eligible, although this strategy would have been probably advisable to reduce the incentives for short-term projects.

Returning to our model, the objective function if the agent incorporates all benefits can be written as:

$$Max V = \int_0^\infty \prod(t) \cdot e^{-rt} dt \qquad (5.a)$$

$$\begin{aligned} \prod(t) = \int_0^{a(t)} A(x)dx + \int_0^{f_1} F_1(x)dx - k_1(u_1) \cdot u_1(t) + \\ \int_0^{L-a-f_1} F_2(x)dx - k_2(u_2) \cdot u_2(t) \end{aligned} \qquad (5.b)$$

And first order conditions for the optimal control model can be shown to be (Caparrós and Jacquemont 2003):

$$\begin{aligned} \frac{P_w W_2(L-a-f_1) + P_c C_2(L-a-f_1) + P_b B_2(L-a-f_1)}{r} - k_2(0) = \\ \frac{P_w W_1(f_1) + P_c C_1(f_1) + P_b B_1(f_1)}{r} - k_1(0) = \frac{A(a)}{r} \end{aligned} \qquad (6)$$

The interpretation of equation (6) follows conventional lines. In the steady-state equilibrium, the revenues of reforesting one additional hectare of forest type 2 have to be equal to the revenues associated to one additional hectare reforested of forest type 1, and to the revenues associated to the agricultural use of that hectare. Nevertheless, the problem hidden in equation (6) is again that only P_w and the commercial part of A are actually provided by existing markets and that the proposed market for carbon will internalize only P_cC. If this is done regardless of the influence on P_bB, the social optimum will be reached only by coincidence. The issue is more relevant since C and B will generally go in opposite directions, so that an alternative established in order to maximize C would reduce B, or even make it negative. If only P_cC is internalized, and with the assumptions made for forests types 1 and 2, the expected outcome is a suboptimal over-plantation of forest type 1 (fast growing alien species). That is, in equilibrium, the agent will equalize equation (6) regardless of the biodiversity term, and since $P_bB_1 < P_bB_2$, the last unit would have gone to reforest with forest type 2 (natural indigenous species) if all benefits were considered. Depending on the form of the carbon and biodiversity functions, exclusive internalization of carbon would lead to an even worse situation from the point of view of society than the current one (if the loss in biodiversity is higher than the gain due to the increase in carbon sequestration).

The conventional solution to this deviation requires the economic valuation of biodiversity, with the problems already mentioned. An alternative approach is to work with the constraints imposed to assess whether or not this could bring the agent's strategy closer to social optimality. As described above, the Marrakech Accords have set an absolute cap for afforestation and reforestation in the CDM (but not for afforestation and reforestation in Annex I countries). However, this alternative provides no incentive to favor forest type 2. This kind of restriction implies that the choice between species is done regardless of the constraint until it becomes binding, favoring therefore forest type 1. Once the restriction becomes effective, no additional forestation would occur at all.

Finally, another alternative would be to limit the total amount of carbon sequestration per unit of land to which an economic value may be given. This is a straightforward strategy to reduce or remove the influence of the different amounts of carbon sequestered by the two types of forests. If the maximum amount of carbon per unit of land taken into account (C_m) is set such that $C_m = C_2$, the differences in carbon sequestration between the two species disappear and the agent will choose one to set up focusing on timber (since biodiversity is not internalized). Nevertheless, both types of forest would be favored compared to agricultural land due to the internalization of carbon sequestration, so that the total area reforested would increase (maintaining the current proportions between forest type 1 and 2). This kind of restriction is justified as long as the difference in biodiversity values is higher than the values internalized by the market and therefore considered by the private agent (timber and carbon sequestration). The IPCC (2000) timidly proposed this option, acknowledging at the same time the difficulties to incorporate non-carbon environmental and social concerns into quantitative limits on carbon credits. However, a conservative value for the major types of regions could reduce the incentives to plant fast growing alien species. In any case, no per hectare cap was included in the Marrakech Accords.

5 The Convention on Biological Diversity: an Alternative Protection for Forests?

Two different environmental agreements, coexisting in the international legal order, are related to forests. As shown in the previous section, the choices in some forest practices with the view to enhance carbon sinks may have an adverse impact on biodiversity. Such a result may conflict with the aims of the CBD or other international conservation instruments related to forests[61]. The question arises as to which instrument should prevail when two overlapping instruments differ with regard to their objectives or the actions to be undertaken. In this context there are two overriding issues. First, whether the CBD or the Kyoto Protocol takes precedence in situations of conflict. Second, whether international legal rules governing these instruments (as codified in the 1969 Vienna Convention) indicate how conflict situations may be resolved. Before answering these questions, an assessment of the CBD's aims and requirements is needed.

5.1 CBD Aims and Requirements

First of all, the CBD is aimed at the conservation of biological diversity, the sustainable use of its components, and the fair and equitable sharing of the benefits arising from the utilization of genetic resources[62]. Thus, the Convention[63] applies to all processes and activities which have or are likely to have significant impacts on the conservation and sustainable use of biological diversity undertaken within each Party's jurisdiction or control, and also beyond their national jurisdictions (in relation to forests it applies only within a Party's jurisdiction)[64]. To do so, the CBD calls upon Parties to adopt national policies consistent with the conservation and sustainable use of biological diversity[65]. It requires Parties to: (i) regulate or manage biological resources relevant to the

[61] See supra note 10.
[62] CBD, Article 1.
[63] On the Convention on Biological Diversity, see Wolfrum (1996).
[64] CBD, Article 4.
[65] Ibid, Article 6.

conservation of biological diversity; (ii) promote the protection of ecosystems by establishing systems of protected areas, (iii) rehabilitate and restore degraded ecosystems; and, (iv) prevent or eradicate alien species that threaten ecosystems[66].

In addition, the CBD requires Parties to identify and monitor processes and activities likely to have significant adverse impacts on biodiversity and to regulate such activities whose significant impacts have been ascertained by the Party in question[67]. In order to accomplish these tasks, each Party, 'as far as possible and as appropriate' shall introduce national procedures requiring environmental impact assessments (EIAs) for activities that may cause significant adverse impacts with a view to avoiding or minimizing those impacts[68]. The CBD requires that the traditional knowledge of indigenous peoples relevant to the conservation and sustainable use of biodiversity should be respected and maintained through national legislation[69]. It aims to integrate biodiversity concerns into the conduct of every human activity. Concerning forests, the CBD COP 3 recognized forests as playing crucial roles in maintaining global biological diversity, such as combating desertification, enhancing habitats, and promoting indigenous species for the re-establishment of native forests[70]. Thus, a role may exist for LULUCF forestry projects that follow the CBD's requirements to contribute to the restoration of degraded ecosystems (Cullet and Kameri-Mbote 1998).

Are these requirements strong enough in protecting forest biodiversity? The wording used in Articles 5 and 6 of the CBD to attain the CBD's objectives is weak[71]. It results from the attempt of the CBD's drafters to link two opposing principles: that States have sovereign rights to exploit their natural resources; and, that States must not cause damage to the environment resulting in a loss of biodiversity[72]. The concept of sustainable development could reconcile these two

---

[66] Ibid., Articles 8 (c), (d), (f), and (h).

[67] Ibid, Article 7(c).

[68] Ibid, Article 14.1 (a).

[69] Ibid, Article 8 (j).

[70] UNEP/CBD/COP/3/16.

[71] 'Each Contracting Party shall, as far as possible and as appropriate ...' and 'each Contracting Party shall in accordance with its particular conditions and capabilities ...' is used respectively in Articles 5 and 6 of the CBD. See Swanson (1999).

[72] This is clearly stated in the Preamble of the CBD: 'Affirming the that States have sovereign rights over their own biological resources, (...) also that States are responsible for conserving their biological resources.' Moreover, CBD Article 3 expressively recalls that:

competing principles by moderating one with the other. The idea that development should meet 'the needs of the present without compromising the ability of future generations to meet their own needs' (WCED 1987) provides a middle position at which sovereign rights and the protection of biodiversity can both be accommodated. However, the notion of sustainable development is not a concept of customary international law from which legal rules governing State behavior can be derived (Beyerlin 1996). Such a consequence is due to the fact that the notion of sustainable development lacks reliable criteria on how States should balance their development with the protection of the environment. International agreements and declarations, which endorse the sustainable development concept, remain elusive in defining it[73], resulting in the evolution of sustainable development simply as a malleable political objective that States aim to achieve (Sands 1999). The legal definition of 'sustainable use' in the CBD (which is derived from the principle of sustainable development) is also abstract and fails to provide binding standards of behavior[74] (especially if interpreted in conjunction with the weak wording of the CBD's other provisions). Article 2 of the CBD states that 'sustainable use means the use of components of biological diversity in a way and at a rate that does not lead to the long-term decline of biological diversity, thereby maintaining its potential to meet the needs and aspirations of present and future generations.' The definition declares how Parties *should* act rather than

---

'States have, in accordance with the Charter of the United Nations and the principles of international law, the sovereign right to exploit their own resources pursuant their own environmental policies, and the responsibility to ensure that activities within their jurisdiction or control do not cause damage to the environment of other States or of areas beyond the limits of national jurisdiction.' It reflects literally Principle 2 of the Rio Declaration on Environment and Development 1992 (UN Doc. A/CONF.151/26/Rev.1, Vol.1), and Principle 21 of the 1972 Stockholm Declaration (UN Doc. A/CONF/48/14/Rev.1).

[73] UNEP Governing Council Decision 15/2 of May 1989 added that sustainable development requires 'the maintenance, rational use and enhancement of the natural resource base that underpins ecological resilience and economic growth (...) and implies progress towards international equity'. (Annex II, GAOR, 44th Session Suppl. No.25 (A/44/25). See also (non exhaustive list): UNFCCC, Articles 2 and 3; the 1992 Convention on the Protection of the Marine Environment of the Baltic Sea Area, (Helsinki) 9 April 1992, Article 15; General Agreement on Tariffs and Trade: Multilateral Trade Negotiations Final Act Embodying the Results of the Uruguay Round of Trade Negotiations, (Marrakech), 15 April 1994, Preamble. In terms of non-legally binding instruments, see: Forests Principle, note 10 above.

[74] CBD Article 2.

how they *must* act in terms of the sustainable use of their respective biological resources. Therefore, the CDB fails to provide strong binding rules and its implementation only relies on the good faith of its Parties (as long as they have sovereign rights over their own biological resources). Thus, states retain significant freedom of action to take measures dealing with economic and social aspects of forests, under the condition that the results shall be the conservation or the sustainable use of those resources. Finally, the CBD lacks real financial incentives, by which developing countries can balance the economic benefits resulting from timber exploitation and carbon sequestration maximization strategies (Wolfrum 1996, Sand 1996, Bothe 1996). However, the CBD does recognize biodiversity loss as a global environmental problem and promotes the conservation of biodiversity a common concern of humankind[75]. Furthermore, the CBD has established a quasi-universal regime, encompassing an ecosystem approach, which takes into account the biological and ecological interactions between species of an ecosystem[76]. The ecosystem approach aims at protecting the complex relations among living and non-living components of ecosystems by limiting negative impacts from human exploitation of one component from impacting on other components in an ecosystem. In this regard, the CBD obliges Parties to adopt and apply rules for the sustainable use of resources. As seen above, it lays down a procedural approach directed at minimizing the negative impacts of human activities on biodiversity[77]. Thus, it enunciates the importance of integrating socio-economic and environmental goals as well as inter-generational considerations. In the end, it is the responsibility of the Parties to the CBD to translate this objective into national regulations.

5.2 The Compatibility Problems

Assessing the adequacy of the international law of treaties in dealing with compatibility problems among environmental international instruments is of relevance. As discussed above, the Kyoto Protocol and the CBD are both concerned with forests, however, their provisions may either conflict with, or supplement each other. How these instruments interact is therefore of critical importance in determining their effects. As there is no hierarchical structure of

[75] CBD Preamble, Paragraph 3.

[76] CBD Article 2.

[77] In October 2001, the CBD had 183 Parties (http://www.biodiv.org). Concerning the ecosystem approach endorsed by the CBD, see CBD Decision V/6, UNEP/CBD/5/23.

international law and because treaties are equally binding[78], when the Kyoto Protocol enters into force, it will be on equal standing with the CBD[79]. Determinations of the compatibility of these instruments therefore rests in the rules of international law as codified by the 1969 Vienna Convention on the Law of the Treaties.

The 1969 Vienna Convention on the Law of the Treaties (1969 Vienna Convention) applies to written agreements that are governed by international law and concluded between States[80]. In cases of conflicts between two international agreements, reliance is placed on Article 30 of the 1969 Vienna Convention. However, with regard to conflicts between the Kyoto Protocol and the CBD, the 1969 Vienna Convention is inapplicable because the 1969 Vienna Convention only applies to successive treaties on the same subject matter[81]. The CBD and the Kyoto Protocol do not govern the same environmental subject matter. They merely overlap on one issue, while their primary aims are different. This feature excludes the application of the *lex specialis* principle, which states that where two treaties govern the same subject matter, the one that adds some specification in comparison to the other prevails to the extent that its provisions do not conflict with the aim of the general instrument (a perfect example is the Kyoto Protocol to the UNFCCC, which lays down more specific rules to reduce GHG emissions than the UNFCCC). Finally, the applicability of the *lex posterior* principle, as set out in Article 30.3 of the 1969 Vienna Convention, depends on the status of the States as Parties or non-Parties to the conflicting agreements. It provides that when all Parties to an earlier agreement are Parties also to a later agreement, the earlier is applicable so far as it is consistent with the later agreement. Thus, these rules do not help to define the relation between two overlapping instruments on one issue.

---

[78] With the exception of a treaty endorsing rules of *jus cogens* with an *erga omnes* effect, or conflicting with such a norm. See Nguyen, Dailler and Pellet (1987). Articles 53 and 64 of the 1969 Vienna Convention established a hierarchy of norms by providing that in case of conflict between a treaty and a norm of *jus cogens,* the treaty is void. Neither the CBD nor the Protocol contains such a norm. Another exception is Article 103 of the United Nations Charter, which establishes a hierarchical structure with regard to the Charter of the United Nations only.

[79] At the time of writing the Protocol had not yet entered into force.

[80] Convention on the Law of Treaties (1969 Vienna Convention), (Vienna) 23 May 1969, Articles 1 and 2.1.

[81] The Convention entered into force on 27 January 1980 and is considered as representing international customary law in its scope. See Wolfrum and Matz (2000) and Nguyen, Dailler and Pellet (1987).

A solution could be found in special treaty clauses, which purport to prevent or solve conflicts between agreements. Article 2.1(a) of the Kyoto Protocol expresses the willingness of the Parties to consider the obligations arising from other international environmental instruments[82]. However, the Protocol lacks a specific compatibility clause, which would prevent conflicts of obligations by establishing the precedence of those instruments adopted before it (Wolfrum and Matz 2000). In addressing the use of sinks, Article 2.1 of the Kyoto Protocol states that Parties should 'take into account' commitments arising from other relevant environmental agreements; however, it does not specify which agreements or obligations should be 'taken into account'[83]. In short, Article 2.1 is too generic to be considered as an equivalent to a compatibility clause as defined under Article 30.2 of the 1969 Vienna Convention, which applies when an agreement 'specifies that it is subject to, or that is not to be considered as incompatible with, . . .' another agreement on the same subject matter. Article 2.1 of the Protocol is more or less a reminder for the Parties to implement sinks policies in a manner consistent with relevant international environmental agreements to promote sustainable development[84]. In contrast, the CBD contains a clause (Article 22), which allows for the precedence of rights and obligations that bind the contracting Parties at the time of ratification[85]. This precedence has a limit: the exercise of those rights under previous agreements shall not threaten or cause serious damage to biodiversity. However, this exemption only applies to obligations and rights of Parties that existed before the CBD was ratified (at the time of writing, the Kyoto Protocol has not yet entered into force). Further, this exemption is strictly limited to the extent that the exercise of these obligations and rights will not threaten or cause serious damage to biological diversity, which leaves a wide margin of interpretation by the Parties to the CBD, and which could lead to the de facto precedence of the CBD in relation to other agreements (Wolfrum and Matz 2000). In principle, Parties to the CBD that ratify the Protocol will have to apply their national policies with the view to implementing

---

82 Protocol Article 2.1(a)(ii) states: 'Each Party included in Annex I [shall implement the] ... protection and enhancement of sinks and reservoirs ... taking into account its commitments under relevant international environmental agreements ...'.

83 Regarding the clause on relevant international agreements, the European Union proposal made reference to the linkage between the UNFCCC, the Convention on Biological Diversity, and the UN Conference on Environment and Development Forest Principles on the matter of sinks and forestry policy. However, the issue was not raised in any subsequent text (Depledge 2000). Neither the UNFCCC nor the Protocol contains such a clause.

84 Protocol Article 2.1 states that each Annex I Party 'in achieving its quantified emission limitation, ... in order to promote sustainable development: shall implement an/or further elaborate policies and measures in accordance with its national circumstances ...'.

85 CBD Article 22.1.

the Kyoto Protocol consistently with the CBD[86]. As such, a conflict between the Kyoto Protocol and the CBD would appear if the Kyoto Protocol imposes any obligations, which violate the rules of the CBD. In this case, the rules of international law are inadequate to resolve this conflict. However, in the present case, the Kyoto Protocol provides for the possibility to use LULUCF activities and gives incentives to do so. The Protocol lays down an obligation for Parties to perform such activities. Therefore, there are no conflicting obligations between the two instruments as such. The only legal question left is how far Parties to both instruments are allowed to follow those incentives, as provided by the Kyoto Protocol, without violating requirements of the CBD. Thus, new solutions have to be found in order to avoid, for the Parties to both conventions, a behavior that follows one instrument's incentives, which may result in a contradiction with their obligations under the other instrument.

5.3 Reconciliation at the National Level

What should be the attitude to be adopted by a State that is a Party to more than one international instrument focusing on a related issue? Based on the *pacta sunt servanda* principle of international law[87], Parties to an international agreement are required to fulfil their commitments ensuing from it in good faith without violating existing obligations from previous international instruments. Starting from this principle, Pontecorvo (1999) and others argue that States Parties to environmental agreements of global relevance, in taking advantage of certain incentives for fulfilling their obligations under one convention that are potentially conflicting with other international requirements, are under the duty to adopt a 'harmonizing approach'. This approach consists of taking into account pre-existing commitments. The CBD and the Kyoto Protocol are 'common interest treaties'. The common interests are that they deal with global environmental problems, establish general regimes, principles and standards, and prohibit Parties from making reservations to their provisions[88], which indirectly highlights the importance given to the principles embodied in these instruments. This implies that Parties to these instruments are under a 'moral obligation', in consideration of the common interest of the international community in the effective preservation of the global environment, to reconcile the provisions of

[86] This is consistent with the fact that the CBD encourages State Parties to promote national policies with the view to enhance the protection and sustainable use of biodiversity, and the fact that Protocol Article 2.1(a)(ii) requires that national policies for enhancing sinks take into account relevant international instruments. States are left to reconcile the policy objectives.

[87] 1969 Vienna Convention, Article 26.

[88] CBD Article 34; and UNFCCC Article 24.

these treaties in the light of a 'common interest clause'[89]. The harmonizing approach should encompass non-legally binding instruments relative to forests, such as the Forest Principle[90], which express general practice recognized by States as rules of international law (*opinio juris*). The harmonization approach seems to be endorsed by the Kyoto Protocol, which contains a 'common interest clause' in Article 2.1 requiring each Party to 'take into account its commitment under relevant international environmental agreements'. However, the fact that countries have avoided the negotiation of a binding regime for forests and delayed the adoption of a forestry protocol under the CBD highlights the fact that States are not prepared to undergo stringent standards for the sustainable use of forestry (Tarasofsky 1996, Henne and Fakir 1999). Rather, States have preferred to maintain their own criteria for forestry and for sustainable development. The Marrakech Accords reflect this idea by leaving the host party to determine whether JI and CDM projects (including forestry projects) contribute to its sustainable development. Attempts to introduce international criteria for sustainable projects have been rejected, as developing countries have been very sensitive to ceding their powers to define domestic sustainable development objectives[91]. This tendency to avoid international criteria is exemplified by the discretion that is inherent to the operation of the Kyoto Protocol's flexible mechanisms. During the process of validation of JI and CDM projects, an EIA (as recommended in the CBD) is only undertaken when a project participant or the host country considers negative impacts of these projects as being 'significant'[92]. No discussions within the international forum were held to define with more precision on what 'significant' negative impacts are. Even if EIAs are deemed as necessary, they are to be conducted in accordance with the procedures required by the host Party. This means that the interpretation of the significance of the impact depends on the Parties concerned with the project. As well, in cases where the host party has not legislated an EIA procedure, no EIA will be undertaken[93]. EIA (or other criteria for sustainable projects) is not therefore a requirement for the processes of

---

[89] Moreover, the interpretation by States of sinks-related provisions should encompass non-legally binding instruments relevant to forests, such as the Forest Principle, as they express *opinio juris* (Agenda 21, chapter XI, 'IPF Recommendations').

[90] See Forests Principle, note 10 above. Forest principles refer to a whole range of resources and services provided by forest ecosystems and call for their sustainable management.

[91] The EU proposal to list eligible activities that are sustainable under CDM projects was strongly rejected by the G77 and China and the Umbrella Group during COP 6 at The Hague in November 2000 (Fernandez 2000).

[92] Annex (Guidelines for the implementation of Article 6 of the Kyoto Protocol), Draft Decision-/CMP.1 (Article 6), FCCC/CP/2001/13/Add.2, Paragraph 33(d), and Annex (Guidelines for the implementation of Article 12 of the Kyoto Protocol), Draft decision-/CMP.1 (Article 12), FCCC/CP/2001/13/Add.2, Paragraph 37(c).

[93] Ibid.

validation and registration of project activities under the Protocol[94]. Without binding sustainable development-related requirements, no real incentives for the preservation of biodiversity exist under the Protocol's flexible mechanisms. The implementation of rules for the sustainable use of forest projects to protect biodiversity relies simply on the good faith of the Parties.

5.4 Inter-institutional Co-operation

Article 5 of the CBD invites Parties to cooperate where appropriate through competent international organizations on matters of mutual interest for the conservation and sustainable use of biodiversity. A synergy between conventions is necessary to ensure a coherent and integrated approach to global environmental problems[95]. Co-operation at the institutional level should enhance the implementation of coordinated measures on common areas of action such as forestry. Usually, efficient inter-institutional cooperation is necessary on three levels: (i) among inter-governmental bodies in relation to decisions on policies and work programmes; (ii) among scientific and other expert groups in information and assessment; (iii) and among secretariats in relation to developing policy, information resources, programmes planning and execution (Kimball 1997). This approach was endorsed by the Secretariat of the CBD, when issuing the Note to the UNFCCC at UNFCCC COP 6, where the Secretariat of the CBD distinguished two groups of collaboration activities[96]. One group is concerned with the analysis of the impacts of climate change on biological diversity and possible response measures. The other group explores the possibility of using the Protocol's incentive measures as vehicles to integrate biodiversity concerns into the policies and measures under the Kyoto regime. As discussed above, further work on these issues is being conducted by the SBSTA for recommendation to UNFCCC COP 9.

Indeed, the Kyoto Protocol offers good incentives to protect biodiversity by giving to Annex I Parties the opportunities to use sinks activities to gain credits. It sets detailed rules for implementation and lays down consistent enforcement

---

[94] As seen above, the requirement to report national legislation or administrative procedures that aim to protect biodiversity and sustainable use of national resources is not an eligibility requirement for participation in the Kyoto flexible mechanisms.

[95] Commission on Sustainable Development, Fifth Session (CDS-5), (New York) 8–25 April 1997, UN Doc. A/S-19/14 E/1997/60, at para. 109. The Nineteenth Special Session of the UN General Assembly recommended that the COPs of the UNCED Conventions co-operate in exploring ways of collaboration in order to have effective implementation of these conventions. See, 'Programme for further implementation of Agenda 21', UN Doc. A/RES/S-19/2, 19 September 1997.

[96] Annex 1, UNEP/CBD/SBSTTA/6/11, Paragraphs 3(a) and (b).

rules and procedures. The CBD's Subsidiary Body on Scientific, Technical and Technology Advice (SBSTTA) will need to develop criteria for the conservation and sustainable use of biological diversity to be used in the design of activities, or to be applied in monitoring and evaluating their implementation. The SBSTTA also will need to establish a list of negative activities to determine which activities should be subject to an EIA. Such work should be undertaken in collaboration with the relevant bodies of the UNFCCC[97]. At Bonn in July 2001, the SBSTA endorsed the formation of a joint liaison group between the CBD and UNFCCC secretariats, and called for the secretariat of the UN Convention to Combat Desertification to join this group in order to enhance cooperation[98]. The Ministerial Declaration issued at the end of UNFCCC COP 7 at Marrakech welcomed such cooperation, thus, encouraging Parties to explore synergies between the two conventions in order to achieve sustainable development[99].

The Marrakech Accords seem favorable towards the integration of the climate regime's policies and measures with those of the CBD. The Marrakech Accords state that the implementation of LULUCF activities must contribute to the conservation of biodiversity and the sustainable use of natural resources[100]. The Marrakech Accords also call upon the SBSTA to elaborate on the definitions and the modalities for afforestation and reforestation projects under the CDM 'taking into account ... the issues of non-permanence ... [and] environmental impacts, including impacts on biodiversity and natural ecosystems'[101]. Moreover, the SBSTA is required to investigate the application of biome-specific forest definitions for future commitment periods[102]. Here, the elaboration of the definition and modalities for sinks projects under the CDM seems to adopt the concerns for biodiversity protection. As previously stated this concern was flawed[103]. However, these concerns for activities under Articles 3.3 and 3.4 and under JI projects are vague (the wording used is 'contribute'), leading to further questions. How will these activities contribute to biodiversity? Which standards of sustainable use will be utilized for these activities in order to ascertain their contribution to biodiversity protection? Further, reporting requirements on biodiversity protection in the Accords is inefficient since it has been decided that

---

97 Ibid., Paragraphs 26(a), (b) and 27.

98 Annex, FCCC/SBSTA/2001/INF.3, Paragraph 9.

99 Decision 1/CP.7, Marrakech Ministerial Declaration, FCCC/CP2001/13/Add.1, Paragraph 3.

100 Draft Decision-/CMP1 (LULUCF), FCCC/CP/2001/13/Add.1, Paragraph 1(e).

101 Decision 17/CP.7 (Article 12), FCCC/CP/2001/13/Add.2, Paragraph 10(b).

102 Decision 11/CP7 (LULUCF), FCCC/CP/2001/13/Add.1, Paragraph 2(b).

103 These definitions will be forwarded to the Ninth Conference of the Parties of the UNFCCC. The definitions should introduce some disincentives relative to forest related-activities that have a significant impact in terms of greenhouse gas emissions such as forest burning, conversion of natural forests into secondary forests, primary forests harvesting.

the sinks reporting requirement under Article 7.2 of the Protocol is not an eligibility condition for the use of flexible mechanisms during the first commitment period[104]. In short, the Marrakech Accords postpone biodiversity concerns about sinks projects to a later date and avoid the issues at stake.

A real synergy between the conventions could be reached if the sustainable approach provided by the CBD was sanctioned and promoted by the Protocol. This approach should require certification that sinks activities are not only assessed for their capacities to remove GHG emissions, but also for their contributions to biodiversity conservation and other environmental needs.

6 Conclusion

We have shown in this chapter that putting an economic value on carbon sequestered by means of *forest management* is not expected to have a great negative influence on biodiversity. However, creating economic incentives for carbon sequestration by *afforestation and reforestation* is expected to yield a suboptimal over-plantation of fast growing alien species with a potential negative impact on biodiversity.

Our discussion on the KP's decisions has shown that the limits to LULUCF activities set out are mainly overall caps in terms of quantity of carbon, with almost no influence on the situation discussed above. Direct references to biodiversity have also been included in the Marrakech Accords, but they are general guidelines and cannot be seen as effective limitations.

On the other hand, the CBD lacks economic incentives to ensure that forest activities follow the Convention's ecosystem approach strategy. Whereas the Kyoto Protocol contains economic incentives, it does not advocate the use of an ecosystem approach, which could be used to protect biodiversity needs. The CBD does not provide stringent rules that allow Parties to see the conservation of biodiversity as a management constraint. However, the CBD does provide relevant guidelines for the sustainable use of forests. For this reason alone, the provisions of the CBD should be further considered when elaborating rules under the Protocol. In particular, attention should be paid to limitations on the use of alien plants and tree species to avoid conflicts between the conventions and to ensure that no incentive measures under the Protocol will result in violations of the CBD (as discussed above, quantitative per hectare cap could reduce the suboptimal incentives created to establish fast growing plantations). This is especially important because international rules governing treaties are not drafted to prevent or respond to such conflicts. The result is that if problems are to be

[104] Annex (Guidelines for the preparation of the information required under Article 7 of the Kyoto Protocol/Reporting of supplementary information under Article7, paragraph 2), Draft Decision-/CMP.1 (Article 7), CCC/CP/2001/13/Add.3, Paragraphs 26.

addressed, the Parties to both the CBD and the Kyoto Protocol must adopt a harmonization approach to their regimes. However, the implementation of this harmonization approach depends on the good will of States. As a result, discrepancies among States on the application of these rules may arise. To address this situation, the adoption of international standards of sustainable use for forestry may be the answer. Finally, our discussion has shown that an integrated approach, using the synergies of both regimes at an institutional level, offers an opportunity to apply harmonized and coordinated biodiversity concerns together with greenhouse gas mitigation.

References

Beyerlin, U. (1996). 'The Concept of Sustainable Development', in R. Wolfrum (ed.), *Enforcing Environmental Standards: Economic Mechanisms as Viable Means?* Berlin: Springer, 95–121.

Bothe, M. (1996). 'The Evaluation of Enforcement Mechanisms in International Environmental Law', in R. Wolfrum (ed.), *Enforcing Environmental Standards: Economic Mechanisms as Viable Means?* Berlin: Springer, 13–38.

Campos, P. and Caparrós, A. (1999). 'Análisis Económico de la Fijación de Carbono por el Pino Silvestre', in F. Hernández (ed.), *El Calentamiento Global en España: un Análisis de sus Efectos Económicos y Ambientales.* Madrid: CSIC, 141–162.

Caparrós, A., Campos, P. and Martín, D. (2004). 'Influence of Carbon Dioxide Abatement and Recreational Services on Optimal Forest Rotation'. *International Journal of Sustainable Development*, 6 (3): 345–358.

Caparrós, A. and Jacquemont, F. (2003, forthcoming). 'Conflicts between Biodiversity and Carbon Sequestration Programs: Economic and Legal Implications'. *Ecological Economics.*

Convention on Biological Diversity (2001). *Overview of the Interlinkages between Biological Diversity and Climate Change – the Climate Change Phenomenon.* http://www.biodiv.org

Cullet, P. and Kameri-Mbote, A. (1998). 'Joint Implementation and Forestry Projects: Conceptual and Operational Fallacies'. *International Affairs,* 74: 393–408.

Depledge, J. (2000). *Tracing the Origins of the Kyoto Protocol: an Article-by-Article Textual History.* FCCC/TP/2000/2, http://www.unfccc.int.

Englin, J. and Callaway, J.M. (1993). 'Global Climate Change and Optimal Forest Management'. *Natural Resource Modeling*, 7 (3): 191–202.

Englin, J. and Callaway, J.M. (1995). 'Environmental Impacts of Sequestering Carbon Through Forestation'. *Climate Change*, 31: 67–78.

Executive Secretary of the Convention on Biological Diversity (ESCBD) (2000). *Climate Change and Biological Diversity: Cooperation between the Convention on Biological Diversity and the United Nations Framework Convention on Climate Change.* http://www.cbd.org

Fernandez, M. (2000). 'The Negotiation of the Clean Development Mechanisms at COP6: Precautionary versus Cost-effective Policies'. *ELNI Newsletter*, 2: 2–11.

Friends of the Earth International (FOEI), World Rainforest Movement, FERN and Future in Our Hands (2000). *Tree Trouble.* FOEI, 1–76. http://www.foei.org

Hartman, R. (1976). 'The Harvesting Decision When a Standing Forest Has Value'. *Economic Inquiry*, 14: 52–58.

Henne, G. and Fakir, S. (1999). 'The Regime of the Convention on Biological Diversity on the Road to Nairobi'. *Max Planck Yearbook of United Nations Law*, 3: 315–361.

Intergovernmental Panel on Climate Change (IPCC) (2000). *IPCC Special Report: Land Use, Land Use Change and Forestry*. Geneva: WMO-UNEP.

Jacquemont, F. and Caparrós, A. (2002). 'The Convention on Biological Diversity and the Climate Change Convention 10 Years After Rio: Towards a Synergy of the Two Regimes?' *Review of European Community and International Environmental Law*, 11(2): 169–180.

Kimball, L.A. (1997). 'Institutional Linkages between the Convention on Biological Diversity and Other International Conventions'. *Review of European Community and International Environmental Law*, 6 (3): 239–248.

Matthews, S., O'Connor, R. and Plantinga, A.J. (2002). 'Quantifying the Impacts on Biodiversity of Policies for Carbon Sequestration in Forests'. *Ecological Economics* 40 (1): 71–87.

Nguyen, D., Dailler, D. and Pellet, P. (1987). 'Application des Traités' in Droit International Public. *Librairie Générale de Droit et de Jurisprudence (LGDJ) 3e Ed.*: 200–264.

Pontecorvo, C.M. (1999). 'Interdependence Between Global Environmental Regimes: The Kyoto Protocol on Climate Change and Forest Protection'. *Zeitschrift für Ausländisches Öffentliches Recht und Völkerrecht*, 59: 705–748.

Samuelson, P.A. (1976). 'Economics of Forestry in an Evolving Society'. *Economic Inquiry*, 14: 466–492.

Sand, P.H. (1996). 'The Potential Impact of the Global Facility of the World, Bank', in R. Wolfrum (ed.), *Enforcing Environmental Standards: Economic Mechanisms as Viable Means?* Berlin: Springer, 479–499.

Sands, P. (1999). 'International Court and the Application of the Concept of Sustainable Development', *Max Planck Yearbook of United Nations Law,* 3: 389–405.

Swanson, T. (1999). 'Why is there a Biodiversity Convention? The International Interest in Centralized Development Planning'. *International Affairs*, 75: 307–331.

Tarasofsky, R. (1996). 'The Global Regime for the Conservation and Sustainable Use of Forests: An Assessment of Progress to Date'. *Zeitschrift für Ausländisches Öffentliches Recht und Völkerrecht*, 56 (II): 668–683.

United Nations (1992a). *United Nations Framework Convention on Climate Change (UNFCCC)*. http://www.unfccc.int.

United Nations (1992b). *Convention on Biological Diversity (CBD)*, http://www.biodiv.org

UNFCCC (1997). *The Kyoto Protocol to the UNFCCC*. http://www.unfccc.int

UNFCCC (2001). *Report of the Conference of the Parties on its Seventh Session (FCCC/CP/2001/13 and Addendum 1 to 4)*, http://www.unfccc.int

Van Kooten, G.C. (2000). 'Economic Dynamics of Tree Planting for Carbon Uptake on Marginal Agricultural Lands'. *Canadian Journal of Agricultural Economics*, 48: 51–65.

Van Kooten, G.C., Binkley, C.S. and Delcourt, G. (1995). 'Effects of Carbon Taxes and Subsidies on Optimal Forest Rotation Age and Supply of Carbon Services'. *American Journal of Agricultural Economics*, 77: 365–374.

Wolfrum, R. (1996). The Convention on Biological Diversity: Using State Jurisdiction as a Means of Ensuring Compliance, in R. Wolfrum (ed.), *Enforcing Environmental Standards: Economic Mechanisms as Viable Means?* Berlin: Springer, 373–411.

Wolfrum, R. and Matz, N. (2000). 'The Interplay of the United Nations Convention on the Law of the Sea and the Convention on Biological Diversity', *Max Planck Yearbook of United Nations Law*, 4: 445–480.

World Commission on Environment and Development (WCED) (1987). *Our Common Future*. Oxford: Oxford University Press.

Chapter 7

SUSAN NICOLE KROHN[1]

Twenty Thousand Leagues Under the Sea: On the Legal Admissibility of Strategies to Mitigate Climate Change by Ocean Sequestration

1 Introduction

In 1977 an article on global warming suggested tackling the problem of climate change by an injection of large amounts of carbon dioxide (CO_2) into the Mediterranean Sea where sinking thermohaline currents could carry and spread them into the deep parts of the Atlantic Ocean[2]. This kind of geo-engineering, as the author put it, was considered as an option to avoid a further increase of greenhouse gas concentrations in the atmosphere.

What reads like an extract from a publication of Jules Verne may – with some modifications – turn out to be reality within the next decades. With their primary objective to stabilize greenhouse gas concentrations on a level that would prevent dangerous anthropogenic interference with the climate system, the United Nations Framework Convention on Climate Change[3] and the Kyoto Protocol[4] open the way to combat global warming by the use of sinks. While the latter agreement explicitly recognizes the sequestration of carbon dioxide in forests and agricultural lands as a means to mitigate climate change[5], the option of carbon sequestration in the oceans has, so far,

---

[1] Dr. iur; lawyer, diploma in management sciences (FH).

[2] Marchetti (1977).

[3] Art. 2 sentence 1 of the United Nations Framework Convention on Climate Change of 9 May 1992 (Framework Convention on Climate Change).

[4] Art. 3 para. 3 and 4 of the Kyoto Protocol of 11 December 1997.

[5] Ibid.

MICHAEL BOTHE AND ECKARD REHBINDER (EDS.), Climate Change Policy, 183–216.

not gained considerable attention in international politics. Nonetheless, several governments have agreed to carry out an international CO_2 field experiment and industry has increased its involvement in the research of ocean sequestration techniques[6]. Notwithstanding the need for cost-reduction and for a decrease in energy consumption[7], the capture and storage of carbon dioxide is, at present, technically feasible.

Recent figures on climate change seem to attach more importance to the concept of disposing carbon dioxide in marine ecosystems. According to the Intergovernmental Panel on Climate Change, an expert body on issues of global warming[8], a worldwide reduction of 850 billion tonnes of carbon (GtC) over the next 100 years will be necessary to stabilize atmospheric CO_2 at an adequate level[9]. It is fairly undisputed that such a goal can only be reached by a significant change in present energy consumption and production patterns. The US Office of Fossil Energy assesses the necessary period for the establishment of generally accepted and economically viable alternatives to today's use of fossil fuels to be between 50 to 100 years[10]. Therefore, mankind may run out of time to find an adequate solution to the apparently daunting task of climate change. In order to avoid a dangerous anthropogenic influence on the climate, the development and implementation of a full set of greenhouse response actions including mitigation strategies will be required.

Whether marine carbon sequestration techniques should become an element of a comprehensive strategy to stabilize greenhouse gas concentrations is the subject of an increasingly controversial discussion among scientists. Proponents of ocean sequestration advocate it on the grounds of its capacity to influence the near-term transient peak of atmospheric CO_2 concentrations[11]. Others add that ocean sequestration may open the opportunity for a continued use of fossil fuels in a more climate-friendly way, enabling societies, especially in developing countries, to choose the rate at which they switch to other sources of energy more freely[12]. In particular the latter argument has brought critics into the arena who

---

6 For details on the International Project see <http://www.co2experiment.org>. A description of further activities on carbon sequestration is included in a report of the British Government Panel on Sustainable Development (2000) Annex C, section 3. For commercial proposals see also Adhiya and Chrisholm (2001).

7 British Government Panel on Sustainable Development (2000), Annex C, section 10.

8 The Intergovernmental Panel on Climate Change was established by the World Meteorological Organisation and the United Nations Environmental Programme based on United Nations General Assembly Resolution 43/53 of 6 December 1988, para. 5.

9 Freund (1996), Table 1.

10 US Department of Energy, Carbon Sequestration: State of Science, Draft February 1999, cited from Amatistova (2001), 2.

11 Amestitova (2001), 12 and Johnston *et al.* (1999), 11 *et seq.*

12 British Government Panel on Sustainable Development (2000), Annex C, section 6.

highlight that ocean sequestration gives no permanent solution to the problem of global warming[13]. The carbon stored in the oceans will eventually be re-exposed to the atmosphere. Thus, there exists a high risk that the problem of climate change will simply be deferred and passed on to future generations. Instead of disposing CO_2 in the oceans, opponents claim, much more efforts should be made to increase energy efficiency and reduce mankind's reliance on fossil fuels[14].

This chapter examines the legality of ocean sequestration mechanisms under existing international environmental law. It summarizes the range of ocean sequestration options and provides insight into the environmental impacts of CO_2-storage in marine ecosystems.

2 Technological Background of Carbon Sequestration

2.1 The Relevance of the World's Oceans in the Global Carbon Circle

To understand the concept of ocean sequestration, one has to start from the fact that the oceans are the largest global "reservoir" of carbon. With about 40 000 GtC in organic and inorganic form they contain more than 50 times as much carbon as the atmosphere (750 GtC)[15]. Before the Industrial Revolution, when no significant anthropogenic releases of CO_2 took place, global carbon reservoirs were in equilibrium. Carbon fluxes occurred between the ocean and the atmosphere with no net gains or losses from one reservoir to another. The discharge of carbon dioxide resulting from the use of fossil fuels as a major source of energy has disturbed this balance. At present, the atmospheric concentration of CO_2 amounts to 335 ppm compared to a pre-industrial level of 280 ppm[16]. This increase is, however, only half as high as one would expect from the combustion of fossil fuels and significant land-use changes such as deforestation. Today, worldwide anthropogenic emissions of carbon amount to roughly 7 GtC/year of which only 3.3 GtC/y remain in the atmosphere[17]. The outstanding fraction is taken up by the oceans and other ecosystems thereby functioning as "carbon sinks" and mitigating the negative effects of greenhouse gas emissions on the global climate. Scientists suggest that the annual net carbon uptake of the

13 Greenpeace (1999).
14 Union of Concerned Scientists on Marine Carbon Sequestration (2000).
15 IPCC, Houghton *et al.* (2001), 197.
16 IPCC, Houghton *et al.* (2001), 39.
17 IPCC, Houghton *et al.* (2001), 208.

oceans is around 2 GtC[18]. Hence, about 85% of present-day anthropogenic CO_2 emissions will enter the oceans indirectly within the next centuries or millennia[19].

The absorption and storage of CO_2 in the sea is driven by physio-chemical and biological processes referred to as the solubility pump and the biological pump. In simple terms, the oceans consist of three different layers of water: an upper mixed layer of about 100 m in depth, a thermocline region extending to a depth of 1000 m and a deep sea region[20]. In the upper mixed layer, equilibrium of carbon concentration between the ocean and the atmosphere exists. It is caused by a rapid transfer of CO_2 at the air-sea interface due to wave action that mixes the surface water down to approximately 100 m[21]. A vertical mixing of the oceans is generally inhibited by the thermocline due to the existing temperature and density gradients[22]. Because of the higher pH of seawater, carbon dioxide is more soluble in the ocean than in freshwater. As the solubility of CO_2 increases with decreasing temperatures, colder waters, particularly in high latitudes, have the capacity to uptake carbon dioxide from the atmosphere. In polar regions, the surface waters are saltier because of ice formation and therefore denser. As a result, they sink to the sea-bottom and transfer carbon to the deeper parts of the ocean. The water is carried to warmer climates and resurfaces mainly in the equatorial regions where during its upwelling to the surface carbon dioxide is degassed to the atmosphere. As the surface waters travel gradually to the poles and cool again, they increasingly absorb CO_2. The process of uptake and release of carbon dioxide is called the solubility pump[23]. Due to the slow transfer of water between the different layers of the oceans, this cycle can take up to 1000 years to complete[24].

The second driving force of the carbon circulation in the oceans is referred to as the biological pump based on phytoplankton activity[25]. In the upper mixed layer of the oceans, phytoplankton converts CO_2 into organic carbon by photosynthesis using sunlight and inorganic nutrients. As the carbon passes through the food-web in the surface waters, most of it is converted back to CO_2 and released to the atmosphere[26]. About 20% of the

---

18 Ibid.

19 US Department of Energy Report (1999), section 3.

20 Herzog (1998), 1.

21 Ibid.

22 Ibid.

23 Falkowski *et al.* (2000), 292; Sarmiento (1993), 37; Ormerod, Freund and Smith (2002), 5.

24 Wong and Hirai (1997), cited from Johnston *et al.* (1999), 6 and Ormerod, Freund, and Smith (2002), 6.

25 Chrisholm (2000), 685; Boyd et al (2000), 695; Sarmiento (1993), 37 *et seq.*; Ormerod, Freund and Smith (2002), ibid.

26 Sarmiento (1993), ibid.

organic carbon, however, is exported to deeper parts of the oceans by sedimentation where it is broken down by bacteria and re-mineralized back into CO_2 that resurfaces after a period of up to a millennium[27].

2.2 Technical Methods of Ocean Sequestration

The idea of ocean sequestration is to accelerate the slow, natural process of carbon uptake and storage artificially[28] either by enhancing the efficiency of the solubility pump or of the biological pump.

a) Direct Injection of CO_2 into the Ocean

Most scientific discussion focuses on the proposal to modify the solubility pump by shortcutting the vertical slow mixing of the ocean through the direct injection of CO_2 into the deeper layers of seawater. After the capture of CO_2 from large single sources, especially coal-fired power stations or certain industrial processes, and its transport to the sequestration site, it could be disposed into the sea. There are several storage methods still under discussion, a number of which are reviewed below[29].

- *Direct injection of CO_2 by pipe or pipeline:* at present, it is technically feasible to disperse carbon dioxide into the mid-water column by a pipeline or a pipe from a moving ship. After its dissolution in the seawater, the carbon dioxide would be carried to the deeper parts of the sea by water circulation and could be kept away from the atmosphere for centuries. The retention time of the disposed CO_2 is regarded as a function of the depths at which it is discharged. Model simulations predict, for example, that 46% of the carbon dioxide injected at a depth of 1500 m would remain in the oceans for at least 500 years[30]. Simulations like these are, however, of limited validity because the water movements in the deep oceans and the rapidity of their vertical mixing are only partially understood and unlikely to be uniform[31].
- *Formation of an artificial lake of liquid CO_2 on the seabed:* at typical pressures and temperatures in the oceans, CO_2 is transformed to a liquid below a line of 500 m whose density exceeds that of seawater below

---

27 Adhiya and Chrisholm (2001), 4.

28 Herzog (1998), ibid; Johnston *et al.* (1999), 11 *et seq.*

29 For further details on the main technical options under discussion see Herzog (1996), passim; Johnston et al (1999), 29 *et seq.*; Caldeira, Herzog, and Wickett (2001), 2.

30 Caldeira, Herzog and Wickett (2001), 7.

31 Johnston *et al.* (1999), 14 *et seq.*; Ormerod, Freund and Smith (2002), 31 *et seq.*; Union of Concerned Scientists on Marine Carbon Sequestration (2000), 3.

3000 m[32]. Due to these properties, it would be possible to form a submarine CO_2-lake on the seabed by using a vertical pipe. Such a deep sea lake would probably be covered by a hydrate crust building a barrier for the dissolution of the injected liquid.

- *Disposal of solid blocks:* another sequestration technique under discussion is the release of solid carbon dioxide blocks at the ocean surface. Since solid CO_2 is denser than seawater, the blocks would sink to deep-water layers or even the seabed. Because of the high expenses connected with the production of solidified carbon dioxide[33] this method, however, seems to be too costly.

b) Ocean Fertilisation

An alternative approach to ocean sequestration is to enhance the net carbon uptake of the biological pump by ocean fertilisation[34]. Photosynthesis of phytoplankton is thought to be controlled by the availability of nutrients and trace elements such as iron. Hence, additions of small amounts of iron to large areas of the sea could promote the growth of phytoplankton and increase the drawdown of CO_2 from the atmosphere[35]. Sedimentation would export a fraction of carbon to deeper layers of the ocean where it could be stored for several centuries. It is argued that the rise of biological productivity in the surface waters by ocean fertilisation may also have effects on fish populations with prospects of better fish catches[36].

At the time of writing, four ocean fertilisation field experiments – although not intended to prove the feasibility of fertilisation for the purpose of climate mitigation – have been conducted on a meso-scale in the Equatorial Pacific and the Southern Ocean[37]. The projects confirmed an iron-induced growth of planktonic species and an increased uptake of carbon dioxide[38]. A long-lived bloom of plankton and a domination of large celled species of phytoplankton was noticed[39]. However, considerable export production of carbon to deeper parts of the ocean could not be observed[40],

---

32 Ohsumi (1995), 59.

33 Herzog (1998), 1.

34 British Government Panel on Sustainable Development (2000), Annex B, section 1 *et seq.*; Adhiya and Chrisholm (2001), 4; Johnston *et al.* (1999), 14; Ormerod, Freund and Smith (2002), 23 *et seq.*

35 Boyd (2000), 695, 697 *et seq.*

36 Ormerod, Freund and Smith (2002), 22.

37 Martin *et al.* (1994); Coale *et al.* (1996); Boyd (2000); Smetacek (2000).

38 Martin *et al.* (1994); Coale *et al.* (1996), 499; Adhiya and Chrisholm (2001), 9.

39 Ibid.; Boyd (2000), 697 *et seq.*

40 Boyd (2000), 695, 699.

which raises doubts on the efficiency of the concept in the context of climate protection policies.

Generally, the transfer of carbon dioxide to deeper sea regions depends upon the local water circulation and hydrography. Because of its importance for intermediate and deep water formulation and the high occurrence of unused nutrients in the surface waters needed for phyto-planktonic growth, the Southern Ocean appears to be well suited for ocean fertilisation[41]. So far, assessments of the potential change in the global carbon cycle by large-scale ocean fertilisation can only be drawn from computer models. The available literature on modelling of fertilising the Southern Ocean shows a decrease in the atmospheric CO_2 concentrations by 6–21%[42]. Since the field experiments revealed a rather short-lived effect from iron addition, fertilisation would need to be maintained over a long period of time exceeding 100 years to show significant impacts[43].

A similar scheme on ocean fertilisation envisages the utilisation of nitrate, phosphate or silicate as fertilizers in areas of the sea where their shortage is currently limiting algae growth[44].

c) Injection of CO_2 into Sub-sea Aquifers and Formations

A third storage option is the disposal of CO_2 in depleted oil or gas fields or deep saline reservoirs. This alternative assumes that the risk of leakage from fractures or seismic activities could be minimized by a careful choice of sites and by their monitoring. As regards Europe, the capacity for CO_2-storage in geological reservoirs is estimated to be about 220 GtC, which is equivalent to 200 years of the continent's greenhouse gas emissions[45]. Substantial amounts of CO_2 are already being transferred into underground reservoirs. For example, in 1996, an operation for the storage of CO_2 purely for the purpose of climate protection began in the Norwegian Part of the North Sea. As an element of the Sleipner Vest gas production project, nearly 1 Mt/y of CO_2 is being disposed in a deep saline reservoir beneath the bed of the ocean[46].

---

41 Boyd (2000), 695.

42 British Government Panel on Sustainable Development (2000), Annex B, section 2.2; Adhiya and Chrisholm (2001), 10.

43 Adhiya and Chrisholm (2001), 9 *et seq.*

44 Johnston *et al.* (1999), 23; British Government Panel on Sustainable Development (2000), Annex B, section 3.

45 Holloway, Heederik *et al.* (1996), 3.

46 British Council on Sustainable Development (2000), Annex C, section 3.

2.3 Environmental Impacts

Most proponents of ocean sequestration acknowledge its potential to cause environmental damage. On the global scale, the overall impact of carbon sequestration on the oceans' ecology is regarded as being proportionally limited. Even if the oceans with an amount of 40 000 GtC were to gain all carbon present in the atmosphere, this would only cause an increase of carbon concentrations by less than 2%[47]. On the regional and local levels, however, the environmental consequences could be severe, but advocates of ocean sequestration strategies consider that these impacts will be offset against the damage avoided by a reduction of the temperature increase over the next centuries[48].

A confident prediction of the nature and scale of environmental impacts of ocean sequestration is extremely difficult since scientists lack sufficient understanding of ocean circulation patterns, deep ocean ecosystems and their biodiversity, lifecycles and habitats[49]. Furthermore, the capacity of the oceans to uptake CO_2 and the movements of water may change as a result of natural variability or of global warming[50].

a) Direct Injection

Despite significant scientific uncertainties, it can be anticipated that an injection of carbon dioxide into the water column will cause acidification, a decline in the pH of seawater[51]. To a certain extent, the effect on the biochemical properties of the oceans will depend upon the technique of CO_2 release[52]. While some models found a reduced pH of about 1 unit from an ambient level of 7.8 within tenths of kilometres from the release

[47] US Department of Energy (1999), section 3.

[48] See Herzog (1998), 2.

[49] US Department of Energy (1999), section 3.4, concluded: "We lack sufficient knowledge of the consequences of ocean sequestration on the biosphere and on natural biogeochemical cycles. Such knowledge is critical to the responsible use of the oceans as a carbon sequestration option … The oceans play an important role in sustaining the biosphere, so any change in the ocean ecosystem function must be viewed with extreme caution."

[50] Johnston *et al.* (1999), 30 *et seq.*

[51] Auerbach *et al.* (1997); Union of Concerned Scientists, Marine Carbon Sequestration (2000), 4; US Department of Energy Report (1999), section 3.2; Johnston *et al.* (1999), 33 *et seq.*; Seibel and Walsh (2001), 319.

[52] A disposal of solid CO_2 blocks, for instance, could minimize the overall environmental impacts because of the greater dispersion of CO_2 in the water column during the sinking of the blocks. A contact with the biologically richest part of the oceans in the surface waters above 500 m could only be avoided by an injection of CO_2 into deeper parts of the ocean, though.

mechanism[53], other methods are expected to cause a reduction in seawater pH by solely 0.1 units within a volume of 0.5 cubic km[54]. A sustained pH of less than 6.5 is, in any case, lethal to many coastal marine organisms and a value of 7–7.5 pH units is likely to be the lowest limit of tolerance for many species[55].

Initially, the discussion started from the assumption that carbon sequestration will almost exclusively affect small organisms with a limited mobility living in the vicinity of the release point[56]. More recently, the results of an in-situ experiment predicted the danger of a more extensive loss of even higher organisms[57]. Scientists found out that several invertebrate and vertebrate species do not necessarily leave more acidic areas and avoid dissolving flocculent carbon hydrates when allured by the scent of food. In regions of low pH, these animals appear to suffer from metabolic suppression due to increased CO_2 concentrations[58]. Apart from that, animals with external calcareous surfaces could be harmed if close enough to an injection point since carbonates may be dissolved in the acidic environment[59].

Even if changes in pH perturbations can be minimized through the proper design of release mechanisms and the dispersion of carbon dioxide over a larger volume of water, the danger to cause severe damages to marine ecology will possibly remain. Despite the very limited scientific knowledge of the deep-sea environment and the inhabiting organisms, general agreement exists on the high susceptibility of deep-living animals to pH excursions[60]. PH concentrations have been stable in the deep sea for a very long period of time so that benthic organisms[61] are highly adapted to these conditions and are expected to have a very limited capacity to pH changes. This capacity may have already been exhausted by the decrease in the pH of the oceans due to changes of the global carbon cycle caused by rising CO_2 emissions. Furthermore, layers of low pH water could build a physiological barrier for some species that migrate vertically in the ocean[62].

In the case of seabed-storage, environmental impacts may be concentrated to a small fraction of the global seabed as long as an outspread

53 Report of the Twenty-Seventh Session of GESAMP (United Nations Joint Group of Experts on the Scientific Aspects of Marine Environmental Protection), Nairobi / Kenya, 14 – 18 April 1997, cited from: Johnston *et al.* (1999), 33 *et seq.*

54 Drange, Alendal and Johannessen (2001).

55 Johnston *et al.* (1999), 34.

56 Herzog (1998), 3.

57 Tamburri *et al.* (2000).

58 Seibel and Walsh (2001).

59 Ibid.

60 Ibid.

61 Organisms that live at the bottom of the estuaries and the sea.

62 Ormerod, Freund and Smith (2002), 17.

of CO_2 can be avoided. Nevertheless, the effects of increased CO_2 concentrations on the marine life on the seabed are largely unknown and need to be further investigated. Although the greatest density of biomass occurs in depths between 500–1000 m and decreases exponentially below that level, there is a significant population of organisms on or near the ocean floor[63]. Keeping in mind their sensitivity to pH changes, the storage of carbon is expected to have severe environmental impacts in disposal areas.

b) Ocean Fertilisation

As mentioned above, a long time-frame of ocean fertilisation, probably more than 100 years, is regarded as a necessity in order to achieve significant changes in atmospheric carbon dioxide concentrations[64]. The consequences of such long-running projects are difficult to anticipate[65]. In spite of the significant uncertainties surrounding the concept of iron fertilisation, agreement exists that it will increase the abundance of large-celled phytoplankton species and will thereby change the phytoplankton community's composition[66]. This is very likely to have negative consequences throughout the food chain and may lead to a decrease in marine biodiversity[67] and an appearance of nuisance species[68].

The explosive growth of algae caused by ocean fertilisation is also expected to reduce the oxygen concentration of oceanic waters[69]. Although the amount of this deoxygenation depends upon different factors like the timeframe, the extent of fertilisation and the depth at which the organic matter will be distributed[70], models suggest a widespread deoxygenation of 10–25% of the Southern Ocean in response to large-scale fertilisation[71]. Deoxygenation of the water column will be followed by high mortalities across the biota and the development of so-called "dead zones" devoid of all living beings[72]. A lower oxygen level could lead to a significant reduction in biodiversity with a potential shift in species composition[73]. The same effect is expected from the use of macronutrients like nitrates and phosphates that

---

63 Ibid.

64 British Government Panel on Sustainable Development, Annex C, section 2.2; Adhiya and Chrisholm (2001), 9 *et seq.*

65 Watson *et al.* (2000); Schmitz (2000).

66 De Baar and Boyd (2000), 61 *et seq.*

67 Coale *et al.* (1996), 500 and Dodson, Arnott and Cottingham (2000), 2662 *et seq.*

68 Johnston *et al.* (1999), 25.

69 Johnston *et al.* (1999), 24 and British Government Panel on Sustainable Development (1999), Annex B, section 2.4.

70 Fuhrmann and Capone (1991), 1951 *et seq.*

71 Johnston *et al.* (1999), 24; Sarmiento and Orr (1991), 1928.

72 Fuhrman and Capone (1991), 1952.

73 British Government Panel on Sustainable Development (2000), Annex B, section 6.

are predicted to cause eutrophication[74] depriving the ocean of oxygen[75]. This process is well known from the enrichment of lakes and coastal waters with nutrients from sewage or agricultural runoff.

Another area of concern relates to the possible release of methane and nitrous oxide by increased primary production due to anoxic conditions[76]. The release of both gases, each classified as a greenhouse gas with a higher warming potential than CO_2, might offset the net gains from reducing carbon dioxide concentrations by ocean fertilisation[77].

Finally, the impact of fertilizers on deep-ocean organisms is still a subject of considerable uncertainty.

3 Legal Implications

Policies to mitigate climate change have to comply with the requirements of international environmental law. At present, no legal instrument directly addresses carbon sequestration in the oceans. Nonetheless, several conventions exist that will be of relevance to the topic.

3.1 Framework Convention on Climate Change and its Kyoto-Protocol

The Framework Convention on Climate Change is not clear on the legal admissibility of carbon sequestration in the oceans. While some rules of the Convention seem to encourage the utilisation of corresponding technologies, others are likely to prohibit it. According to Article 4, paragraph 1 (b) of the Convention, all Parties are obligated to formulate, implement, publish and update national and regional programmes to mitigate climate change by addressing anthropogenic emissions by sources and removals by sinks. Additionally, Article 4, paragraph 1 (d) stipulates the duty to enhance sinks. Since oceans fall under the definition of sinks enshrined in Article 1, paragraph 8 of the Convention[78] and are recognized as contributors to climate stability in paragraph 4 of the preamble[79], the obligations mentioned

[74] Eutrophication is defined as a condition in an aquatic ecosystem where high nutrient concentrations stimulate blooms of algae.

[75] British Government Panel on Sustainable Development (2000), Annex B, section 3.4.

[76] Owens *et al.* (1991), 293 *et seq.*

[77] Fuhrmann and Capone (1991), 1951 *et seq.*

[78] This provision states: “Sink means any process, activity or mechanism which removes a greenhouse gas, an aerosol or a precursor of a greenhouse gas from the atmosphere”.

[79] “Aware of the role and importance in terrestrial and marine ecosystems of sinks and reservoirs of greenhouse gases”.

above appear to justify the use of ocean sequestration strategies[80]. A further argument for the legality of such activities can be drawn from one of the Convention's principles laid down in Article 3, paragraph 3 sentence 3, which calls for a comprehensive policy formulation covering *all* relevant greenhouse gas sources and sinks[81].

On the other hand, Article 4, paragraph 1 (d) of the Convention does not only bind Parties to enhance their sinks, but also to sustainably manage and protect them. In contrast to projects in the field of agriculture and forestry, mainly expected to cause positive outcomes for other environmental goods than climate stability, large-scale ocean sequestration is likely to cause severe negative impacts on marine ecosystems, in particular on their biological diversity. Climate change mitigation by means of such a technique may therefore be inconsistent with the duty to promote sustainable management and sink protection, assuming that this rule requires respect for the aspect of biodiversity preservation. One might doubt whether – in the context of a climate change regime – the obligation to protect and sustainably manage sinks asks for anything else than upholding their climate-stabilizing function, which will not be impaired by sequestration projects. The definition of sinks laid down in Article 1, paragraph 8 solely refers to their capacity for greenhouse gas storage and, consequently, favours a narrow interpretation of Article 4, paragraph 1 (d). There are, however, more convincing arguments to support a broader understanding. According to the rules of treaty interpretation enshrined in Article 31 of the Vienna Convention on the Law of Treaties, an agreement has to be interpreted by taking into account any relevant rules of international law applicable in the relations between the Parties.[82]

A global environmental agreement to give thought to in this context is the Convention on Biological Diversity, which obligates Parties to develop and implement policies for the conservation of biodiversity and the sustainable use of its components[83]. A more holistic interpretation of the notion of "sinks" by paying regard to their importance for biological diversity preservation will not only conform to these duties. It will also correspond to several decisions of the Conference of the Parties under the Framework Convention of Climate Change[84] and the Biodiversity Convention[85] that aim

[80] This point of view is followed by the Scientific Committee on Ocean Research – Intergovernmental Oceanographic Commission Advisory Panel on Ocean CO_2, Watching Brief: Ocean Carbon Sequestration.

[81] Heinrich (2002), 3, who highlights that this norm is somehow qualified by the requirement of cost-effectiveness laid down in Art. 3 para. 3 sentence 2.

[82] Art. 31 para. 3 (c) of the Convention of 23 May 1969.

[83] Art. 8 10 of the Convention of Biological Diversity of 5 June 1992.

[84] With Dec. 11, para. 1 (e) of the 7th Conference of the Parties, Marrakech/Morocco, 29 October–9 November 2001, the Conference of the Parties called upon the Parties to report on

at a coherent approach in formulating and implementing policies for biodiversity and climate protection. Finally, an interpretation of the obligation to protect sinks in a more comprehensive manner will give respect to the interdependence of biological and physio-chemical processes that occur within ecosystems.

Existing scientific uncertainties surrounding carbon sequestration techniques will not free states from their obligation to make efforts for the preservation of marine biological systems under Article 4, paragraph 1 (d). Although the provision to sustainably manage and protect sinks gives no guidance on state conduct in cases of missing scientific evidence, the precautionary principle will demand action in this regard. Whether its present recognition on the international level suffices to consider the principle as a general element of customary international law is the subject of considerable debate among scholars[86]. In the context of marine environmental protection, however, the precautionary principle has been endorsed by "virtually all recent international agreements"[87], so that it is justified to assume its normative character in this area. Despite the very general character of the obligation of Article 4, paragraph 1 (d), which leaves a high degree of discretion to Parties, it cannot be concluded from the Framework Convention on Climate Change that the precautionary principle enshrined in specific international environmental law can be ignored when implementing the Convention's provisions in the relevant context[88].

Not only the duty of Article 4, paragraph 1 (d) to sustainably manage and protect sinks, but also one of the treaty's principles may stand in the way of ocean sequestration activities. Under Article 3, paragraph 5 of the Framework Convention, Parties are called upon to promote sustainable

---

their legislative and administrate measures to ensure that sink-related activities under Art. 3 para. 3 and 4 of the Kyoto Protocol contribute to the conservation of biodiversity and the sustainable use of natural resources.

[85] E.g. Dec. V/3, para. I 4 of the 5th Conference of Parties, Nairobi, Kenya, 15–26 May 2000, "Progress report on the implementation of the programme of work on marine and coastal biodiversity"; Dec. V/4 para. 11, 16 *et seq.*, "Progress report on the implementation of the programme of work for forest biological diversity"; Dec. V/21, para. 3, "Cooperation with other bodies" highlighting the general importance to strengthen the cooperation with the Climate Protection Regime on relevant issues.

[86] The following authors regard the precautionary principle as an element of customary international law: Sands (1995), 213; Primosch (1996), 232; Epiney and Scheyli (1998), 107. This point of view is not followed by: Odendahl (2000), 222 *et seq.* and Beyerlin (1994), 139 *et seq.*

[87] United Nations (1990), 20, para. 60.

[88] The Framework Convention itself refers, although in some qualified way, to this principle in Art. 3 para. 3.

development, which includes efforts for sustainable resource management[89]. The World Commission on Environment and Sustainable Development defined sustainable development as "development that meets the needs of present generations without comprising the ability of future generations to meet their own needs"[90]. The concept's underlying idea, intergenerational equity, has not been implemented into precise guidelines to influence states' behaviour yet. Despite this, it seems appropriate to draw from it an obligation of states to give some guarantees that today's human activities will not cause unsolvable problems for future generations[91]. From short- and medium-term perspectives, carbon sequestration in the oceans may help to mitigate climate change. Little consideration, however, has been given to the question of the perils arising from the leakage of CO_2 to the atmosphere during the upwelling of CO_2-enriched deepwater in the centuries to come. No sufficient assurances can presently be given that ocean sequestration will not simply shift the burden of climate change to future generations, thereby imposing upon them possibly unmanageable risks. Consequently, proposals to mitigate climate change by carbon storage in the sea conflict with the principle of intergenerational equity[92].

Apart from the issue of intergenerational equity, the admission of ocean sequestration will lead to problems concerning states entitled to use the amount of sequestered carbon as an offset against emission reduction targets. In other words, it touches upon the issue of intragenerational equity[93]. If sequestration activities take place in areas under state jurisdiction, the assignment will raise no problems – but carbon dioxide injection or ocean fertilisation activities are also expected to be carried out on the high seas including the deep seabed. A state's right to use this common space for its own purposes, which could be derived from the internationally accepted principle of the "freedom of the high seas" laid down in the United Nations Convention of the Law of the Sea[94], is not accepted in the context of the

---

[89] Sustainable resource management is generally accepted as an element of the concept. See, e.g. Sands (1995), 59; Boyle (1999), 9 and Rehbinder (1994), 96 *et seq.*

[90] World Commission on Environment and Development (1987), 43.

[91] Bartholomäi (1997), 320 *et seq.* For further details on the concept see also Brown Weiss (1992).

[92] Johnston *et al.* (1999), 12.

[93] The aspect of intragenerational equity is widely recognized as an element of the concept of sustainable development. See, for example, Sands (1995), 60 *et seq.* and Handl (1990), 27.

[94] See Art. 87 para. 1 of the United Nations Convention of the Law of the Sea (UNCLOS) of 10 December 1982, which states: "The high seas are open to all States, whether coastal or land-locked. Freedom of the high seas is exercised under the conditions laid down by this Convention and by other rules of international law. It comprises, *inter alia*, both for coastal and land-locked States: (a) freedom of navigation; (b) freedom of overflight; (c) freedom to lay submarine cables and pipelines, subject to Part VI; (d) freedom to construct artificial islands and other installations permitted under international law, subject to Part VI; (e)

Framework Convention's provisions on strategies to combat global warming. Article 4, paragraph 2 (a) of the Framework Convention on Climate Change obligates every Annex I Party – mainly developed nations are listed in this Annex – to adopt mitigation policies by limiting its anthropogenic emissions and protecting and enhancing "its" greenhouse gas sinks and reservoirs. Hence, activities suitable to comply with the Convention's provisions must refer to sinks under state jurisdiction. Article 3, paragraph 3 sentence 4 and Article 4, paragraph 2 sentence 3 permit cooperation between Parties and a more flexible approach in this regard, but no authorization can be drawn from these norms to use activities in areas beyond national jurisdiction for treaty fulfilment. With the adoption of the Kyoto Protocol, the provisions mentioned above were specified by permitting Annex I states to get involved in projects to be carried out in other developed or even in developing countries in return for emission reduction units usable to comply with commitment targets[95]. These regulations and the decisions of the Conference of the Parties on how to implement them[96] express the maximum achievable consensus of states on instruments that allow to comply with emission reduction obligations by carrying out climate protection measures outside the states' own territories. The admission of cooperative approaches for treaty implementation was a controversial issue during the negotiations of the Kyoto Protocol because many countries, especially developing states, regarded these schemes as a means for industrialized nations to free themselves from the development of effective national climate protection policies[97]. The compromise finally achieved and expressed in the Protocol would be undermined if climate mitigation operations in areas beyond national jurisdiction were permitted to be used as carbon dioxide offsets. Consequently, activities on the high seas cannot be used to count towards the emission reduction targets as long as no specific legal basis allows states to do so. Against the background of the Protocol's drafting history, it seems rather unrealistic that a corresponding proposal will be met with approval permitting technologically equipped countries to impose environmental risks on all coastal nation states in order to fulfil their treaty obligations.

---

freedom of fishing, subject to the conditions laid down in section 2; (f) freedom of scientific research, subject to Parts VI and XIII".

[95] See the so-called "joint implementation" under Art. 3 para. 10, 11 and Art. 6 para. 1 and the "clean development mechanism" of Art. 3 para. 12 and Art. 12. Apart from these mechanisms, a "joint fulfilment" of the obligations under Art. 3 para. 1 and the emission trading scheme of Art. 3 para. 10, 11 and Art. 17 provide for a flexible implementation.

[96] See especially the Marrakech Accords, Draft Decision-/CMP.1, adopted at the 7th Conference of Parties, 29 October–9 November 2001, Marrakech, Morocco.

[97] Baker Röben (2000), 214 *et seq.*

In the context of proposals for CO_2-injection into the deep seabed, the seabed's status as "common heritage of mankind" has to be taken into consideration[98]. According to Article 140, paragraph 1 of UNCLOS, activities in this area shall be carried out for the benefit of mankind as a whole, which demands an equitable sharing of financial and other benefits derived from such actions. Although the relevant regulations of UNCLOS primarily aim at governing the exploitation of deep seabed resources, their wording permits a broader interpretation including other activities in this area. Hence, states will not be able to obtain CO_2 reduction units at the expense of this common property belonging to all nations. Instead, states will be obligated to give some kind of compensation for the benefits derived from the seabed's utilisation, which will create economic constraints for ocean sequestration operations in this area.

In sum, ocean sequestration proposals will fall under different provisions of the international regime on climate change, some favouring this technology, others acting as an obstacle. It has to be noted that every norm of the Framework Convention mentioned above does not contain an absolute standard[99]: While the relevant obligations are qualified either by their high degree of generality or by their soft wording[100], the principles in question solely provide for a restricted amount of guidance. In comparison to a rule that generally necessitates a special outcome if its requirements are fulfilled, a principle operates more like a guideline: it has to be implemented as efficiently as possible, but divergences are permitted on reasonable grounds[101]. Hence, the conflicting norms must be reconciled when deciding upon the legality of ocean sequestration.

Until the end of the first commitment period from 2008 to 2012, the Kyoto Protocol will not permit Parties to use ocean sequestration activities as a means to comply with the climate regime's obligations. Article 3, paragraph 3 and 4 of the Protocol solely allows greenhouse gas removals resulting from human-induced land use, land-use change and some kind of forestry to be used as an offset against emission targets.

As the duty to promote, research and develop carbon sequestration technologies in Article 2, paragraph 1 (a) (iv) of the Kyoto Protocol indicates, the drafters of the agreement foresaw the possibility to include further sink-related activities into the strategy to combat global warming in the future. Such an inclusion could be done by an amendment to the Kyoto Protocol under its Article 19, but it seems more realistic to assume the

---

98 Art. 136 UNCLOS. For the concept of "common heritage of mankind" see also: Fitschen (1995).

99 See also Bodansky (1993), 501, 504, 518.

100 Art. 4 para. 1 (b): to formulate, implement "policies"...; Art. 4 para. 1 (d): to "promote" sustainable management, enhancement and protection of sinks.

101 Dworkin (1978), 24 *et seq.*

admission of further sink enhancement activities within later commitment periods.

How the conflict between the rules and principles of the Framework Convention in the context of carbon sequestration will be resolved in the future depends upon the question to what extent they will be compromised by different policy options. The availability of alternative greenhouse gas reduction or mitigation strategies, the gravity of ecological damages to the oceans and the status of scientific knowledge will be crucial factors in balancing the interests concerned. Additionally, the environmental standards enshrined in international marine law will set the framework for the admission of ocean mitigation proposals. Such an understanding is already indicated in the Kyoto Protocol, which states in Article 2, paragraph 1 (a) (ii):

> Each Party in Annex I ... shall ... implement and / or further elaborate policies and measures in accordance with its national circumstances, such as ... protection and enhancement of sinks and reservoirs of greenhouse gases ... *taking into account its commitments under relevant international environmental agreements*. [italics added]

Although it is generic, this provision shows that the Kyoto Protocol does not intend to overrule environmental standards existing in other areas.

3.2 The London Convention 1972 and its 1996 Protocol

A universal agreement in the field of marine ecosystem protection to be taken into consideration when deciding upon the legality of ocean sequestration proposals is the Convention on the Prevention of Marine Pollution by Dumping of Wastes and other Matter, adopted in London in 1972 (London Dumping Convention). The treaty aims to control and reduce sea pollution caused by dumping of wastes and other matter[102]. By the December 2003, 80 states had ratified the agreement, among them all nations presently involved in scientific research on ocean sequestration. As regards dumping, the London Convention imposes different restrictions on Parties depending on the harm the respective material is expected to cause to marine life or human health. It distinguishes between three categories of pollutants. Wastes and other matter listed in Annex I of the Convention may not be discharged into the sea[103]; while dumping of matter on Annex II is not generally prohibited, but requires special permission[104]. As far as substances are not enumerated in one of the Annexes, they are allowed to be dumped on

[102] Art. 1 of the London Dumping Convention of 29 December 1972.
[103] Art. IV para. 1 (a).
[104] Art. IV para. 1 (b).

the basis of a prior general permit[105] issued under less specific conditions than those required for a disposal of Annex II material[106]. When deciding upon granting one of the permissions, the factors set forth in Annex III of the treaty[107], among them the nature of the material, the characteristics of the dumping site, the methods of disposal and the possible effects of the discharge on marine life, have to be taken into account[108]. Hence, although the Convention does not generally prohibit the disposal of certain matter, it requires an informed decision-making process with due regard given to the environmental impacts of the proposed activity[109].

Whether the Convention governs carbon sequestration projects depends on how the term "dumping" is defined. Article III, paragraph I (a) of the Convention gives the following definition:

> Dumping means any deliberate disposal at sea of wastes or other matter from vessels, aircraft, platforms or other man-made structures at sea.

Apart from carbon injection from land-based sources every proposed method of ocean sequestration falls under this definition as a consequence of the use of the broad term "any deliberate disposal of other matter"[110].

Article III, paragraph 1 (b), however, excludes some activities from the scope of the term "dumping". In relation to carbon sequestration, subparagraph (ii) will be of importance. It states that dumping does not include a placement of matter "for a purpose other than the mere disposal thereof, provided that such placement is not contrary to the aims of this Convention". As the phrase "for a purpose other than the mere disposal" demonstrates, the intention of the action being undertaken is decisive for the application of the Convention. As long as actions are not only carried out directed towards the matter's disposal[111], the drafters of the Convention, in principle, meant to exclude them from the treaty's regulatory regime[112].

A direct injection of CO_2 into the oceans pursues a limitation of greenhouse gas concentrations in the atmosphere. Because of the expected positive effect on the climate, one may assert that the intention of CO_2 storage goes beyond a "mere disposal" of the substance. However, the creation of environmental benefits is no sufficient criterion to exclude an operation from the scope of the treaty: It is not unusual for the disposal of a

---

105 Art. IV para. 1 (c).

106 This conclusion can be drawn from a comparison of the definitions of a "special permit" and a "general permit" under Art. III paras. 5 and 6; see Leitzell (1973), 510.

107 Art. IV para. 2.

108 Art. IV para. 2 in connection with Annex III of the treaty.

109 Timagenis (1980), 216.

110 According to Art. III para. 4 the notion "matter" covers all kind of substances.

111 Such as in cases of the placement of fishing gear, scientific research instruments or military installations on the seabed which are possibly not intended for recovery.

112 Timagenis (1980), 134, 200; Leitzell (1973), 505.

matter to be carried out for the sake of other ecological goods like the integrity of terrestrial ecosystems or the cleanness of the air or of groundwater. In this regard, the injection of CO_2 does not distinguish itself from other dumping activities. On the contrary, all these cases have in common that the intended environmental effects are the final purposes of the disposal and not additional ones to it. Hence, the disposal of CO_2 into the sea from a ship or a platform will fall out of the scope of the exemption clause of Article III, paragraph 1 (b) and into the purview of the London Dumping Convention.

The consequence of the treaty's application to direct carbon injection depends upon the question whether CO_2 falls under one of the categories of substances listed in the Convention's Annexes. CO_2 itself is not named in any of the attachments, but Annex I prohibits the discharge of "industrial wastes" into the oceans defined as wastes "generated by manufacturing or processing operations"[113]. The fraction of carbon dioxide intended to be disposed into the oceans will stem from such industrial processes[114]. Hence, if CO_2 can be regarded as waste, it will not be allowed to dump it at sea. The London Dumping Convention gives no guidance on this issue[115], but the following definition of the notion of "waste" is set out in many treaties on waste management and can be regarded as being generally accepted under international law:

> substances or objects which are disposed of or are intended to be disposed of or are required to be disposed of by national law[116].

As the broad term "substance or object" indicates, the material's state of aggregation is not essential for the distinction between "waste" and "non-waste". In relation to gaseous matter a wide interpretation will, however, lead to overlaps with "emissions", which are generally governed by other regulatory regimes[117]. To differ between these two, it will be crucial whether the material is delimitable. If so, it can be regarded as waste. CO_2, captured from industrial sources and loaded on ships to be carried to a disposal site, fulfils this requirement irrespective of its solid, liquid or even gaseous form.

Some doubts remain whether a naturally occurring matter can be regarded as waste. International law gives no specific answer to this

---

[113] Annex I, para. 11.

[114] See above.

[115] Art. III para. 4 solely interprets the phrase "waste and other matter".

[116] See, e.g. Art. 2 para. 1 of the Basel Convention on the Control of Transboundary Movements of Hazardous Wastes and their Disposal of 22 March 1989; Art. 1 para. 1 of the Bamako Convention on the Ban of the Import Into Africa and the Control of Transboundary Movement and Management of Hazardous Wastes Within Africa of 29 January 1998.

[117] See, e.g. the Convention on Long-Range Transboundary Air Pollution of 13 November 1979.

question. Although space here does not permit a detailed discussion on this issue, it appears that national and even supranational law supports such a classification[118]. There is no apparent reason not to transfer this way of thinking into the context of the London Dumping Convention. On the contrary, the inclusion of naturally occurring substances into the definition of waste accords with the Convention's objectives since these substances can also harm living resources and marine life, damage amenities, or interfere with other legitimate uses of the sea[119].

To sum up, CO_2 from industrial processes intended to be injected into the ocean constitute "industrial waste", the disposal of which violates the London Dumping Convention[120].

Whether open-ocean fertilisation projects are also governed by the London Dumping Convention[121] is another challenging question. In this regard, the scope of the exemption clause enshrined in Article III, paragraph 1 (b) (ii) needs some further consideration. During ocean fertilisation activities, chemical substances are added to the ocean not with the aim to discard them, but to cause specific bio-chemical reactions. Thus, introduction of the material is carried out for a "purpose other than its (mere) disposal". However, to be excluded from the scope of the Dumping Convention, Article III, paragraph 1 (b) (ii) demands that the material's placement does not run contrary to the aims of the Convention. Article I defines the promotion of an effective control regime for all sources of

---

[118] For US case law see: Heinrich (2002), 4. For European case law see: European Court of Justice, Case C-9/00 Palin Granit, judgment of 18 April 2002, according to which leftover stones from a granite quarry can constitute waste.

[119] See Art. I of the Convention, stating: "Contracting Parties shall individually and collectively promote the effective control of all sources of pollution of the marine environment, and pledge themselves especially to take all practicable steps to prevent the pollution of the sea by the dumping of waste and other matter that is liable to create hazards to human health, to harm living resources and marine life, to damage amenities or to interfere with other legitimate uses of the sea.

[120] Consenting: GESAMP, Report of the Twenty-Seventh Session of GESAMP (United Nations Joint Group of Experts on the Scientific Aspects of Marine Environmental Protection), Nairobi/Kenya, 14–18 April 1997, cited from Johnston *et al.* (1999), 40: "For technical and financial reasons CO_2 (dry ice) disposal appears to be an an attractive option unless it is to be dumped from vessels. Dumping from ships, however, comes under the aegis of the Convention on the preservation of Marine Pollution by Dumping of Wastes and Other Matter (London Convention, 1972). In 1993 Contracting Parties to this Convention adopted a prohibition on dumping of industrial wastes … at sea that took effect on January 1st, 1996. It therefore seems unlikely unless the Convention can be amended to permit the dumping of CO_2, from ships, that any of the current Parties to the Convention … could give approval to such a practise. It should be further noted that the same conclusion would apply to liquid CO_2 disposal from vessels and platforms which would also fall within the purview of the London Convention 1972". See also Amestitova (2001), 9 and Heinrich (2002), 4.

[121] In this direction, see Jones and Young (2001), 4.

marine environmental pollution and pollution prevention itself as the Convention's objectives.

If ocean fertilisation constitutes such pollution, it cannot be covered by the exemption clause. Otherwise, matter with potential to cause environmental harm would be excluded from the international control regime established by the treaty, which would run contrary to its objectives. Unfortunately, the London Convention does not define the meaning of the term "pollution"; but many global or regional marine agreements are based on the following definition:

> Introduction, directly or indirectly, by human activities of wastes and other matter into the sea which results or is likely to result in such deleterious effects as harm to living resources and marine ecosystems, hazards to human health, hindrance of marine activities including fishing and other legitimate uses of the sea, impairment of quality for the use of seawater and the reduction of amenities[122].

To constitute pollution, the likeliness of a material to cause substantial harm is sufficient[123]. Such an approach corresponds to the recognition of the precautionary principle in the context of the London Dumping Convention, which was formally endorsed by a resolution of the Conference of Parties in 1991[124]. Due to their potential deleterious effect on the marine environment, ocean fertilisers fall under the category of pollutants. Their exclusion from the scope of the treaty must, therefore, be dismissed since it would be "contrary to the aims of the Convention".

At present, under the London Dumping Convention, ocean fertilisation would solely need authorization by a national body in the form of a general

---

[122] Art. 1 para. 1 (4) of the UNCLOS of 19 December 1982; Art. 2 (a) of the Convention of the Marine Environment and Coastal Area of the South-East Pacific of 12 November 1981; Art. 2 (f) of the Convention for the Protection of the Natural Resources and Environment of the South Pacific Region of 24 November 1986; Art. 3 para. 1 (d) of the Convention for Cooperation in the Protection and Sustainable Development of the Marine and Coastal Environment of the Northeast Pacific of 18 February 2002; Art. 10 of the 1996 Protocol to the Convention on the Preservation of Marine Pollution by Dumping of Wastes and other Matter, 1972 and the Resolutions Adopted by the Special Meeting of 7 November 1996.

[123] Some older treaties seem to require evidence of causal links between the discard of matter and environmental degradation as a constituting element of the definition of "pollution". On the grounds of the wide acceptance of the precautionary principle in international marine law, it must be assumed that this requirement was modified to include situations in which deleterious harm may probably arise.

[124] Resolution 44/14 of 6 September 1991. A precautionary approach can also be derived from some regulations enshrined in the Convention's annexes: Annex I, Part II, para. 8. 1(a) urges states to develop criteria to determine the potential impact of pollutants in order to select appropriate incineration sites. Annex III B para. 9 calls for Parties to consider whether an adequate scientific basis exists for assessing the characteristics and the composition of the matter to be dumped and its impact on marine life and human health.

permit. Because of the criteria to be taken into consideration, the treaty will lead to an informed decision-making process. Nonetheless, a high margin of discretion will be left to Parties on how to balance environmental issues against other interests.

Upon the entry into force of the 1996 Protocol to the 1972 Convention on the Prevention of Marine Pollution by Dumping of Wastes and Other Matter and Resolutions Adopted by the Special Meeting, ocean sequestration will become generally prohibited. The Protocol, which will supersede the London Dumping Convention[125], embodies a structural revision of the Convention. Instead of banning the discharge of certain matter, the Protocol obligates Parties to refrain from dumping any substance as long as it is not listed in the Protocol's Annex[126]. Carbon dioxide or materials expected to serve as ocean fertilisers are not enumerated on this list and will fall under the category of prohibitive materials[127]. The agreement will further restrict the potential use of the oceans as a sequestration site because of the Protocol's application to the storage of materials in the seabed and the subsoil thereof[128].

3.3 United Nations Convention on the Law of the Sea

UNCLOS sets up a comprehensive legal regime for the oceans and, thus, provides a framework for the protection of the marine environment and the sustainable use of its resources. Most of its environmental provisions are to be found in Part XII, which starts from the general obligation to protect and preserve the marine environment[129] and subsequently elaborates more detailed provisions. The first regulation to be of specific importance in the context of carbon sequestration is Article 194, paragraph 2. It obligates states to take all measures to ensure that activities under their jurisdiction or control are so conducted as not to cause damage by pollution to other states and their environment. Furthermore, states have to ensure that pollution arising from activities under their jurisdiction or control does not spread beyond the areas where they exercise sovereign rights.

This duty is based on the customary rule of *sic utere tuo ut alienum non laedas* expressed in the well-known Trail Smelter Case in which a tribunal concluded that no state has the right to use or permit the use of its territory in such a manner as to cause injury to another state[130]. The scope of this rule

---

[125] Art. 25 of the Protocol.
[126] Art. 4 para. 1 subpara. 1 of the Protocol in connection with Annex I of the treaty.
[127] Labour-Green (2001), 2.
[128] Art. I para. 4 subpara. 1 of the Protocol.
[129] Art. 192 of the Convention.
[130] Report on International Arbitral Awards 1941 III, pp. 1909 *et seq.* For further details on the judgments, see Mickelson (1993).

was extended to areas beyond jurisdiction by a large number of international documents[131] and is generally accepted as a part of customary international law[132].

Due to their potential to cause severe environmental damage to areas beyond national jurisdiction, carbon sequestration activities are likely to contravene this duty. There are, however, widely accepted qualifications of the obligation not to cause transboundary harm which will restrict the norm's utility in governing mitigation policies. In its judgment, the Trail Smelter Tribunal held that the restriction on states' sovereignty is only applicable when the case is of serious consequence and the injury is established by a clear and convincing evidence[133].

There are good reasons to argue that the scope of the obligation, at least in the context of international marine law, has been extended by the precautionary principle from cases with "clear and convincing" proof to activities not necessarily resulting in harmful effects but causing risks of serious damage[134]. Nevertheless, the question remains how the threshold of serious harm should be determined. No criteria can be drawn from the rule itself for circumstances under which a risk exceeds the acceptable level[135]. Hence, the "Trail Smelter" obligation as enshrined in Article 194, paragraph 2 of UNCLOS has often been regarded as a conceptual basis for the development of more specific rules on pollution prevention[136].

The UNCLOS regulations on pollution from land-based sources such as pipelines (Article 207), from seabed activities (Articles 208 and 209) and on pollution by dumping (Article 210) stipulate an obligation to adopt laws and regulations to prevent, reduce and control pollution of the marine environment caused by each respective source. In accordance with the treaty's character as a framework convention, these provisions leave it to the Parties to define more specific environmental standards in international or national consensus-building processes[137]. Thus, they give little guidance to states' behaviour in the context of ocean sequestration policies. Something different, however, applies as far as dumping is concerned. Article 210, paragraph 6 calls upon states, when adopting national legislation, to draw up provisions no less effective in preventing, reducing and controlling pollution

[131] See, e.g. Art. 3 of the Convention of Biological Diversity of 5 June 1992, Principle 2 of the Rio Declaration of 13 June 1992; para. 3 of the Preamble of the Vienna Convention on the Protection of the Ozone Layer of 22 May 1985.

[132] International Court of Justice Advisory Opinion on the Legality of the threat or use of nuclear weapons, 1996 ICJ Reports 241; Kiss (1983), 1075; Birnie and Boyle (1992).

[133] Report of International Arbitral Awards 1941 III, 1938.

[134] Primosch (1996), 233; Freestone (1991), 31.

[135] Wolfrum (2000), 29.

[136] Dahm, Delbrück and Wolfrum (1989), 446.

[137] Yankow (1999), 279.

than globally accepted rules and standards. With this requirement to adhere to international regulations, the Convention exerts the doctrine of *erga omnes*[138]. States will be bound to the environmental rules contained in international dumping treaties, especially the London Dumping Convention, irrespective of their membership to these agreements. Consequently, the entry into force of the 1996 Protocol of the London Convention will lead to a far reaching prohibition of ocean sequestration also covering states that have not ratified the Protocol.

In contrast to this, the UNCLOS rules on pollution from land-based sources, which are important for carbon sequestration through the discharge of CO_2 from pipelines, provide for very limited international influence on states' behaviour. Article 207 entails the duty to adopt laws and regulations to prevent, reduce and control pollution of the marine environment from land-based sources. In doing so, states shall "take into account" internationally accepted rules (Article 207, paragraph 1). UNCLOS does not obligate them to develop such rules. States acting through competent international organisations or diplomatic conferences are solely required to "endeavour" to do so (Article 207, paragraph 4). This weak wording is the result of the lack of readiness of many states to create international standards on issues covered by their territorial sovereignty, thereby possibly compromising economic interests[139]. Presently, only very general and weak regulations exist on the international level dealing with pollution from land-based sources. As a result of the "Montreal Guidelines for the Protection of the Marine Environment against Pollution from Land-based Sources"[140] and chapter 17 of Agenda 21[141], the "Global Program of Action for the Protection of the Marine Environment from Land-based Activities"[142] was developed. The Program intends to provide for guidance in the development of national, regional or sub-regional programs[143], but contains no targets, timetables or material standards to influence states' behaviour more intensively. As a consequence, it is for each national government to determine which substances and operations require regulation and control[144].

Article 26 of the Vienna Convention of the Law of Treaties necessitates states to perform treaties in good faith. Starting from this obligation and the

138 Wolfrum, Röben and Morrison (2000), 258, 270; Birnie and Boyle (1992), 353; Boyle (1986), 355 *et seq.*; Timagenis (1980), 505 *et seq.* For another perspective see Vukas (1990), 417 *et seq.*

139 Mensah (1999), 312 and Yankow (1999), 280.

140 Adopted by the United Nations Environmental Program on 24 May 1985.

141 Section 17.24 *et seq.*

142 The Program was adopted on 3 November 1995.

143 D 14.

144 The Program, however, names some priority areas for action: See Section V A-I on sewage, persistent organic pollutants, heavy metals, radioactive substances or oils.

general duty to protect the environment under Article 192 of UNCLOS, one may argue that states have to react upon environmental hazards to the oceans linked with new strategies like ocean sequestration. Regardless, a great margin of discretion on the extent and the political level to do so will continue to exist and limit the capacity of UNCLOS to govern the issue.

3.4 Regional Sea Agreements

Several regional agreements give effect to UNCLOS by specifying and implementing its general provisions. Such regional treaties allow for higher environmental standards because they help to accommodate the special circumstances and needs to be found in certain marine areas[145].

Provisions on the prevention and reduction of pollution can be found in almost every regional sea agreement. The following description is limited to conventions covering parts of the Atlantic and the Pacific Ocean that are of special interest for carbon sequestration activities due to their substantial contribution to deep water formation and circulation. At present, six regional sea agreements cover these marine areas.

The Convention for the Protection of the Marine Environment of the North-East Atlantic (OSPAR Convention) is applicable to parts of the Atlantic and Arctic Oceans between the east coast of Greenland, the North Pole, the continental North Sea coast and the Strait of Gibraltar[146]. In comparison to other regional sea treaties, its obligations are very far-reaching. As regards dumping of wastes from vessels, a general prohibition of these activities with enumerated exceptions similar to those in the 1996 Protocol to the London Dumping Convention exists[147]. Carbon dioxide is not listed in any of the annexes so that a direct disposal of CO_2 from a ship will contravene the obligations of the OSPAR Convention.

In relation to marine pollution from land-based sources, especially pipelines, the OSPAR Convention goes a little beyond the standards set by UNCLOS. Apart from the general duty to take all possible steps to prevent and eliminate pollution from such sources[148], states are called upon to use the best available techniques and environmental standards for point sources. The OSPAR Convention places point source discharges, which may affect marine areas, under national authorisation or regulation[149].

---

145 Birnie and Boyle (1992), 259.

146 Art. 1 (a) (i).

147 Art. 3 in connection with Annex II, Art. 3 para. 1 and 2.

148 Art. 3 of the Treaty and the regulations of Annex I.

149 Art. 2 para. 1 of Annex I, which allows for some international influence through legally binding decisions of the OSPAR Commission.

In contrast to discharges from land-based sources, the agreement prohibits any dumping of wastes from offshore installations[150]. A disposal of CO_2 from platforms at sea will, therefore, be inconsistent with the treaty.

During the meeting of the OSPAR Commission in June 2003, some delegations expressed their concern on the fact that the legal compatibility of ocean sequestration activities will depend upon the method of CO_2 release[151]. They highlighted that this could lead to unwelcome situations. After Greenpeace International had informed the Commission at the previous meeting that an experimental release of CO_2 in Norwegian waters was planned and had alleged that the project would violate the regulations of the OSPAR Convention, the Commission[152] had asked the Group of Jurists and Linguists to undertake a detailed evaluation in this regard[153]. In their preliminary report, the experts took the view that the placement of CO_2 by ship could constitute dumping and could therefore be prohibited[154]. Discharges of carbon dioxide from land through pipelines, however, would solely be subject to regulation and authorisation.

Ocean fertilisation is excluded from the scope of the Convention according to Article I, paragraph (g) (ii), which states that dumping does not include a

> placement of a matter for a purpose other than the mere disposal thereof, provided that, if the placement is for a purpose other than that for which the matter was originally designed or constructed, it is in accordance with the relevant provision of the Convention.

As mentioned above, ocean fertilisation operations are undertaken for a purpose other than the mere disposal of the substances in question. Normally, these substances were produced with the aim to use them as fertilizers, so that ocean fertilisation activities fall under the exemption clause and are not governed by the regulatory regime of the treaty.

Other international agreements on the preservation of the Atlantic are the Convention for Cooperation in the Protection and Development of the Marine and Coastal Environment of the West and Central African Region (Abidjan Convention)[155] and the Convention for the Protection and Development of the Marine Environment of the Wider Caribbean Region[156] (Cartagena Convention). The former applies to the coastal zones from Mauritania to Namibia falling within national jurisdictions[157], while the latter

[150] Art. 5 in conjunction with Art. 2 para. 1 of Annex III.
[151] OSPAR (2003), 4.47.
[152] It is made up by the representatives of the Parties according to Art. 11.
[153] OSPAR (2002).
[154] OSPAR (2003), 4.46.
[155] Adopted in Abidjan on 23 March 1981.
[156] Adopted in Cartagena de Indias on 24 March 1983.
[157] Art. 1 of the Abidjan Convention.

covers the marine environment of the Gulf of Mexico, the Caribbean Sea and the Atlantic areas adjacent thereto[158]. In accordance with UNCLOS, both agreements require Parties to ensure the effective application of internationally accepted standards in the context of dumping from ships, which will give effect to the provisions of the London Dumping Convention and its 1996 Protocol[159].

Some detailed rules referring to the prevention of sea pollution from land-based sources can be found in the Cartagena Convention's Protocol Concerning Pollution from Land-based Sources and Activities to the Convention for the Protection and Development of the Marine Environment of the Wider Caribbean Region[160]. Starting from the general obligation to take appropriate measures to prevent, reduce and control pollution from land-based sources, Parties are asked to develop more precise rules[161] on the international level for "source categories, activities and associated pollutants of concern"[162]. Annex I of the Protocol lists criteria to identify such pollutants, among them the potential of the substances to cause undesirable changes in the marine ecosystem, irreversible and durable effects, negative impacts on marine life, eutrophication or migration[163]. Although the Protocol contains solely an obligation to elaborate more specific rules on pollutants of concern, the criteria of Annex I, if fulfilled, will put pressure on the Parties to move towards a stronger regime for the prevention of pollution from land-based sources. Thus, in the case of proposals for carbon sequestration by pipeline injection, the Protocol may hinder Parties to withdraw from enacting more stringent protection measures. A prohibition of carbon injection activities, however, cannot be derived from the treaty.

As regards the Pacific, three regional sea agreements presently exist: the Convention for the Protection of the Marine Environment and Coastal Area of the South-East Pacific comprises coastal zones of all South-American states and the high sea up to a distance within which pollution may affect the area under national jurisdiction[164]. The 200 mile zones around the small island states of the Southern Pacific are governed by the Convention for the Protection of the Natural Resources and Environment of the South Pacific

[158] Art. 2 of the Cartagena Convention.

[159] Art. 6 of the Abidjan Convention, Art. 6 of the Cartagena Convention.

[160] Adopted at Aruba on 6 October 1999.

[161] Including effluent and emission limitations, best management practises and timetables for achieving the set limits; see Art. IV para. 1 sentence 2.

[162] Art. IV para. 1.

[163] Annex I of the Cartagena Protocol, C, 2, e), g), i), j).

[164] Art. 1 of the Convention for the Protection of the Marine Environment and Coastal Area of the South-East Pacific, adopted at Lima on 12 November 1981(Lima Convention).

Region[165]. Furthermore, on 18 February 2002, the Convention for Cooperation in the Protection and Sustainable Development of the Marine and Coastal Environment of the Northeast Pacific was adopted in Antigua, but has, so far, not entered into force[166]. All three agreements contain rather general regulations on marine pollution prevention[167]. It is, however, noteworthy that the Convention covering the South Pacific Region generally prohibits waste disposal in the seabed[168]. Furthermore, it contains provisions similar to those of the London Dumping Convention under which carbon sequestration activities would require a general permit[169].

3.5 Antarctic Treaty and its Madrid Protocol

As the Southern Ocean is the area of main interest for ocean fertilisation, corresponding operations will be governed by the Antarctic Treaty Regime. According to its Article IV, the Antarctic Treaty is applicable to the area south of 60 degrees south latitude[170]. Environmental provisions for human activities within its jurisdiction can be found in the Protocol on Environmental Protection to the Antarctic Treaty (Madrid Protocol) serving as its supplement[171]. As regards the question of the admissibility of ocean fertilisation, Article 4 of Annex IV will be of special importance. It states:

> The discharge into the sea of any noxious liquid substance, and any other chemical or other substance, in quantities or concentrations that are harmful to the marine environment, shall be prohibited[172].

Dissolved iron-compounds do not fall into the category of noxious liquid substances only covering matter listed in Annex II of the International Convention for the Prevention of Pollution from Ships, 1973 as amended in 1978[173]. Due to their potential to cause negative environmental impacts, they will, however, most certainly be regarded as substances harmful to the

[165] Art. 1 para. 1, Art. 2 (a) (i) of the Convention for the Protection of the Natural Resources and Environment of the South Pacific Region, Noumea, 24 November 1986 (Noumea Convention).

[166] It is applicable to the marine areas of the North-East Pacific at the coasts of the Latin American countries.

[167] Art. 4 (a) (i) (iii) of the Lima Convention, Art. 6 para. 1 (a) (i), (iii) of the Antigua Convention, Art. 6 of the Noumea Convention.

[168] Art. 11 para. 1 sentence 3 of the Noumea Convention.

[169] Art. 6 of the Protocol for the Prevention of Pollution of the South Pacific Region by Dumping, adopted at Noumea on 25 November 1986.

[170] Antarctic Treaty of 1 December 1959.

[171] Art. 4 para. 1 of the Protocol of 4 October 1991.

[172] Art. 2 of the Protocol. The provisions of Annex IV are solely applicable to discharges from ships not owned or operated by states (Art. 11 para. 1).

[173] Art. I (c) of Annex IV of the Madrid Protocol.

marine environment. The wording of Article 4 may suggest that the norm is only applicable in the case of proof of a causal link between the discharge of a substance and the expected environmental harm. Because of the wide acceptance of the precautionary principle in the context of the preservation of Antarctica[174], it seems legitimate to assume that the discharge of substances causing significant risks for the polar environment is not allowed either[175]. Such an approach accords with the environmental principles set forth in Article 3 of the Madrid Protocol[176]. According to Article 3, paragraph 2, activities shall be planned and conducted so as to limit negative environmental impacts. Significant adverse effects on environmental goods shall be avoided[177]. Thus, ocean fertilisation projects with their high potential to cause substantial harm will most certainly be regarded as being inconsistent with the Antarctic Treaty Regime.

4 Conclusion

The proposal to use the oceans in order to mitigate climate change raises questions of a technical, ecological, economic, legal, social and political nature with many issues not yet thoroughly discussed. Presently, no regulation on the international level explicitly refers to ocean sequestration. Hence, the statement that carbon sequestration has emerged within a "grey area of existing regulatory regimes"[178] that were designed before the issue of carbon storage in the ocean has come to public attention, is a good one to describe the situation. Nevertheless, and in spite of its degree of ambiguity, existing international environmental law has the capacity to provide a general framework for ocean sequestration issues. This chapter has attempted to demonstrate that most carbon sequestration techniques under discussion today cannot be used to offset emission reduction targets. With its entry into force, the Kyoto Protocol will hinder states to do so until the end of 2012. Apart from that, international marine law prohibits ocean sequestration in many respects. The London Dumping Convention prevents a direct injection of CO_2 from ships or platforms. Ocean fertilisation is

---

174 See, for example, the Final Report of the XVIIIth Treaty Consultative Meeting, Kyoto, Japan, 11–22 April 1994, para. 88, stating that: "Providing technical advice including – minimum monitoring needs to meet the requirements of the Protocol, based on a precautionary approach"; see also Redgwell (1994), 608.

175 Every activitiy to be carried out in Antarctica is subject to an environmental impact assessment; see Wolfrum (2000), 339.

176 According to Redgwell (1994), 607, the principles of Art. 3 form a "safety net" applicable to all operations not specifically addressed in the Protocol.

177 Art. 3 para. 2 (b).

178 Heinrich (2002), 2.

presently not generally forbidden under international environmental law regarding marine ecosystems, but it will become illegal with the entry into force of the Convention's 1996 Protocol. Furthermore, fertilisation of the most attractive area, the Southern Ocean, is prohibited under the Antarctic Treaty. As far as the injection of CO_2 into the deep seabed is concerned, some regional treaties that are already in force disallow states to carry out these activities. The 1996 Protocol to the London Dumping Convention will set global standards in this regard.

As a consequence of the very rudimentary development of legal norms on pollution from land-based sources, an injection of CO_2 by pipeline may be permitted in the future, assuming that the general obligation set out in the "Trail Smelter" Case is complied with and substantial risks to the marine environment are avoided.

The elaboration of ocean sequestration techniques may go ahead, but existing international environmental law will, to a large extent, prevent research results to be put into practice.

References

Adhiya, J. and Chrisholm, S.W. (2001). *Is Ocean Fertilisation a Good Carbon Sequestration Option?*, White Paper Prepared for the Laboratory for Energy and Environment at MIT, Massachusetts, USA.

Amatistova, L. (2001). *Storing Carbon in the Oceans: An Option for Addressing Global Climate Change*, GEE Discussion Paper 2001. http://www.gee21.org/publications.html.

Auerbach, D.I. *et al.* (1997). 'Impacts on Marine Life: I. A Toxological Assessment Integrating Constant-Concentration Laboratory Assay Data with Variable-Concentration Field Exposure'. *Environmental Modelling and Assessment* 2: 333–343.

Baker Röben, B. (2000). 'Protection of Global Atmospheric Components', in Morrison, F.L. and Wolfrum, R. (eds.), *International, Regional and National Environmental Law*. The Hague, London, Boston: Aspen Publishers, 201–224.

Bartholomäi, R. (1997). *Sustainable Development und Völkerrecht*, Baden-Baden.

Beyerlin, U. (1994). 'Rio-Konferenz '92: Beginn einer neuen Umweltrechtsordnung?'. *Zeitschrift für ausländisches öffentliches Recht und Völkerrecht*, 54: 124–145.

Birnie, A. and Boyle, P. (1992). *International Law and the Environment*. Oxford: Oxford University Press.

Bodansky, D. (1993). 'The United Nations Framework Convention on Climate Change: A Commentary', *The Yale Journal of International Law*, 18: 451–558.

Boyd, P.W. *et al.* (2000). 'A Mesoscale Phytoplankton Bloom in the Polar Southern Ocean Stimulated by Iron Fertilisation'. *Nature*, 407: 695–702.

Boyle, A. (1986). 'Marine Pollution under the Law of the Sea Convention'. *American Journal of International Law*, 79: 347–372.

Boyle, A. (1999). 'Introduction', in Boyle, A. and Freestone, D. (eds.), *International Law and Sustainable Development*. Oxford: Oxford University Press, 1–18.

British Government Panel on Sustainable Development (1999). *Sequestration of Carbon Dioxide*; Annex B: Sequestration of Carbon Dioxide by Ocean Fertilisation; Annex C: Capture and Storage of CO_2 from Large Industrial Sources. London : British Government Panel on Sustainable Development, http://www.sd-commission.gov.uk/panel-sd/ position/ co2

Brown Weiss, E. (1992). *In Fairness to Future Generations: International Law, Common Patrimony and Intergenerational Equity*. Tokyo: Transnational Publishers, United Nations University.

Caldeira, K., Herzog, H. and Wickett, M.E. (2001). *Predicting and Evaluating Effectiveness of Ocean Sequestration by Direct Injection*. Paper presented at the 1st National Conference on Carbon Sequestration, Washington DC.

Chrisholm, S. W. (2000). 'Stirring Times in the Southern Ocean'. *Nature*, 407: 685–687.

Coale, K. *et al.* (1996). 'A Massive Phytoplankton Bloom Induced by an Ecosystem-Scale Iron Fertilisation in the Equatorial Pacific Ocean'. *Nature*, 383: 495–501.

Dahm, G., Delbrück, J. and Wolfrum, R. (1989). *Völkerrecht – Volume I/1*. 2nd ed., Berlin, New York: de Gruyter.

De Baar, H. and Boyd, P.W. (2000). 'The Role of Iron in Plankton Ecology and Carbon Dioxide Transfer of the Global Oceans', in Hanson, R., Ducklow, H, Field, J. (eds.), *The Dynamic Carbon Cycle: A midterm synthesis of the joint global ocean flux study*. Cambridge: Cambridge University Press, 61–140.

Dodson, S.I., Arnott, S.E. and Cottingham, K.L. (2000). 'The Relationship in Lake Communities Between Primary Production and Species Richness'. *Ecology*, 81: 2662–2679.

Drange, A., Alendal, G. and Johannessen, O.M. (2001). 'Ocean Release of Fossil Fuels: A Case Study'. *Geophysical Research Letter*, 28: 2637–2640.

Dworkin, R.M. (1978). *Taking Rights Seriously*. Cambridge: Cambridge University Press.

Epiney, A. and Scheyli, M. (1998). *Strukturprinzipien des Umweltvölkerrechts*. Baden-Baden: Nomos Verlag.

Falkowski, P. *et al.* (2000). The Global Carbon Cycle: A Test of Knowledge of Earth as a System, Science 290 (13 October 2000), pp. 291–296.

Fitschen, T. (1995). 'Common Heritage of Mankind', in Wolfrum, R. and Philipp, C. (eds.), *United Nations: Law, Politics and Practise*. München, Dordrecht, London, Boston: CH Beck, 211–220.

Freestone, D. (1991). 'The Precautionary Principle', in Churchill, R. and Freestone, D. (eds.), *International Law and Global Climate Change*. London: Kluwer Academic Publishers, 21–39.

Freund, P. (1996). 'International Initiative to Combat Climate Change, Targets for Development of Greenhouse Gas Control Technology', in IEA Greenhouse Gas R&D Programme. http://www.iaegreen.org.uk/pfghgt4b.htm.

Fuhrmann, J. and Capone, D. (1991). 'Possible Biochemical Consequences of Ocean Fertilisation'. *Limnology and Oceanography*, 36: 1951–1959.

Greenpeace (1999). *Ocean Dumping of Carbon Dioxide – No Solution to Climate Change*. Greenpeace Briefing. http:// www.greenpeace.org/multimedia/download/1/17073/0/final_ backgrounder_co2_dumping.pdf.

Handl, G. (1990). 'Environmental Security and Global Change: The Challenge to International Law'. *Yearbook of International Environmental Law*, 1: 3–33.

Heinrich, J. (2002). *Implications of CO_2 Storage*. Laboratory for Energy and the Environment. Massachusetts: Massachusetts Institute of Technology.

Herzog, H.J. (1996). *Proceedings of the 3rd International Conference on Carbon Dioxide Removal*. Cambridge, UK.

Herzog, H.J. (1998). *Ocean Sequestration of CO_2 – An Overview*. Paper presented at the Fourth Conference on Greenhouse Gas Control Technologies, 30 August–2 September 1998, Interlaken, Switzerland.

Holloway, S. and Heederik, J.P. (1996). 'The Underground Disposal of Carbon Dioxide'. *British Geological Survey*, Nottingham, UK.

Houghton, J.T. *et al.* (2001). *Climate Change 2001: The Scientific Basis*. Cambridge: Intergovernmental Panel on Climate Change (IPCC).

Johnston, P. *et al.* (1999). *Ocean Disposal / Sequestration of Carbon Dioxide from Fossil Fuel Production and Use: An Overview of Rationale, Techniques and Implications*. Exeter/Amsterdam: Greenpeace Research Laboratories, Technical Note 01:99.

Jones, I.S.F. and Young, H.E. (2001). *The short and land term role of the ocean in Greenhouse Gas Mitigation*. Paper presented at the First National Conference on Carbon Sequestration. Washington: US Department of Energy, National Energy Technology Laboratory, 14–17 May 2001. http://www.netl.doe.gov/proceedings/01/carbon_seq/p44.pdf.

KAHEA The Hawaiian Environmental Alliance (2000). *Union of Concerned Scientists on Marine Carbon Sequestration*, Position Statement. http://www.kahea.org/ocean/co2/unionsct.html.

Kiss, A. (1983). 'The International Protection of the Environment', in MacDonald, R., St. John, J., Douglas M. (eds.), *The Structure and Process of International Law*. The Hague, Boston, London: Martinus Nijhoff Publishers, 1069–1093.

Labour-Green, J. (2001). *Legal and Political Aspects to Enhance Natural Carbon Sequestration in the Southern Ocean*. Paper Presented at the Symposium of the Australian Academy of Technological Science and Engineering, Looking South – Managing Technology, Opportunities and the Global Environment, November 2001.

Leitzell, T.L. (1973). 'The Ocean Dumping Convention – A Hopeful Beginning'. *The San Diego Law Review*, 10: 502–513.

Marchetti, C. (1977). 'On Geoengineering the CO_2 Problem'. *Climatic Change*, 1: 59–68.

Martin, J. *et al.* (1994). 'Testing the Iron Hypothesis in the Ecosystems of the Equatorial Pacific Ocean'. *Nature*, 371: 123–129.

Mensah, T. (1999). 'The International Law Regime for the Protection and Preservation of the Marine Environment from Land-based Sources of Pollution', in Boyle, A. and Freestone, D. (eds.), *International Law and Sustainable Development*. Oxford: Oxford University Press, 297–324.

Mickelson, K. (1993). 'Rereading Trail Smelter'. *Canadian Yearbook on International Law*, 31: 219–234.

Odendahl, K. (2000). *Die Umweltpflichtigkeit der Souveränität. Reichweite und Schranken territorialer Souveränitätsrechte über die Umwelt und die Notwendigkeit eines veränderten Verständnisses staatlicher Souveränität*. Berlin: Band 88.

Ohsumi, T. (1995). 'CO_2 Disposal Options in the Deep Sea'. *Marine Technological Society Journal*, 29/3: 58–66.

Ormerod, W.G., Freund, P. and Smith, A. (2002). *Ocean Storage of* CO_2. IEA Greenhouse Gas R & D Programme, 2nd ed.

OSPAR (2002). *Convention for the Protection of the Marine Environment of the North-East Atlantic*. Meeting of the OSPAR Commission, Amsterdam, 24–28 June 2002, Summary Record : OSPAR 02/21/1/E, 9.24 (b).

OSPAR (2003). *Convention for the Protection of the Marine Environment of the North-East Atlantic*. Meeting of the OSPAR Commission, Bremen, 23–27 June 2003, Summary Record: OSPAR 03/17/1-(A-B)- E.

Owens, N.P.J. *et al.* (1991). 'Methane Flux to the Atmosphere from the Arabian Sea'. *Nature* 354 : 293–296.

Primosch, E.G. (1996). 'Das Vorsorgeprinzip im Internationalen Umweltrecht'. *Austrian Journal of Public and International Law*, 51: 227–241.

Redgwell, C. (1994). 'Environmental Protection in Antarctica: The 1991 Protocol'. *International & Comparative Law Quarterly*, 43: 599–634.

Rehbinder, E. (1994). 'Precaution and Sustainability: Two Sides of the Same Coin?', in Kiss, A. and Burhenne-Guilmin, F. (eds.), *A Law for the Environment*. Cambridge: Cambridge University Press, 93–100.

Sands, P. (1995). *The Principles of International Environmental Law, Volume I*. Cambridge: Cambridge University Press.

Sands, P. (1995). 'International Law in the Field of Sustainable Development: Emerging Legal Principles', in Lang, W. (ed.), *Sustainable Development and International Law*. London: Aspen Publisher, 53–66.

Sarmiento, J.L. (1993). 'Ocean Carbon Cycle'. *C & E News*, 31 May 1993, 30–43.

Sarmiento, J.L. and Orr, J.C. (1991). 'Three-dimensional Simulations of the Impact of Southern Ocean Nutrient Depletion on Atmospheric CO_2 and Ocean Chemistry'. *Limnology and Oceanography*, 36: 1928–1950.

Schmitz, B. (2000). 'Plankton Cooled a Greenhouse'. *Nature*, 407: 143–144.

Scientific Committee on Ocean Research – Intergovernmental Oceanographic Commission Advisory Panel on Ocean CO_2, Watching Brief: Ocean Carbon Sequestration; http://www/ioc.unesco.org/iocweb/co2panel/LegalPolicy.htm

Seibel, B.A. and Walsh, P.J. (2001). 'Potential Impacts of CO_2 Injection on Deep-sea Biota'. *Science*, 294: 319–320.

Smetacek, V. (2000). 'RV Polarstern Cruise ANT XVIII/2 EISENEX'. *AWI weekly report 1*. http://www.awi-bremerhaven.de/Pelagic/eisenex.html.

Tamburri, M.N. *et al.* (2000). 'A Field Study of the Effects of CO_2 Ocean Disposal on Mobile Deep-Sea Animals'. *Marine Chemistry*, 72: 95–101.

Timagenis, Gr.J. (1980). *International Control of Marine Pollution*. Dordrecht, Boston, Alphen aan den Rijn: Kluwer Academic Publishers.

United Nations (1990). *Report on the Law of the Sea*. New York: UN Doc. A/45/721, 19 November 1990.

US Department of Energy Report (1999). *Carbon Sequestration: State of Science*. Washington: Office of Science, Carbon Sequestration Research and Development. http://www.fe.doe.gov/coal_power/sequestration/reports/rd/index.html.

Vukas, B. (1990). 'Generally accepted International Rules and Standards', in Soons, A. A. (ed.), *Implementation of the Law of the Sea Convention through International Institutions*. Honolulu: Law of the Sea Institute, 405–422.

Watson, A.J. *et al.* (2000). 'Effect of Iron Supply on Southern Ocean CO_2 Uptake and Implications for Global Atmospheric CO_2'. *Nature*, 407: 730–733.

Wolfrum, R. (2000). 'International Law and Environmental Law', in Morrison, F.L. and Wolfrum, R. (eds.), *International, Regional and National Environmental Law*. The Hague, London, Boston: Aspen Publishers, 1–70.

Wolfrum, R., Röben, V. and Morrison, F.L. (2000). 'Preservation of the Marine Environment', in Morrison, F.L. and Wolfrum, R. (eds.), *International, Regional and National Environmental Law*. The Hague, London, Boston: Aspen Publishers, 225–284.

Wolfrum, R. (2000). 'Environmental Protection of Ice-Covered Regions', in Morrison, F.L. and Wolfrum, R. (eds.), *International, Regional and National Environmental Law*. The Hague, London, Boston: Aspen Publishers, 329–341.

Wong, C.S. and Hirai, S. (1997). *Ocean Storage of Carbon Dioxide: A Review of Oceanic Carbonate and CO_2 Hydrate Chemistry*. Cheltenham.

World Commission on Environment and Development (1987). *Our Common Future*. New York: Oxford University Press.

Yankow, A. (1999). 'The Law of Sea Convention and Agenda 21: Marine Environmental Implications', in Boyle, A. and Freestone, D. (eds.), *International Law and Sustainable Development*. Oxford, New York: Kluwer Academic Publishers, 271–295.

PART II

EU and Climate Change Policies

Chapter 8

MERCEDES FERNÁNDEZ ARMENTEROS

State Aid Issues Raised by the Implementation of Climate Change Policy Instruments

Introduction

The introduction of the new Community Guidelines on State Aid for Environmental Protection in February 2001 has so far not raised significant reactions[1]. This is probably due to the fact that the main concerns in the field of state aid are generally discussed in sectors of horizontal aid other than environmental aid, such as agriculture and regional aid.

Nevertheless, the panorama of state aid in favour of the environment is changing as Member States make more frequent use of this kind of subsidies. Not only has the quantity of environmental state aid been increased considerably, but over the years Member States have also adopted increasingly innovative environmental policies and the environmental subsidies they provide is often in forms which are very different from those traditionally included in the European regulatory framework of state aid[2].

But if state aid in favour of the environment is acquiring special significance this is largely due to the impact of national policies related to both climate change as well as renewable sources of energy. In this regard, DG Competition of the European Commission is very conscious of the increasing role that this type of aid will play in the future. It is important to note that the Commission already had the opportunity to share its views about the compatibility of two national emissions trading systems (in the UK and Denmark respectively) with EU rules. Equally, the ECJ ruled last year in

---

[1] Community Guidelines on State Aid for Environmental Protection, OJ C 37, 3 February 2001, 3.

[2] A comparative analysis of the yearly reports submitted by the Commission allows to appreciate the significant increase of state aid for environmental protection in the last years.

MICHAEL BOTHE AND ECKARD REHBINDER (EDS.), Climate Change Policy, 219–258.

favour of the German system of subsidized prices for renewable energy by rejecting the interpretation of the Commission regarding state aid[3]. By the same token, decisions on the compatibility of fiscal regimes favouring alternative energy and climate policies are also a common object of Commission decisions.

Emanating from the above considerations and the emergence of climate policy issues, this article offers an analysis of the compatibility between the novel economic instruments in the field of climate change and energy policy with the Commission's current regulatory approach to state aid for the protection of the environment. A short overview will be given firstly of the compatibility between competition policy and environmental policy and secondly of the scope of the integration principle in the new Guidelines. A general assessment of the current EU framework of state aid for environmental protection will then be offered, including a comparison with the preceding Guideline on State Aid for environmental protection. The focus will then be on the compatibility of national measures adopted to combat greenhouse gas emissions with the EU regulatory regime for state aid. Given the great variety of instruments that Member States can introduce to reduce greenhouse gas emissions, the analysis will concentrate on the most novel instruments recently implemented by Member States, as well as on advance issues that might be the object of concern in the years to come.

1 Compatibility between competition and environmental policy rules

The European regulatory body for state aid, enshrined in Chapter I Title VI of the EC Treaty, relating to "Competition Rules" – articles 87-89 – is characterised not only by the absence of formal legislation, but also by its sensitive political character. Those two aspects distinguish state aid from the competition rules relating to enterprises, where the Treaty – articles 81 and 82 – provides a precise and rigid framework. Both characteristics explain why in the area of state aid the EU Commission enjoys a large discretionary power to guarantee that state aid does not jeopardise competition in view of ensuring the single market objectives.

Nevertheless, the relationship between competition and environmental policies has often been tense. Public authorities as well as academics have most of the times emphasised the differences separating both fields rather than their common elements. From this angle, it should not be ignored that the legal basis for the integration of both policies rests on the polluter-pays principle (PPP), introduced for the first time in 1974 in the Council Decision

3 Case C-379/98, *PreußenElektra*, 13 March 2001, ECJ Rep. 2001 I, 2099.

on the inclusion of environmental costs by public authorities[4] and later in the Treaty itself by the Single European Act, currently article 174 (2). Indeed, although the polluter-pays principle has traditionally been considered an environmental principle, there is no doubt that the objective of this principle fundamentally is the internalisation of environmental costs. Thus, it aims at achieving economic efficiency inherent in market based competition principles. It can thus be concluded that the polluter-pays principle would have been applied, sooner or later, by the EU, even if it had not been explicitly embodied in the environmental chapter, as it is both an environmental and an economic principle. Thus, the fact that the polluter-pays principle forms part of the logic of the competition and environmental policies is a guarantee, at least in principle, of the compatibility of both objectives.

The analysis of the regulatory framework of state aid for environmental protection leads us to a more profound analysis of the content of the polluter-pays principle.[5] The Treaty does not offer any definition of the principle, and the same is valid for other environmental principles. Nevertheless, environmental policies must be in line with both the PPP as well as the criteria set out in article 174 (3), especially "the economic and social development of the Community as a whole and the balanced development of its regions" (article 174 (3) and (4)). This implies that the PPP would be balanced by economic and social criteria in such a way that, when the PPP is applied to environmental state aid policy, the Commission should consider the economic and social consequences of the imposition of environmental costs. It is a different question whether – as suggested by certain authors – the Community Guidelines on State Aid for Environmental Protection constitute a proper instrument to strike this balance of interests[6].

The relationship between competition and environmental policies, however, is far from irreconcilable. Over the years, environmental and competition policies have rather converged. First of all, from an environmental point of view, both the Member States and the Commission have repeatedly preferred economic instruments to the classical environmental measures, known as the "command and control approach"[7]. Regarding the single market and competition norms, we have also witnessed a fundamental change in regulatory philosophy which started to accept the

---

4 Council Recommendation, 3 March 1975, OJ L 194, 1. In other legal systems, the PPP is not expressly approved, but it is commonly used, for instance in the American system.

5 See Jans (1995), 291.

6 In this respect, Grabitz and Zacker (1989) have suggested that the Council should assume its legislative task to concretise the PPP as the Guidelines did not represent an adequate instrument.

7 For an overview of the origin of economic instruments at the Community level, see Rehbinder (1993).

active role of the private sector in contributing to the protection of the environment. From this perspective, a telling example is the recent controversial CECED decision in which the Commission approved for the first time an agreement between several enterprises to end the production of high energy consuming washing machines with the intention of saving energy[8]. This case confirms the flexibility of the Commission towards the initiatives of the private sector which favour the environment. Equally, the case is an example of the integration of environmental policy objectives into competition policy as required by article 6 of the Treaty[9]. Along the same lines, by including new kinds of environmental subsidies, the new regulatory framework of environmental state aid embraces the spirit of integrating the requirements of environmental protection into competition policy. And, as a matter of fact, it is quite significant that the new Guidelines recognise the integration principle as the new guiding principle of the regulatory framework.

2 The Integration Principle and the New Regulatory Framework for Environmental State Aid

The new normative framework for environmental state aid (the Guidelines) that entered into force in February 2001 echoes the normative developments occurring in the last years. As pointed out above, the integration of environmental considerations prescribed by article 6 of the Treaty guides the development of the new rules. In this respect, by emphasising that environmental and competition policies are not mutually antagonistic, point 3 of the Guidelines emphasizes the duty of integration with the purpose of promoting sustainable development. However, while the new rules show an unambiguous willingness of recognising the environment as a factor to be

---

[8] F-1/36.718; OJ L 187, 26 July 2000. See also European Commission (2001). In this decision, the Commission considered that the agreement at stake, although restricting competition in the sense of article 81.1, could be exempted pursuant to article 81.3, as it diminished energy consumption by also helping consumers to save energy.

[9] The changing role of private actors regarding environmental policies and their effect on competition policy is emphasised by Jans (2000): "… In this context, the traditional black and white approach on which European competition policy and law were founded is becoming increasingly inappropriate. This approach assumed that it was up to the public authorities to establish and pursue environmental policy by means of legislation and its application through prohibitions and permits, and up to "the market" to respond by operating as profitably as possible within these constraints. In this climate, competition policies which allowed only marginal room for environmental considerations were appropriate. However today, where the market is very much expected to act responsibly on its own account, environmental objectives should play a greater part in competition policy than is currently the case …".

considered in the design and assessment of state aid, this will not entail total integration. It is clear that the Commission has expanded the cases of accepted environmental state aid in order to motivate Member States to take new environmental measures[10]. But it is also true that the Commission continues to maintain the basic concepts of state aid policy, namely, that environmental state aid "can not ensure the development of activities whose economic viability is not possible"[11].

It should also be clear that the application of the integration principle in relation to state aid is not entirely a novelty of the recent regulatory framework, since, apart from having been incorporated in the Single European Act in 1986 and later in article 130R, the integration principles already inspired the normative text on environmental state aid in 1994[12]. Yet, the fact that the integration principle is nowadays included in article 6 - which forms part of the chapter dedicated to the principles on which EU policies must be based - could in future tip the balance even more in favour of environmental considerations when cases of state aid are evaluated by the Commission. As suggested by certain authors, it is not inconceivable that in the years to come the Commission could exempt certain categories of environmental state aid through the mechanism included in article 89[13]. Indeed, Regulation 994/98[14] authorises the Commission to grant an exemption for certain categories of state aid, the environment being one of them (article 1(1)(iii)). That would imply that once a category is exempted, environmental state aid could be considered compatible with the internal market rules and, consequently, would not be subject to the notification requirement of article 88 (3). This same line of thought appears in the comments submitted by the German government to the provisional text of the Guidelines on state aid for environmental protection. As a matter of fact, the German government lamented that the Commission had not made use of the power given to it by the Council in 1998 to exempt environmental state aid from the state aid regime, which would correspond to a real and effective integration of environmental concerns into other policies as provided by article 6 of the Treaty. It is interesting to observe, however, that in the

---

[10] Focusing on the integration principle as far as the Guidelines on State Aid for environmental protection are concerned , see Vedder (2001).

[11] Alexis (2001), 26.

[12] OJ C 72, 10 March 1994, 3.

[13] Article 89 of the Treaty reads: "The Council, acting by a qualified majority on a proposal from the Commission and after consulting the European Parliament, may make any appropriate regulations for the application of Article 87 and 88 and may in particular determine the conditions in which article 88.3 shall apply and the categories of aid exempted from this procedure". Similar observations regarding the 1994 normative framework of state aid for environmental protection were already advanced in the past, see Jans (1995), 108.

[14] OJ 1998 L 142.

previous version of the environmental state aid framework the Commission left the door open to the possibility of a future general exemption for environmental aid. However, it can be argued that despite possible advantages of a future exemption by category, such an exception would not be considered reasonable nowadays. Considering the experimental character of most state aid modalities which will soon be implemented by Member States, it seems premature to elaborate a precise and clear category to be exempted. In fact, an exempted category would most likely include only those state aid forms already identified, deterring Member States from testing other types of aid not contained in the list of exemptions – especially if this list is interpreted in a restrictive way. This would be a real obstacle to further development, especially in the climate change domain in which new instruments are constantly tested by Member States.

Furthermore, the content of the integration principle suggests that it should influence the very concept of internalisation of costs, and, in turn, the content of the polluter-pays principle. Certainly, even though the objective of the internalisation of costs appears to be one of the aspects of the integration principle, it is also clear that a thorough internalisation has not been achieved in the framework of the new rules on environmental state aid nor in the general environmental policies. First of all, it should be noted that in the case of environmental aid, the uncertainties concerning the internalisation of costs affect not only the cost side, but also the benefit side. As to the costs, the *raison d'être* of environmental subsidies is precisely not to internalise them. Subsidies are only an imperfect solution as the issue is not only to allocate all costs of measures protecting the environment to those undertakings causing the environmental harm, but particularly to eliminate subsidies benefiting sectors like the carbon or nuclear energy sector which prevent the development of renewable energy sources. Admittedly there are political difficulties to eliminate state aid to those polluting sectors, and environmental subsidies appear the second best option. As regards benefits, it seems that the new normative framework remains limited. Point 6 of the Guidelines defines internalisation of costs as taking into consideration, when calculating the production costs of a firm, all costs associated with the protection of the environment. Nevertheless, the positive costs, in other words, all environmental benefits derived from certain activities – for instance the benefits derived from renewable energy sources – should also be considered in any cost–benefit analysis. However, it should be acknowledged that the difficulties in defining 'environmental costs' are not a specific problem of state aid for environmental protection. They rather affect all domains of environmental law where economic valuations have to be made[15].

If the integration principle, understood as integration of environmental costs and benefits, shows considerable gaps, this does not hold true for the material scope of the principle: it appears positive. At least in principle, it is a positive novelty introduced by the new Guidelines that the Commission is required to take into account environmental aspects when adopting or reviewing a state aid framework, and that Member States may be compelled to provide for environmental impact studies whenever "they will notify [the Commission] of an important aid project, irrespective of the sector involved". Hence, the integration requirement affects community institutions (the Commission when framing other regulations on state aid) inasmuch as national authorities (when elaborating environmental impact assessment studies or when notifying relevant projects on state aid). This will become one of the most significant aspects of the integration principle.

3 Analysis of the New Regulatory Regime of Environmental State Aid

3.1 The Previous Regulatory Framework

Before analysing the preceding regulatory framework on state aid, some words must be devoted to the Guidelines as a regulatory instrument. In the area of state aid, the EU Commission, as the main decision maker, develops general substantive policy either through formal legislation or through informal rule-making. Indeed, in most cases the Commission has developed policy through guidelines and communications rather than through formal legislation. The rules and policy framework has been established for particular sectors as well as for other matters including the environment. The ECJ has held that it is lawful for the Commission to structure its discretion through such guidelines, provided that they do not depart from the Treaty rules[16]. Indeed, the reasons for using such guidelines are of a practical nature. They reduce the discretion and the arbitrariness of the Member States, they facilitate transparency and legal security and make the work of the Commission easier by eliminating the need to treat many cases individually[17].

---

15 Generally on the rationality and economic efficiency of the state aid regime at Community level, see Niccolaides and Bilal (1999).

16 Case C-288/96, *Germany v. Commission*, ECR 2000 I, 8237, para. 62.

17 For the legal issues raised by the value of this kind of rule-making see Craig and de Búrca (2003).

The recognition of the necessity of exceptions to the PPP and also of state aid gave rise in 1974 to the adoption by the Commission of a Memorandum on environmental state aid, which was modified in 1980 and 1986.[18] The most significant trait of this first normative framework was the transitional character of environmental state aid. For the future, the text envisaged an integral application of the polluter-pays principle[19].

In 1994, the Commission, with the experience acquired during almost two decades, developed a more complete regulatory framework on environmental state aid.[20] Although the 1994 framework provided for substantial modifications, the Commission continued to be headstrong *vis-à-vis* environmental state aid and hardly showed any intention to encourage Member States to adopt measures which could fall under the state aid regime. It is only with the promulgation of the regulatory framework of 2000 that differences in conceptions of environmental state aid become noticeable.

While the first and second phase are characterised by a desire to clarify actual cases of environmental aid without encouraging Member States to adopt this kind of subsidies – a concept which corresponded to the harmonisation of environmental policy practised at the community level – this concept disappeared over the years. In the new framework, a clear encouragement is given to Member States to experiment with new forms of environmental subsidies, a fact which should not be underestimated from a political point of view. Regarding the content, the primary difference between the old and new regime lies in the increase of operating aid cases brought into the text and in the decline in investment aid cases[21]. The importance of operating aid in the entirety of Member States is expressly recognised by the Commission in point 2 of the new Guidelines, stating that "new forms of operating aid are also on the increase".

In the following sections, a detailed analysis of the most relevant legislative modifications regarding the different categories of environmental state aid will be given.

---

[18] Memorandum of 6 November 1974, Letter to Member States SEC(74) 4264 ; Fourth Report on Competition Policy, points 175 to 182. Letter to Member States SG(80) D/8287, 7 July 1980, and letter SG (87), D/3795, 23 March 1987.

[19] This practice of the Commission *vis-à-vis* State Aid for environmental protection has been considered as being very restrictive. In this sense, see Van Calster (1998).

[20] For an analysis of the juridical evolution of state aid until 1995, see Jans (1995), 108 et seq.

[21] The weight given to operating aid can be seen by the sheer extension of the text, not only in comparison with the former text but also compared with the paragraphs on investment aid: whereas investment aid is covered by 11 points, the text dealing with operating aid extends over 25 points. In the former text it was only covered by one provision, namely point 3.4.

3.2 Modifications in the investment aid measures

Investment aid is the first element where changes in the new regulatory setting of environmental state aid can be observed. According to the 1994 framework, there are three modalities for this kind of aid:

1. aid aiming at enabling existing undertakings to adapt to new environmental standards or at encouraging them to adopt them in advance,
2. aid encouraging the adoption of standards higher than those required by the norm,
3. aid aiming at achieving environmental standards derived from environmental agreements, unilateral programs, i.e. standards which are not or not yet legally compulsory.

Within the new framework, investment aid disappears in the first case. While in the past undertakings could benefit from a subsidy of 15% of eligible costs for a limited period of time to be able to adapt to EU environmental standards, from now on undertakings will no longer be able to benefit from this kind of aid to fulfil their obligations. Still, small and medium size enterprises (SME) represent an exception to the rule: according to point 28 of the current Guidelines, they can benefit from state aid in order to meet EU standards during three years from the moment those standards become compulsory and up to a maximum of 15% of eligible costs. Logically, paragraph 3 of point 3.2.2 A of the former regime regulating the relocation of enterprises has been deleted from the text. According to the former regime, enterprises existing at least two years were not considered 'new enterprises' if they opted for replacing their installations by a new establishment rather than adapting these to meet obligatory standards. As a consequence, they also had the right to state aid not exceeding the adaptation costs of the old installation. The new framework, instead, considers that relocation is not necessarily beneficial to the environment and, therefore, there is no basis for granting environmental aid, except in cases when a firm established in an urban area or in a Nature 2000 designated area lawfully carries on an activity that creates major pollution and must, on account of this location, move from its place of establishment to a more suitable area (point 39).

Similar to the previous norm, the new rules consider it appropriate to grant special treatment to those cases in which Community environmental standards are improved. Points 20 and 29 of the Guideline provide for the possibility of granting investment aid if it serves as an incentive to achieve levels of protection which are higher than those required by Community standards, or where no Community standards exist.

It is clear from the above that the Commission has considerably extended the internalisation of costs concept, by limiting the state aid to specific cases of SME. On this point, it is quite significant that the Commission insists on expressing this change of concept throughout the text. In point 19 of the Guidelines the Commission states, for instance, that after the Fifth Action Programme on the environment – based on the polluter-pays principle – firms already have had several years to adapt to the gradual application of the principle. Equally, in point 21 the Commission states that environmental standards constitute the ordinary law with which enterprises have to comply, and that there is no need to provide financial incentives to fulfil these legal requirements, *inter alia* because it has not been proven that such a kind of aid really serves as an incentive for firms to comply with the legal requirements.

As to the policies for climate protection, the new regulatory setting also provides for innovations. First of all, the new Guidelines adopt a broad definition of environmental protection (point 6), which does not only include remedies or prevention of damage to natural resources, but also the efficient use of those natural resources. Certain basic climate policies fall in this last category, namely measures for saving energy and the use of renewable energy. Whereas in the 1994 environmental guidelines state aid for energy saving policies and for renewable energy were simply accepted as investment aid, in the 2001 framework the Commission is more flexible regarding aid to those sectors. It even allows to go beyond the normal limit of eligible costs which is fixed at 40%, up to 100% in certain cases (point 32). Another innovation of the current regime which is highly relevant for climate policy is the inclusion within the category of investment state aid of state aid to the combined production of electric power and heat (point 31).

3.3 Modifications of the Horizontal Aid Measures

A second category of environmental state aid which is modified by the new regime is horizontal environmental state aid. This kind of aid provides support for disseminating knowledge, for audits and for consulting and training services as well as for disseminating information. Pursuant to the new environmental guidelines, the only kind of authorised aid will be advisory/consultation aid for SME.

3.4 Modifications of the Operating Aid Measures

Another relevant branch of the 2001 environmental state aid framework is represented by what is known as 'operating aid'. Under the previous regulatory regime, operating aid was acceptable only as aid concerning waste management and as exemption to environmental taxes. As pointed out above, the 1994 regime (point 2.3) stated that aid for renewable energy could fall under the investment aid category, whereas the Commission would judge on an ad hoc basis in the case of operating aid for the production of renewable energy, After the new rules came into force, specific operating aid rules were introduced for the case of combined production of electric power and heat (points 66 and 67), renewable energy (points 54-65) and for the reduction of, or exemption to, taxes (points 47-53). As these categories are of particular importance in the context of climate change policy, they will be analysed in greater depth in the following paragraphs.

3.5 Conclusion

Generally, three elements of evolution can be discerned as a result of the comparative study of the environmental state aid regulatory framework in the last years:

1. The energy sector and, more specifically, renewable energy acquired more significance in relation to both investment and operating aid.
2. The new Guidelines shift the emphasis by increasing the possibilities of operating aid, which has a direct effect on the production costs of the beneficiary of the aid. This change of spirit is somewhat surprising given the traditional distrust of the Commission as regards this kind of aid.
3. Contrary to the past, the Commission leaves the door open to environmental aid triggering market mechanisms, while aid for compensating for economic and social burdens derived from the implementation of environmental policies remain an exception. The favourable treatment of aid in the renewable energy sector and the limitation of aid for the small and medium enterprises are proof of this evolution.

4 National Instruments for the Reduction of Greenhouse Gas Emissions and Their Compatibility with the Normative Regime of State Aid

4.1 The New Chapter F "Policies, Measures, and Instruments for Reducing Greenhouse Gases"

Undoubtedly one of the salient aspects of the new regime of environmental state aid is the inclusion by the Commission of a chapter devoted to measures which Member States might adopt with the aim of complying with the objectives to be attained under EU, national and international greenhouse gases reductions policies. However, chapter F entitled "Policies, Measures and Instruments for reducing Greenhouse Gases" is interesting not for what it says, but rather for what it does not say. Although the consequences of the introduction of this chapter will be analysed below, it can already be stated here that the Commission was aware of the fact that future climate change measures adopted by a Member State might fall under the regime of state aid[22]. Nevertheless for the time being the Commission will avoid to prejudge the content of such measures and that is why the four paragraphs under Chapter F do not provide particular guidelines regarding how climate change policies would be assessed in respect of the state aid regime. By doing that, the Commission is giving Member States ample room to manoeuvre to introduce different and original climate change instruments. Point 71 of the Guidelines expressly states that despite the fact that certain measures might be subject to the state aid regime, "it is still too early to lay down the conditions for authorising any such aid".

In the past the Commission already anticipated the tension between national climate policy measures and EC state aid rules[23]. The increasing importance of climate change policies appears for instance in the 8th Survey on State Aid in the European Union[24]. Interestingly, this Survey for the very first time included a title devoted to the fiscal regime in favour of CO_2 emission reductions, stating that

> the category of state aid for environmental protection comprises a subcategory, regarding the regimes implemented by certain Member States in order to encourage CO_2 emission reductions and others for sustaining the use of renewable energy, being the main reason justifying those regimes, the necessity for the EU and each Member State to abide by the engagements derived from the Kyoto Protocol and to attain the general objective of sustainable development[25].

It is particularly interesting to attentively observe how the Commission expresses its views on this topic. From the wording of the new Guidelines it seems as if, despite all future difficulties derived from its application, DG Competition is reluctant to interfere with the support and pressure that the DG Environment has put on Member States to carry out as soon as possible

---

22 Already in 1989, the importance of economic instruments, such as subsidies and charges, and their compatibility with the state aid regime was emphasised, see Grabitz and Zacker (1989).

23 Conflicts might also emerge between national climate measures and WTO rules, in particular the Agreement on Subsidies and Countervailing Measures, see Assunçao and Zhang (2002).

24 8th Survey, COM (2000) 205 (01), 11 April 2000.

25 8th Survey, COM (2000) 205 (01), 11 April 2000, point 32.

those policies necessary for achieving the objectives to which the EU as a whole is committed at the international level. The Guidelines thus acknowledge in paragraph 71 that "some of the means adopted by Member States to comply with the objectives of the Protocol could constitute State aid," but a condoning attitude is offered by expressing that "it is still too early to lay down the conditions for authorising any such aid". In the near future, however, the tensions among the different objectives aimed at by different DGs – Environment, Competition and Transport-Energy – regarding climate change policy will unavoidably be felt. From another angle, this wording reflects two practical consequences. First, even though the Commission already had the opportunity to give its opinion about several national climate policy instruments, such as the Danish and UK Emissions Trading program, and even though the Commission already predicts in paragraph 69 that many of those instruments will cause trouble *vis-à-vis* the state aid regime, it recognises the lack of experience regarding these new instruments. Second, there is no doubt about the Commission's intention to allow Member States to behave as laboratories of policies, given on the one hand the existing need of innovation and learning about effective climate policies and on the other hand the absence of legislation at Community level[26].

Thus, the new Guidelines for environmental state aid remain silent regarding the way in which Member States should design the market instruments aimed at fulfilling their climate change reduction obligations. This silence is significant especially if compared to the preceding versions of the guidelines presented by the Commission in which there were even indications about how Member States should distribute the permits in case of a market permit system. But the reluctance of Member States towards a centralised European model led the Commission to withdraw this suggestion from the definitive text in the end. Another possible reason why the Commission preferred to refrain from fixing concrete parameters for the design of new climate protection instruments might be the fact that climate policies are intimately linked to energy policies, which remain to a large extent within the competence of Member States. In this regard, the first Declaration of the Conference of Permanent Government Representatives of 1985 regarding article 130R – current article 174 – states that national energy policies will have priority over EU environmental policies (Paragraph 9): "The Conference confirms that Community action regarding the environment should not interfere with national natural resource policies".

[26] Despite the primary role of the EU as an actor in climate change policies at the international level, all documents issues by DG XI stress the importance of the subsidiarity principle and the support given to Member States for innovating their climate change policies. See for instance COM (1999) 230, 19 May 1999.

Since operating aid rules are particularly relevant in the context of climate change policies, the content of the operating aid regime and its relevance for climate protection instruments will be analysed next.

4.2 Rules on Operating Aid in Favour of Renewable Energy

During the last years, various Member States have implemented legislative frameworks to encourage the development of renewable energy, as traditional energy markets do not offer the conditions which enable renewable energy sources to compete with traditional sources of energies. The main reason for this failure lies in the fact that "the available technical procedures do not allow the production of renewable energy at unitary costs comparable with traditional energies". It is therefore understandable that in the analysis of national legislation involving state aid for renewable energy, the Commission will primarily consider whether the objective is purely to cover the difference between the production cost of the renewable energy in question and the market price of this energy. This implies that the Commission will undertake a detailed examination of the competitive position of the renewable energy that the Member State tries to favour.

As set out above, in the area of state aid for renewable energy the most remarkable developments of the new EU normative framework appear to be the articles dealing with operating aid. The Commission, given the diverse modalities of state aid for renewable energy sources currently applied by Member States, incorporated a chapter on the "Rules applicable to operating aid for renewable energy source" (points 54-65). As indicated by the Commission, there are three types of aid for renewable energy that could be accepted despite being technically deemed state aid.

The first case is regulated by points 58-60 of the Guidelines , namely aid aimed at covering the difference between the production costs and the market price of renewable energy.

In the second option (points 61-62), the Commission endorses the multiple market instruments used by Member States in order to promote renewable energy. The Guidelines give as example the green certificate systems and tenders, more and more employed by Member States. The favourable treatment that the Commission gives to this kind of aid can be appreciated, especially if compared to the general conditions of operating aid. Whereas the normal period for operating aid in favour of waste management is five years (point 45), the Commission will authorise the aid systems for renewable energy for a period of ten years (point 62), to be renewed if the aid continues to appear necessary. The Commission will, however, require an array of conditions which Member States must fulfil for these mechanisms to be exempted from the state aid regime. National authorities should demonstrate that the financial support is essential to ensure the viability of the renewable energy concerned (necessity condition) and also to attest that the aid will not result in an overcompensation or discourage renewable energy producers to become competitive (point 62).

The third option (points 63-65) appears to a certain extent the most coherent with the polluter-pays principle. According to the wording of the Guidelines, Member States can accord aid to firms producing renewable energy, based on the external costs avoided, defining external costs as the environmental costs that society should take on if the same quantity of energy was produced through conventional sources of energy. This option is the most perfect one from the point of view of cost internalisation, as it considers not only the negative aspects – damage to the environment – but also the positive ones – environmental benefits – which are rarely taken into account in the practice of cost-benefit analysis. Yet, it is difficult for Member States to implement this option, at least until national authorities acquire more expertise about how to estimate the costs to be integrated and the method of calculation.

This favourable treatment of renewable energy is consistent with parallel developments in EU legislation. On several occasions, the EU had the opportunity to emphasise the future role of renewable energy. Whereas the Renewable Energy Directive was approved in September 2001[27], the White Book on the Access of Renewable Energy to the Grid ushered a total reversal of the European approach to renewable energy. From being considered as a mere research and development policy, renewable energy has been converted into one of the goals of the Transport and Energy DG. The reason for this conceptual change was the fact that several Member States started in the last years to introduce different market mechanisms in order to promote the development of green energy markets, green certificates standing out as the main one of these new market instruments.

Because of the importance of renewable energy support mechanisms, it seems pertinent to analyse, albeit it briefly, the ECJ ruling in the *PreussenElektra* case. It might be argued that the ruling of the Court in this case amounts to a justification of the different national market support mechanisms for renewable energy in different Member States.

---

[27] OJ L 283, 27 October 2001, 33.

4.3 The *PreussenElektra* Case and the German Feed-in Law for the Promotion of Renewable Energy: its Relevance for the Future of National Renewable Energy Schemes

As pointed out before, *PreussenElektra* represents a landmark ruling on the issue of renewable energy and climate change[28]. The double relevance of the case is due to the fact that it covers the compatibility of the German feed-in law for the promotion of renewable energy with the state aid regime on the one hand and with the free movement of goods on the other. Furthermore, the Court makes, for the first time, remarkable observations about the significance of the climate change obligations of both the EU and its Member States.

As far as the state aid issue is concerned, the question was whether the Law of 1990 on feeding-in electricity from renewable energy sources into the public grid, as amended in 1998, which in its paragraph 2 established the "obligation to purchase", constituted state aid. Put differently, the ECJ had to assess whether the obligation for electricity supply companies to buy green electricity produced in their region and to do so at a guaranteed minimum price fixed by the government, constituted state aid according to article 87.1.

Thus, *PreussenElektra* involves once again the scope of the definition of state aid. Although article 87 does not give any concrete definition of state aid, it contains all elements that the ECJ uses to assess whether a measure constitutes state aid: "aid granted by a Member State or through State resources in any form whatsoever which distorts or threatens to distort competition by favouring certain undertakings or the production of certain goods insofar as it affects trade between Member States".

The EU Commission as plaintiff in the case argued for a broad interpretation of the state aid concept regarding the existence of a transfer of state resources, reading article 87.1 in conjunction with article 10 of the Treaty which affirms in its second paragraph that "Member States shall abstain from any measure which could jeopardise the attainment of the objectives of the Treaty." The Commission claimed then that the ECJ should consider the German scheme as constituting a "transfer of state resources". On the other hand, the German government supported by renewable energy companies maintained the view that the German legislation did not constitute state aid since there was no economic value granted by the state either directly or indirectly.

The Court held that despite the fact that the purchase obligation entailed an economic advantage for a certain group of producers – i.e. renewable energy producers – it does not constitute state aid "awarded by the State" or

[28] EC Case C-379/98 *PreussenElektra v. Schleswag*, Judgment of 13 March 2001.

"through state resources", as there was no direct or indirect transfer of resources (para. 54). Concerning the question whether the financial burden resulting from the obligation to purchase energy at minimum prices affected in a negative way the economic results of the undertakings which were subject to the purchase obligation, which might in turn reduce the fiscal revenues of the German state, the Court argued that such a consequence was inherent to that kind of regulation and could not be seen as a means to grant renewable energy producers an advantage at the expense of the State (para. 62).

Thus, the Court upheld the German regulation concluding that a State introducing a quantity/price regulation through a general law in pursuing environmental and economic objectives does not grant a subsidy according to article 87. Furthermore, the Court considered that States had a discretion to opt for a purchase obligation system or for other methods.

It can be argued that this jurisprudence will deprive the Commission of any control on renewable energy schemes and, furthermore, of any possibility to create some consistency within the EU renewable energy market. By the same token, the decision undermines the effect that the new Guidelines on State Aid for Environmental Protection could have in the field of renewable energy. Indeed, some of the cases of renewable energy schemes subsequently analysed by the Commission directly refer to the *PreussenElektra* case.

But if the *PreussenElektra* decision provides a general justification for the use of market mechanisms for renewable energy, the second part of the case in which internal market issues and the compatibility of the German law with article 30 of the Treaty are discussed, can be considered as a failure concerning the future of an European renewable energy market. Indeed, Advocate-General Jacobs raised questions regarding the legality of the German law, which prevents utility companies which are subject to the purchase obligation from acquiring renewable energy from sources situated outside the German territory. On this point, the Court declared that the German rule could at least potentially hinder intra-community trade. However, the Court rather relied on political and policy arguments than on legal arguments to justify the legality of the German law, in a way which is not very convincing. First, the case follows the *Wallonian Waste* case in which mandatory requirements of environmental protection were recognised as justifying discriminatory measures. Secondly, the ECJ alludes to the very nature of renewable energy by emphasising the complexity to track the origin of electricity. Nevertheless, the Court departs from the *Outokumpu* case in which it struck down a Finnish law establishing a system of taxation varying according to the method of production of electricity and which was designed to favour renewable energy. In fact, the Finnish system was based on the assumption that given the characteristics of electricity, the origin and the method of production could not be determined once the electricity entered the network and that, therefore, all imported electricity could be subject to an average rate of taxation, which in many instances amounted to a higher rate than the lowest rate applied to Finnish electricity[29]. Here, the ECJ considered that despite the difficulties to clarify the origin of the electricity, the Finnish government would at least have to give the opportunity to importers to prove that that the imported electricity was renewable energy so as to be able to benefit from the same tax applicable to Finnish producers of the energy in cause. The ECJ did not, however, apply the *Outokumpu* jurisprudence in *PreussenElektra*. More surprisingly, the Court made reference to the international obligation of the EU to reduce greenhouse gases emissions in order to justify the validity of the German system, disregarding the obvious fact that the production of renewable energy contributes to the reduction of CO_2 , independently from the country where the renewable energy is produced, as CO_2 has global effects.

Thus, by referring to the absence of EU legislation on renewable energy, to the international obligations of the EU according to the Kyoto Protocol and to a large number of Treaty articles on the environment, especially to article 6 on the "integration of environment into other policies", the Court reaches a pragmatic and politically correct decision. Unfortunately, the Court avoids entering into a more profound and correct legal analysis that might have resulted in the German legislation to be struck down because of contravening the internal market requirements. Thus, if from an environmental point of view this decision can be deemed as a success, the Court, by upholding the German law, missed the opportunity to establish a real internal market for renewable energy.

4.4 Green Certificate Systems: UK and Belgium

Apart from supportive price mechanisms for renewable energy, such as the ones practised in Germany, Spain and Denmark , tradable green certificates are attracting increasing interest in Europe as a potential mechanism for promoting environmental policy goals in a way that is compatible with the demands of an increasingly liberalised European energy market. Still, it is important to note that renewable energy targets are not denominated in carbon units but rather in energy units, a distinction which is relevant for understanding the future relationship between carbon trading and renewable energy policy.

Generally, a green certificate represents the green quality of a unit generated from renewable energy. That means that each unit is divided in

[29] Case C 213/96, *Outokumpu Oy*, ECR 1998 I, 1777.

two parts: the physical electricity and its associated green quality. Both parts can be deemed dissociated and as such commercialised in different markets: the conventional electricity market and the green electricity market.

During 2001 the Commission had the opportunity to evaluate the legality of the Belgian and of the UK green certificate system in light of the state aid regime.

The first case concerned the Flemish Green Certificate System that aimed at increasing sustainable energy and co-generation in view of achieving the international and European obligations binding Belgium concerning CO_2 reductions [30]. According to the Flemish system, a green certificate is granted to producers of renewable energy for a period of ten years. At the end of each year, all Flemish operators and providers of electricity have to be in possession of a minimum quantity of green certificates (quota), to be bought from green energy producers. Producers do not necessarily have to be physically located in the Flemish region, and if that is the case, a specialised organ will control the authenticity of the green certificates. At the end of each year, the certificates will be submitted to the competent organ which means that they are withdrawn from the market. Nevertheless, green certificates can also be used at a later stage as their validity is five years.

Belgium notified this scheme to the EU authorities, but considered that the scheme could not be deemed to constitute state aid, since it did not involve state resources, and even if it did, the conditions of the Community Guidelines for environmental state aid would be met and the scheme could thus be exempted.

In assessing the legality of the Flemish regulation, the Commission had to decide whether the measure in question represented a transfer of state resources. Yet before that, the Commission analysed the other elements of the definition of state aid. First, the Commission held that the obligation imposed upon electricity distributors to acquire green certificates from renewable energy producers amounted to an economic advantage for green energy producers (economic advantage). Furthermore, the Commission considered that green electricity producers constituted a particular group of electricity producers participating in community trade (selectivity character), and that the position of these producers were favoured by the Flemish regulation, a fact which might theoretically entail a modification of market

---

[30] State Aid N. 550/2000, 25 July 2001, Green certificate in the electricity sector. A further Commission decision concerning the Walloon green certificate system - C(2001)3738fin: State Aid N. 415/A/01 – Belgique, Projet d'arrêté du Gouvernement wallon relatif à la promotion de l'électricité verte – repeats the arguments of the decision reached on the Flemish certificate system. For a general review of the green certificate regulatory system in Belgium and its regions see Block and Haverbeke (2001).

conditions for their competitors, in other words, it might affect community trade (effect on competition and trade)[31].

After having analysed these three constitutive elements of state aid, it is interesting to observe that the Commission in assessing the issue of state resources transfer, resorted to the *PreussenElektra* jurisprudence. Hence, the Commission argued that the obligation to acquire a minimum quantity of green certificates is comparable to the obligation to acquire green electricity at minimum prices and that, consequently, the former obligation did not imply a transfer of state resources in the sense of article 87.1. The Commission argued that the provision applicable to producers of renewable energy of allocating green certificates that can be sold in the market is only the official proof that the green electricity was produced. And when green electricity producers sell those certificates such transactions do not involve state resources. However, although the Commission does not consider the measures to constitute state aid, it clearly emphasises its right to decide in the future whether a distribution of the funds created through the fines to be paid by non-compliant distributors constitutes state aid. Given that the rules governing such a fund did not yet exist in the Flemish legislation, the Commission reserved its view until such time as the measures were notified by the Belgian authorities.

Apparently, the Commission had an interest in providing a didactic decision as it was the first case on tradable green certificates. Thus. instead of halting its decision at this point, the Commission continued its analysis as if the measure were a state aid and tested it against the 2001 legal framework. Considering that the Guidelines deal with green certificates in points 61 and 62, the Commission recalls that even if these green certificates fell under the state aid regime, they could be exempted if they fulfilled four conditions: (1) they are indispensable to ensure the viability of renewable energy; (2) they do not overcompensate; (3) they do not dissuade green electricity producers from becoming more competitive and (4) they are limited to ten years. Conditions that, according to the Commission, the Belgian scheme satisfied.

The second Commission decision on the compatibility of green certificates with state aid rules involved the British system known as "renewable obligation"[32]. Although the system also comprises investment

---

[31] Under the European concept of state aid, there is in principle no defined threshold with regard to the distortion of competition. If the application of the system is selective, any distortion of competition, however slight it may be, is sufficient to constitute aid. In giving its reasons for its decision, the Commission is not required to give an exact economic analysis of the effect on trade which constitutes a distortion of competition, Schohe and Arhold (2002).

[32] State Aid N. 504/2000 – United Kingdom Renewable Obligation and Capital Grants for Renewable Technologies, 28 November 2001.

subsidies[33], the analysis was limited to the renewable obligation system, an obligation which can be fulfilled in three different ways: (a) the UK distributor provides renewable energy, (b) the distributor acquires green certificates independently of the energy he provides, or (c) the distributor pays a buy-out price, which will be collected in a fund.

As in Flanders, the UK administration allocates green certificates to renewable energy producers that can be traded between those producers and the suppliers of energy. The system is valid for ten years and, in contrast to the Flemish system, is limited to the production of renewable energy produced within the UK territory, since the UK government considered that the current means to verify the origin of the energy are insufficient. A relevant question for UK authorities was the integration of the green certificate system into the emissions trading system. The question was left to the future and it will necessarily influence the Commission approval since it is possible that it constitutes state aid.

In analysing whether the renewable obligation constituted state aid, the Commission used the same arguments as in the Flemish case. First, the Commission, applying the *PreussenElektra* rule, held that the renewable energy obligation did not constitute state aid since there was no direct or indirect transfer of state resources. In contrast to the Flemish certificate decision, the Commission examined whether the fund fed by resources stemming from the "buy-out-price" represented state aid. In this respect, the Commission applied the three cumulative criteria established by the jurisprudence in order to detect the presence of state resources whenever money is transferred to a fund. These criteria are whether a fund is (a) established by the state, or (b) fed by contributions imposed by the State, or (c) favours certain enterprises.

As to the first criterion: the fund is created by a state organ. Secondly, suppliers will *de facto* not be able to escape from paying the "buy-out price" and will, thus, not have the possibility of choosing between other options. Thus, the payment is "imposed by the State". Third, the fund benefits renewable energy producers. Thus, the Commission concludes that the "buy-out" mechanism constitutes state aid. It then deals with the possibility to exempt it from the prohibition of state aid by applying points 61 and 62 of the Guidelines. In this respect, the Commission considers that the UK system is essential for the viability of renewable energy. Regarding the over-compensation criterion, the Commission observes that overcompensation may occur, but in general terms the system will eliminate it in the future especially because the buy-out mechanism will be reviewed after five years, as declared by the UK authorities. Concerning competition, the Commission believes that the mechanism will contribute to increasing competition among

---

33 Ibid.

renewable energy producers and will encourage them to augment their capacities. Finally, the duration of the scheme will be ten years. Consequently, the Commission concludes that the renewable obligation scheme satisfies the four criteria contained in points 61 and 62 of the Guidelines in order to be exempted from the state aid prohibition, being compatible with article 87.3.(c) EC Treaty.

Although this is not in direct relation with state aid issues, the Commission uses this opportunity to refer to the Directive on renewable energy and the obligation of the UK to recognise, at least two years before the entry into force of the Directive, the certificates issued by other states. The Commission, thus, reserves its competence to determine in the future the legality of the system if restrictions on green certificates from other countries are introduced.

From these two Commission decisions concerning green certificate systems, two conclusions can be drawn. First, the jurisprudence of *PreussenElektra* has been and will be decisive for the numerous support schemes in favour of renewable energy that Member States have on their agenda. Secondly, the interpretation by the Commission of the new Guidelines on State Aid for environmental protection regarding economic instruments for renewable energy seems to be quite favourable for the future of such schemes.

4.5 Rules Relating to Operating Aid in the Form of Tax Reductions and Exemptions

The rules on operating state aid in the form of tax reductions or exemptions take precedence over national measures against climate change. Operating aid in the form of tax reductions or exemptions (points 47-53) is not new. The last paragraph of point 3.5 of the old regulatory framework allowed the temporal reduction of, or exemption from, environmental taxes. Nevertheless, the current framework offers more detailed rules on the modalities in which Member States are able to exempt enterprises from environmental taxes so as not to affect their international competitive position. Certainly, these rules on reduction of or exoneration from taxes apply to any type of environmental tax, but are of special interest in the climate change context, given that most of the Member States are currently establishing fiscal reforms relating to environmental protection that include exemptions from or reductions of the CO_2 tax for certain commercial sectors. In this regard, the Commission has already approved on several occasions the exemption from CO_2 taxes submitted by Member States.

Rules on the exemption from or reduction of taxes are very detailed. Generally, a distinction should be made whether Community tax harmonisation has been carried out or not. In the absence of Community harmonisation concerning CO_2 reduction or limitation measures, the relevant rules are those contained in point 51 where reference is made to the cases in which Member States introduce new taxes or where the tax is superior to the tax fixed by the Community norm. If this is the case, the Guidelines establish two cases of permissible tax reduction or exemption:

1. If firms or associations of firms undertake to achieve environmental protection objectives through voluntary agreements or other kinds of policy means (point 51.1.a), or
2. If certain other conditions are fulfilled in the absence of such environmental agreements (point 51.1.b).

In the case of the existence of a voluntary agreement between the Member State and the beneficiary enterprises, the Commission accepts the aid for a ten year period, provided that environmental agreements aiming at environmental protection objectives are concluded. What is really interesting to notice here is that the Commission in an indirect and subtle way defines what is meant by "voluntary agreement" and the conditions the Commission requires to be met for it being considered a valid environmental agreement. First of all, the agreement must be negotiated by the Member State. This implies *a contrario* that informal agreements unilaterally drafted by enterprises and submitted to state authorities, so-called "gentleman's agreement", remain outside the scope of the environmental agreements for the purposes of this provision. Furthermore, Member States must secure the implementation of these agreements, or in the wording of the Guidelines, Member States are required to "ensure strict monitoring of the commitments entered into by the firms or association of firms". As a last prerequisite, the Commission interestingly requires the inclusion in the agreement of penalty provisions in case the enterprises do not fulfil their commitments. With this last requirement, the Commission intends to prevent the practice common throughout all Member States where voluntary agreements do not have any legal force as there are no legal sanctions in case of non-compliance. It is also interesting to note that although the prerequisites referred to above were already specified in the Communication concerning environmental agreements in 1994, the 2001 Guidelines on State Aid for Environmental Protection, by imposing upon Member States these requirements, give legal force to what then was a mere recommendation expressed by the Commission. Moreover, there is a new element that makes the new Guidelines a real instrument to control the effectiveness of national environmental policies. Indeed, according to the Guidelines, the Commission will have the competence to "*assess the substance of those agreements*" when the aid project is notified by a Member States (point 51.1.a). In other words, the Commission is competent to determine not only their appropriateness, but also their environmental effectiveness. This involves a power of ex-ante control over the content of the agreements, as Member States have to notify state aid projects together with the environmental agreement in question, in order to obtain the approval of the Commission to put in place the tax reduction or exemption – the state aid measure. To a certain extent, it could be argued that the DG Competition, in assessing the state aid and the appropriateness of the environmental agreement plan, might become a fundamental actor to execute environmental policies. The Commission could even refuse to approve a reduction of or exoneration from taxes in case it regards the environmental protection objectives incorporated into the agreements as not ambitious enough. What would be tantamount to saying that the environmental benefits derived from the agreements are inferior to the economic advantage provided by the tax exemption or reduction.

Point 51.1 (b) also provides for two cases where no agreement or commitment is necessary. The first case is if taxes have a community origin and the tax level after reduction is higher than the Community minimum or, in the case of purely national taxes, undertakings must pay a significant portion of that tax.

The second case concerns taxes which are levied as a result of a Community directive (point 49). This provision regulates the way in which state aid questions will be treated if in the future the CO_2 tax is levied. Point 49 (a) regulates the case in which a Member State imposes a rate higher than the minimum rate established by the directive and in which certain firms are exempted from the highest national rate. They have to be taxed at least at the minimum rate established by the directive. Point 49 (b) present more problems as it admits the possibility that Member States apply the minimum rates established by the directive and provide for reductions or exemption in favour of certain firms, which would then pay less than the minimal rate. Even in this case, the Commission can admit the state aid, provided that the measure is necessary, limited in time and not disproportionate in light of the Community objectives pursued.

Finally, a further option concerning tax reductions or exemptions is regulated by point 53.2 which allows Member States to reduce taxes according to the general rules of points 45 and 46. In this case, Member States have two possibilities: either they grant a degressive aid limited to five years, with a maximum intensity of 100% in the first year, falling to zero by the end of the fifth year; or in the case of non-degressive aid, its duration remains limited to five years, not exceeding in intensity over 50%.

At this point, it should be mentioned that the previous versions of the Guidelines on State Aid for Environmental Protection issued by the

Commission differ from the regulatory framework finally adopted[34]. The Commission, in the Guideline proposals, showed a very restrictive attitude regarding state aid in the form of tax exemption or reduction which corresponded to its mistrust of operating aid. The Commission only included the possibility of tax reduction or exoneration in a degressive way and for a five year period, admitting exceptionally the possibility of a period longer than five years for sectors with large energy consumption. These provisions were the object of massive criticism from the part of most governments. For instance, the German government argued that the temporal and degressive requirement were not appropriate in the context of the fiscal reforms for ecological purposes being undertaken by States[35]. The German government severely criticised the time limitation of five years as well as the degressivity, whereas the Commission argued for both conditions that after a certain adaptation time, firms will remain internationally competitive. In rejecting this argument, the German government contended that the competitive situation of firms situated in different Member States and subject to very different rates of taxation did not change simply because of the passage of a certain period of time. As shown above, the Commission finally accepted the objections put forward by Member States, including the tax reduction or exoneration for a ten year period without degressivity. Moreover, according to point 23 of the Guidelines, Member States maintain the capacity to "notify again the measures in question to the Commission" which could prolong them if positive environmental results are proved. This indirectly implies an incentive for Member States to undertake a serious monitoring of the implementation of new market instruments in order to show their effectiveness. These obligations should be seen as causing positive effects, in view of the fact that a large part of the criticism concerning environmental market instruments rests on the absence of data demonstrating their environmental effectiveness and in general on the non-existence of state policies for evaluating these new instruments.

Different examples of tax exemptions and reductions in the context of climate change policies can be found, for instance, in the exoneration of the CO_2 tax for Denmark and the Netherlands (XXIInd Competition Report, points 75 and 451) and recently the approval of the tax exemption granted by Germany to combined-cycle power plants. The approval of the tax exemption for the UK climate change levy is interesting inasmuch as the measure – the tax exemption – is combined with the existence of a voluntary agreement[36]. Indeed, together with the approval of the exemption from the

---

34 This version can be found in <www.wind-energie.de/englisch/eu_guidelines_environment.html>.

35 See the comments of the German government regarding the first proposals for the Environmental Guidelines: <www.wind-energie.de/englisch/c_guidelines_env.html>.

36 State Aid, N. 123/2000.

CO_2 tax in other fields, the Commission endorses the exemption or reduction for companies entering into climate change agreements, considering them to be compatible with article 87.3 ECT[37]. Thus, the sectors that can agree on targets for improving their energy efficiency or for reducing carbon emissions will get a reduction of 80% of the tax. Monitoring and evaluation will show whether the objectives are actually attained.

5 State Aid and Emission Trading Systems

After having analysed a considerable number of operating aid cases included in the new normative framework for state aid for environmental purposes, encompassing some of the more innovative instruments that Member States are currently using in the area of renewable energy and climate change, it is unavoidable to deal with the instrument that will become the centre of gravity of climate change policies, not only for the EU but also for most of its Member States, namely the emission trading system. Not only have certain countries, like the UK and Denmark, already implemented a market trading program, but most importantly, the Commission, after having launched a Green Paper on the creation of an emissions trading system (ETS) in Europe in 2001, issued a Directive proposal in 2001. The Parliament and the Council reached an agreement on the Directive in July 2003.

Within the category of economic instruments for climate change protection, market permits enjoy a vast acceptance on the part of Member States and Community authorities. But if it is true that this instrument is adequate not only from an economic efficiency, but also from an environmental point of view, its design and implementation will provoke numerous points of friction with the state aid regime of the Community. Having anticipated these conflicts, the Commission, in its Green Paper on emission trading,[38] raised the question whether the system of allocating permits to individual firms should be the object of a specific Community agreement or whether the rules on state aid were sufficient to guarantee equitable treatment.

It is precisely in the distribution of pollution permits that the ETS and the state aid regime might collide. For the time being and despite the attempt to include concrete parameters in the Directive, the question will be resolved

---

[37] Apart from the exemptions applicable to the sectors having negotiated climate change agreements, exemptions or reductions were also accepted for:
- electricity, gas and carbon used in public transports and rail transport of goods;
- the use of fuels and electricity from combined heat and power generating installations;
- the use of use of electricity from renewable energy.

[38] COM (2000) 87 final, 8 March 2000.

by the Commission as the new environmental state aid framework remains silent regarding the way in which Member States allocate the market permits. This silence is not accidental. A previous version of the Guidelines gave at least the principles according to which the Commission considered that the allocation of permits should be operated[39]. Such a provision was not included in the definitive text, due to the lack of agreement among Member States on the model on which a market permit system should be based. In this regard, the different Member State positions expressed in reaction to the Green Paper are telling. While some states advocated the necessity of a full harmonisation of the sectors to be covered and of the allocation modalities – as was the case in Belgium – others, led by the UK, relied on the subsidiarity principle and on the freedom of the Member States to decide the affected sectors and the modalities of distribution of permits[40]. In the end, the decentralised solution was retained by the directive proposal. There is no doubt that the preservation of state freedom to decide its own permit market mechanism will unavoidably cause a new concern for Community authorities in charge of state aid control. But trying to reach an agreement on the allocation method would have even more delayed the adoption of the directive, since those issues raised at least the same national reticence and created political obstacles which could be observed in the discussion about the CO_2.

Thus, the legal certainty resulting from the clarification of the rules regarding the allocation of permits on the part of the Commission has been sacrificed for two reasons. First, the reservations Member States have towards the control that the Commission could exercise in a centralised system and second the novelty of the ETS. That is why the Community authorities will have to assess on an *ad hoc* basis the national allocation plan of each Member State with reference to the state aid rules.

Before analysing the points of the national allocation plans that might conflict with the state aid regime, I will address first the two Commission decisions assessing the UK and Danish national emissions trading regimes, as this will shed light on the future approach of the Commission in controlling the national allocation plans.

---

[39] See points 65 and 66 in <www.wind-energie.de/englischer-teil/eu-guidelines-environment.htm#9>.

[40] In support of decentalised allocation systems, see Zhang (1998). For reasons of legal certainty, some authors have instead proposed to incorporate in the Guidelines for State Aid for Environmental Protection rules concerning the potential conflict between state aid and market permits.

5.1 The Danish Emissions Trading Decision

The Commission decision of 29 March 2000 approving the Danish CO_2 quota system in the electricity sector gave the green light to other EU countries aiming to establish a national permit market[41]. The political context should be considered when assessing the content of this decision as the Danish CO_2 quota system represents the first scheme of this type developed in Europe. It would have been illogical to block the first pollution permit market in Europe, as on the one hand, the Commission had insisted on pushing Member States towards the experimentation with, and implementation of, climate change policies and, on the other hand, the Danish CO_2 quota system fits the plans of the Commission to create a EU wide permit trading system. Furthermore, because of the commitments derived from the Kyoto Protocol, by which both the EU and the Member States are bound, the ensuing national climate change policies become one of the major inputs in the EU environmental policies. Both instruments will undoubtedly trigger the positive attitude taken by the Commission towards national permit market systems.

In general terms, the Commission ruled in the Danish decision that despite the fact that granting permits without any compensation constitutes state aid according to article 87.1 EC Treaty, the Danish system can be exempted on the basis of article 87.3(c), as the scheme contributes to the development of environmental protection[42]. Yet the Commission clarified that the approval of the Danish scheme is not prejudicial to future decisions on national schemes regarding the allocation method.

Briefly, the Danish system is based on an annual limit fixing the total quantity allocated to the electricity sector which is responsible for almost 40% of the total gas emissions. The system also envisages its development over a period of three years, fixing a degressive limit for every year: 22 million tons for 2001, 21 million tons for 2002 and 20 million tons for 2003. This represents a significant reduction in view of the 30.3 million tons yearly average during the period 1994-1998. This considerable reduction by 30% is regarded by the Commission as a real achievement in environmental terms.

The initial allocation system establishes that all electricity producers in Denmark – both nationals and foreigners – will receive quotas in proportion to their historic emissions from 1994 to 1998 (grandfathering system). For the new energy producers, the Commission deemed it positive that the Danish government reserved a part of the quota for the new entrants to the

[41] Decision N. 653/99, quota CO_2, 29 March 2000.

[42] This decision was issued before the new Guidelines on State Aid for Environmental Protection were approved.

energy market – in order to preserve the freedom of establishment – as well as to allocate quotas based on objective and non-discriminatory criteria.

The functioning of the Danish permit market system is based on the idea of cost efficiency. Bearing in mind that the fine for electricity producers exceeding their assigned CO_2 amount is 40 DKK per ton, the producer will reduce emissions up to the point that the reduction cost per ton is lower than 40 DKK. In other words, the fine indirectly establishes the market price for CO_2. If the electricity producer manages to remain below its CO_2 emission limit for that year, the difference may be saved and used the following years (banking) or sold to other producers.

The Danish system also provides for exceptions. Producers generating electricity not causing CO_2 emissions – wind energy or waste energy – will not be submitted to the permit system. Furthermore, small-scale combined heat and power plants not exceeding 100,000 ton of CO_2 emissions per year are exempted.

The Commission assessment logically starts by verifying whether the Danish system can be defined as state aid. In this respect, the four conditions described above for qualifying a state measure as state aid must be borne in mind.

1. Transfer of state resources: according to the uniform jurisprudence of the ECJ, the definition of state aid should be interpreted in a broad sense, including both direct and indirect aid. The Commission argues that Denmark obviously, through the free distribution of permits based on historic emissions (grandfathering), provides electricity producers with "an intangible asset for free, which could be sold in the market". In order to ascertain the existence of this resource transfer, the situation must be compared to the case in which the State uses auctioning instead of grandfathering. In the latter case, the State would receive revenues. Thus, the State foregoes certain revenues, and this implies an indirect transfer of resources[43].

2. Advantage for the firm: it is clear that the system favours a sector, the electricity sector, excluding others. Furthermore, this exclusion entails that the categories not submitted to the permit market system are subject to another kind of measure – direct regulation or taxation. This may constitute more intensive charges than the ones imposed by a permit market system.

3. By analysing the advantage criterion, the Commission examines whether the financial position of the firms affected is favoured *vis-à-vis* their

[43] For examples of indirect aid in the context of environmental aid, see Jans (2000), 294 et seq.

competitors[44]. From this point of view, there is not doubt that the financial position of the Danish firms affected by the permit market system improves, not only because they do not have to buy permits, but also because they can sell them, obtaining revenues which can theoretically be invested to increase their competitive position.

4. Regarding the adverse effects on competition and the internal market between States, the Commission determined that the existence of substantial trade in the electricity sector between Member States shows a potential distortion of competition among producers of electricity in Denmark and in other Member States, since a Danish firm is able to sell its permits and can use the profit to increase its competitive position in Denmark and in other Member States[45]. As is commonly known, the Commission in practice presumes, whenever state aid is at stake, the existence of competition distortion affecting Community trade.

After examining the existence of state aid, the Commission exempts it for environmental reasons on the basis of article 87.3.(c). Although state aid for environmental purposes would normally have been tested against the 1994 Guidelines for environmental state aid[46], this framework did not contain this new form of state measure, nor does the current one. Thus, the Commission

---

[44] Jans (2000), 298.

[45] Some authors argue that the allocation of permits through grandfathering will, from the point of view of the market and competition distortion, raise less problems than the exemptions of environmental taxes:

> "It should be pointed out that although grandfathering is thought of as giving implicit subsidies to some actors, it is less trade-distorting than the exemptions from carbon taxes. To understand their difference, it is important to bear in mind that grandfathering itself also implies an opportunity cost for firms receiving permits - what matters here is not how you get your permits, but what you can sell them for, as it is this which determines opportunity cost. Thus, relative prices of products will not be that distorted and switching of demands towards products of those firms whose permits are awarded gratis will not be induced by grandfathering. This makes grandfathering different from the exemptions from carbon taxes. In the later case, there exist substitution effects. The Commission proposed exemptions for the six energy-intensives industries from the carbon and energy tax. This not only reduces the effectiveness of the CEC tax in achieving its objective of reducing CO_2 emissions, but also makes the industries, which are exempt from paying the CEC tax, improve their competitive position in relation to those industries which are not. Therefore, there will be some switching of demand towards the products of these energy-intensive industries which is precisely the reaction that such a tax should avoid." See Zhang (1998), 227.

[46] It should be remembered that this decision was issued in 2000, before the new Guidelines on State Aid for Environmental Protection entered into force. Thus, the relevant normative framework were the 1994 Guidelines.

emphasises the environmental effectiveness of the scheme by considering that the requirement imposed on electricity producers to reduce their emissions constitutes proof that the system contributes to the improvement of the environment. The fact that producers are allocated quotas representing a considerable reduction in comparison to the previous situation, requires them to adopt measures to reduce CO_2. That means that the Commission, in assessing the validity of the measure, takes into account whether the advantage of free allocation is compensated by the environmental objectives pursued by the system. The question which could arise in the future is whether the Commission will approve a permit market system in which emissions reductions are not sufficiently ambitious. What is clear is that the Commission, through its control over state aid, will supervise the implementation of environmental policies.

Finally, the limitation of the scheme to electricity producers is justified according to the Commission because of the importance of this sector for emissions of CO_2. Furthermore, the Commissions insists on the positive aspect of not restricting imports or exports of energy. The Commission accepts an allocation system based on grandfathering because Danish authorities ensure an equal treatment of producers having received quotas and the future market entrants. These internal market aspects are emphasised by the Commission throughout the decision, for instance by pointing out that the approval of the system does not entail the green light for intra-community trade of allowances in case other countries also establish permit market systems. Consequently, any change to allowing an intra-community or international trade of permits, as well as any other revision of the system, must be communicated to the Commission.

5.2 The UK Emissions Trading Scheme

A year after the decision about the Danish CO_2 quota system, the EU Commission also approved the British emission trading scheme[47]. Its design is extremely complicated, but its basic elements are the following.

- Grants are awarded as an incentive to companies in return for absolute emission reductions for which they bid in an auction. These companies enter into the system as "direct participants".
- An emissions trading regime is created in which target holders from different mechanisms ("direct participants" in the emission trading scheme, "participants" in the climate change agreements) may trade

[47] Decision N. 416/2001, 28 November 2001. In close relation to this decision, the Commission also took the decision State Aid N. 123/2000 regarding the climate change levy for the free allocation of allowances for companies entering into climate change agreements.

emission allowances between them and other participants (entering via an emission reduction project or by simply opening a trading account), provided they reduce emissions further below their target. Emission allowances are allocated to participants for free.

The UK authorities emphasise that the scheme is expected to yield annual emission reductions much beyond what would have been achieved by a "business as usual" approach. As to the grants, they are distributed among the direct participants on the basis of the greenhouse gases they emit both directly or indirectly. Moreover, the UK authorities point out that this incentive will not be used to achieve Community standards but rather to go beyond those standards, or to fulfil these standards sooner than would otherwise be required.

Regarding the allowance, it is allocated according to the specific characteristics of each entrance mechanism. Direct participants receive allowances matching their emission cap for the forthcoming compliance period (cap and trade), provided that they held allowances for the previous compliance period which were at least equal to their emissions. Climate change agreement participants will be allocated allowances at the end of each two-year compliance period if they have reduced their energy use or emissions below their target. The amount of allowances issued corresponds to the amount of over-achievement. The UK system also includes the possibility of granting permits for UK based emission reduction projects, as well as for green certificates if individual suppliers over-comply with their obligations. In this case they can convert the amount of their over-compliance into credits measured in CO_2 emissions.

Interestingly, the UK government argued that if the measure constitutes state aid, the aid should be exempted according to article 87.3.(b) as it would promote an important project of common European interest, and if article 87.3.(b) did not apply, the scheme would at least be compatible with article 87.3.(c) as it complies with Chapter F of the environmental aid Guidelines.

In assessing the scheme, the Commission first examined whether it constituted state aid, both in respect of the incentive fee as well as the trading mechanism. As to the first issue, the Commission holds that the grant which the UK government provides to companies which undertake to attain emission reduction targets, confers an advantage that distorts competition and potentially affects trade between the Member States. In other words, the incentive money represents state aid in the sense article 87.1 EC Treaty.

As far as the trading mechanism is concerned, the UK government allocates a limited number of transferable emission permits free of charge to the direct participants. The UK, thus, provide these companies for free with an intangible asset which can be sold on a market yet to be created. The fact that there will be a market is a sign of the value of the asset being allocated, and it represents an advantage for the recipient companies. Furthermore, this advantage distorts competition between companies since the recipient companies could use the profit from the allowances for their business, competing with other companies not having access to such a scheme. Finally, the Commission states, as in the Danish emissions trading decision, that the UK foregoes revenues which could be derived from auctioning the emission permits. As a consequence, the trading mechanism constitutes state aid as well.

As regards the legal basis for the assessment, the Commission acknowledges that point 73 of the Guidelines states that state aid for the protection of the environment could be exempted according to article 87.3.(b) for reasons of "promoting the execution of important projects of common European interest which are an environmental priority and will often have beneficial effects beyond the frontiers of the Member States concerned". While accepting that climate change is a main concern of environmental policy and that emissions trading will help Member States to fulfil their reduction obligations, the Commission tests the system not against article 87.3.(b), but against article 87.3.(c) EC Treaty. Thus, the Commission does not exclude that emissions trading projects could be considered as projects of common European interest. This would mean that although for the time being emissions trading will be exempted according to article 87.3.(c), in the future it could also be exempted according to article 87.3.(b). The difference is that in the latter case, the Commission may authorise aid at higher rates than the limits laid down for aid authorised pursuant to article 87.3.(c). Indeed, no one doubts that the establishment of an emissions trading regime fulfils the criteria contained in point 73 of the Guidelines regarding "projects of common European interest". First, emissions trading has beneficial effects beyond the country that implements it and it makes an exemplary and clearly identifiable contribution to the common EU interest. Nevertheless, the Commission did not consider it opportune to declare it to be of "a common European interest", for two reasons. First, having assessed the Danish system, the Commission used article 87.3.(c) EC Treaty as the legal basis. Second, the Commission noticed in the decision that the UK scheme differs substantially from the EU emissions trading proposal[48].

The last point of interest in this decision refers to the compatibility of the British scheme with chapter F "Policies, Measures and Instruments for Reducing Greenhouse Gases" of the Guidelines (points 68-71). As has been pointed out above, these provisions do not establish specific criteria for the compatibility of economic instruments of climate policy with the state aid

[48] In analysing cases which can be exempted from the State Aid regime according to article 87(3), the Commission will assess a) the necessity to achieve the environmental objective; b) the proportionality of the measure; c) its non-discriminatory character and d) its transparency.

regime. Rather, apart from recognising the desirability of those instruments, the Guidelines acknowledge that it is still too premature to lay down the conditions for authorising this type of aid, despite the fact that some of them could constitute state aid. But in considering the elements that prompt the Commission to exempt the UK system from the state aid regime, it is interesting to observe the weight the Commission attributes to the environmental benefit of the system, as well as to the exemplary effect of the instrument for the rest of the EU countries.

Finally, the procedure by which the Commission exempts the UK and the Danish scheme according to article 87.3.(c) is worth mentioning. The Commission follows the classical path of emphasising the necessity to achieve the environmental objective, the non-discriminatory character and the transparency of both systems. One condition, however, that the Commission uses in exempting state aid, namely proportionality, is not well reasoned in either case. Does the Commission recognise that auctioning constitutes a more proportional means than grandfathering by acknowledging that auctioning would not represent a transfer of state resources contrary to grandfathering? The Commission does not enter into the proportionality and appropriateness of the measure in any of the cases, probably because it admits grandfathering as the only option which is politically feasible.

5.3 The National Allocation Plans of the Emissions Trading Directive and the State Aid Regime

Most of the concerns about state aid in respect of EU tradable systems raised by Member States relate to the allocation issue. As pointed out above and identified by the Commission, grandfathering or attribution of quotas based on historical emissions causes friction with the rules on state aid. Auctioning would be the alternative system to grandfathering, as it would not constitute state aid, at least not at first sight. By paying for the allowances granted by the state, firms would not be the beneficiaries of any transfer of state resource. However, even if certain states defend an auctioning system, most of them prefer grandfathering, as this is politically a more acceptable solution for the firms[49]. As the Commission stated in the Danish trading system decision, grandfathering would amount to an aid provided by the State and having an impact on state resources since the State forgoes revenues, which would not happen in case of selling or auctioning the

[49] Norway suggested in its comments to the Green Paper on the EU Emissions Trading System that auctioning was a better option not only from an efficiency point of view, but also for reasons of distributional order.

allowances. Yet, the Directive on Emissions Trading – in its last version, approved on 22 July 2003 by the Council after the amendments of the European Parliament – stipulates in article 10 that the allocation method for the three year period starting 1 January 2005 will be grandfathering, at least for 95% of the allowances. However, the fact that the EU Directive has recognised grandfathering as the allocation method does not exclude that the implementation of this method practically constitutes state aid, both at the level of the allocation plan and at that of the allocation decision regarding individual enterprises.

At any rate, even if the distribution of permits by grandfathering does not affect the competitive position of products as such, the compatibility between a market of pollution permits and the polluter-pays principle remain unclear[50]. Indeed, grandfathering has been largely criticised and creates major issues in terms of distributive justice. In order to judge a market permit system based on grandfathering, it is necessary to compare it with the classic system of command and control. In this respect it can be argued that grandfathering clauses are not an exclusive characteristic of market permit systems, as they also exist within the classic environmental law system relying on the command and control approach. It is not unusual that environmental regulations stipulate different requirements for different sectors, but also for existing and new firms, granting to the former more favourable conditions to adapt to the requirements of new regulations. Furthermore, command and control regulations fixing upper limits for lawful emissions – for instance in the case of an environmental licence – allow to reach the legal limit without having to pay or suffer sanctions. Bearing all this in mind, it seems that an emissions trading regime is not very different from a traditional environmental licence system. Indeed, even if the free allocation of allowances can be considered a state aid, the important element of the system is its environmental performance, as enterprises placed under the scheme have to reduce their emissions further than their normal target levels. This represents the environmental benefit of the system.

If grandfathering is the main concern regarding the allocation plan in the light of the problem of state aid, it is not excluded that a tradable permit system based on auctioning also involves state aid. This would be the case, for instance, if the state decides to recycle the funds obtained from the allocation process to the firms participating in the auction. Recycling of funds, however, can also take place in a grandfathering system if the revenues gained from the fines for exceeding quotas are channelled back into the scheme.

---

[50] For an excellent criticism of the compatibility between the polluter-pays principle and a permit market systems, see Remy Nash (2000).

While grandfathering appears at first sight as the most problematic allocation method in light of the state aid regime, there are also other issues to be raised in respect of the national allocation plans. For instance, if the method of allocation is discriminatory because it benefits in real terms certain operators or categories of installations, that could also be deemed as a specific measure of state aid. Put differently, if a Member State makes distinctions in applying differently the criteria of Annex III –"Criteria for National Allocation Plans" – of the emissions trading directive, this would amount to a measure by which the State grants a particular advantage to certain enterprises and, as a result, it could constitute state aid. It should be noted that article 9 of the Council common position asserts that the national allocation plan shall be based on objective and transparent criteria, "*including those listed in Annex III*". This could be interpreted in the sense that Member States will have enough flexibility to add other criteria in establishing the national allocation plan, in addition to those of Annex III, for instance, if a Member State claims special national circumstances. It should be noted that the European Parliament Report amended article 9.1 in the sense that the plan should be based on Annex III only, but this proposal has not been included in the last agreement between Council and Parliament[51]. As it stands now, article 9 allows Member States to include further criteria which would favour certain undertakings over others, complicating the Commission's task when assessing the national allocation plan and its compatibility with the state aid regime.

Other criteria contained in Annex III could also raise concerns relating to state aid. This is the case of point 7 of Annex III. This criterion allows Member States to favour early action in their national allocation plans. However, in doing so, a Member State must take into account state aid rules. It is still too early to anticipate the way in which the Commission will assess early action. But one can argue that where early action was achieved in the past by enterprises to which the government also grants favourable treatment in the national allocation plan, it could result in a kind of double aid, which is not acceptable under the state aid regime.

From the early drafting stages, many actors as well as Member States demanded that the Directive on Emissions Trading shed more light on the allocation issues as this would help national authorities in drafting their allocation plans and its compatibility with the state aid regime. Unfortunately, no clear information is provided by the final directive. Article 11.3 merely states that allocation and issuing of allowances should be carried out in conformity with the Treaty and in particular with articles 87 and 82.

---

[51] Report of the European Parliament on the Proposal for a European Parliament and Council Directive establishing a Framework for Greenhouse Gas Emissions Trading within the EC and amending Council Directive 96/61EC, 13 September 2002 (A5-0303/2002).

Equally, Annex III of the Directive declares that the "plan shall not discriminate between companies or sectors in such a way as to unduly favour certain undertaking or activities in accordance with the requirements of the Treaty, in particular article 87 and 88". From the text of articles 9–11 and Annex III, it seems as if the Committee referred to in article 23.1 will be in charge of assessing the national allocation plan in the light of the criteria of Annex III. Considering that one of the criteria included in Annex III is the compatibility with the state aid regime, this Committee will indeed determine if the national allocation plan is in conformity with the rules contained in article 9 and ss. and Annex III, requiring *inter alia* conformity with the state aid regime.

Now – as suggested at the beginning of this chapter – it might seem plausible that state aid issues can emerge not only in the design of the allocation phase, but also in a later phase, during its implementation. Indeed, although the allocation plan will be assessed by the Committee referred to in article 23, and although the relevant criteria include compliance with the state aid regime, nothing ensures that at a later stage, during the allocation and issuing of allowances, Member State authorities will not violate the state aid regime, for instance by allocating permits in a different way than prescribed in the national allocation plan. In this case, any interested party – such as a competing firm – would have the right, in conformity with article 20.2 of Regulation 659/99, to inform the Commission about the existence of a state aid. The Commission would then start a detailed investigation[52]. It should be noted that in this case, DG Competition would not be asked to assess the national allocation plan, but rather the specific way in which the Member State in question allocated and issued the allowances. One could even consider the consequences if the Commission found the issuing of the allowances an infringement of article 87 and article 88 EC Treaty. The Commission may, after allowing the State to comment, take a decision on its compatibility with the common market and provide for a suspension injunction, which in this case could amount to a suspension of the distribution of allowances[53]. Understandably, a decision by the Commission requiring a Member State to stop the allocation of allowances to one or more operators could negatively influence the commencement or continuation of the emissions trading market, especially if one considers that in line with article 11.1 of the Directive on Emissions Trading, the period in which Member States must decide on the allocation of the allowances to the

---

[52] Reg. 659/99, (1999) OJ L 83, 1, regulates the procedural rules that apply to State Aid. It defines interested parties in article 1(h) as "any Member State and any person, undertaking or association of undertakings whose interests might be affected by the granting of aid, in particular the beneficiary of the aid, competing undertakings and trade associations", see also article 20(2).

[53] Article 11.1 Reg. 659/99.

operators of each installation is only three months before the beginning of the first period of the emissions trading that starts in January 2005.

Thus, because of the lack of harmonisation regarding the national allocation plan criteria, and the lack of clarity regarding the criteria contained in Annex III, Member States are likely to violate the state aid provisions in the elaboration and implementation of their national allocation plans. Even if, as required by article 9.1, the Commission elaborated by the end of 2003 guidelines regarding the implementation of the criteria listed in Annex III, Member States will still have enough discretion to circumvent the Community state aid regime[54].

In the explanatory memorandum of its proposal, the Commission explained that state aid issues would be raised if different allocation methods are used for different sectors or undertakings within a single Member State. Thus, if the national allocation plan uses different methods, this could be subject to state aid control. This requirement is also expressed in Annex III point 5 which requires that "national allocation plans shall not discriminate between companies or sectors in such a way as to unduly favour certain undertakings or activities". Thus, discrimination may appear not only between covered sectors, but also between covered and non-covered sectors and between competing sectors located in several Member States.

Based on the text of the directive, other possible violations of the rules on state aid can be anticipated. First, article 29 allows Member States in case of *force majeure* to allocate additional allowances to installations. The distribution of these additional allowances and the reasons for allocating them will have to be attentively scrutinised by the Commission in order to exclude any violation of the state aid regime. One could also think of a Member State neglecting to enforce its domestic climate change policy including its domestic regime of emissions trade for climate protection purposes. Could such a neglect amount to a subsidy for those firms against which the system is not enforced?[55]

---

[54] Competition issues different from those of state aid can also emerge in an ETS. For instance, article 81 could apply to a group of operators which entered into an agreement for distributing a set of emissions allowances. In addition, there might be an abuse of dominant positions violating article 82 in the case where some undertakings retain a large amount of allowances preventing new entrants for instance from getting access to those allowances. On these issues, see METRO (2002).

[55] In this respect, the Sniace Decision, N 118 /97, should be considered in which the Commission stated that a case where Spanish authorities neglected the enforcement of environmental legislation did not constitute state aid, because there was no transfer of public capital.

6 Joint Implementation, Clean Development Mechanism and State Aid Control

Although the national allocation plan is the most problematic feature as far as state aid is concerned, it can be anticipated that other flexible instruments such as Joint Implementation (JI) and Clean Development Mechanism (CDM) might also raise concerns regarding their compatibility with the state aid regime. The implementation of these instruments might lead to a transfer of public resources since Member States can involve, for instance, public money in JI-activities in a number of different ways, such as exempting JI-projects from certain taxes, financing investments in JI, and acquiring Emissions Reduction Units from JI-projects. There is no reason why EC rules on state aid would not apply to JI-projects that take place within the Community. Furthermore, state aid rules – like other EC competition rules – have an extraterritorial effect. Thus, state aid rules could also be applicable outside the Community, which means that the state aid regime could also apply to CDM that will necessarily take place outside the EU. Some other examples where state aid issues would be involved are, for instance, financial assistance by a Member State to JI- or CDM-projects. In the same way, if a EU government bought credits (Emission Reductions Units) derived from JI-projects, for instance, at a price higher than the market price, this could constitute state aid.

References

Alexis, A. (2001). 'Adoption du nouvel encadrement des aides d'Etat en faveur de l'environnement. *Competition Policy Newsletter*, 1: 26.

Assunçao, L. and Zhang, Z.X. (2002). *Domestic Climate Change Policies and WTO.* UNCTAD Discussion Paper Series, August 2002.

Block, G. and Haverbeke, D. (2001). 'Droit d´émission et certificats verts: le nouvel encadrement légal en matière d´électricité'. *Aménagment-Environnement*, 2: 109-122.

Craig, P. and de Búrca, G. (2003). *EU Law 3rd edition.* Oxford : Oxford University Press.

European Commission (2001). *XXX Report on Competition Policy 2000.* SEC (2001) 694 final, 7.5.2001.

Grabitz, E. and Zacker, C. (1989). 'Scope for Action by the EC Member States for the improvement of environmental protection under EEC law : the example of environmental taxes and subsidies'. *Common Market Law Review*, 26: 439.

Jans, J. (1995). 'State Aid and article 92 and 93 of the EC Treaty: Does the polluter really pay?'. *European Environmental Law Review*, April 1995.

Jans, J. (2000). *European Environmental Law.* Groningen : Kluwer Law International.

METRO (2002). *Emissions Trading and Competitive Positions.* Research Report for VNO-NCW.

Niccolaides, P. and Bilal, S. (1999). 'An Appraisal of the State Aid Rule of the European Community, do they promote efficiency?'. *Journal of World Trade*, 33 (2): 97-124.

Rehbinder, E. (1993). 'Environmental Regulation Through Fiscal and Economic Incentives in a Federalist System'. *Ecology Law Quarterly*, 20: 57 et seq.

Remy Nash, J. (2000). 'Too much market? Conflict between tradable pollution allowances and the polluter-pays principle'. *Harvard Environmental Law Review*, 24: 465-536.

Schohe, G. and Arhold, C. (2002). 'The Case-Law of the European Court of Justice and the Court of First Instance on State Aid, 1998-2001'. *European State Aid Quarterly*, 10: 4.

Van Calster, G. (1998). 'State Aid for Environmental Protection : Has the EC Shut the Door?'. *Environmental Taxation and Accounting*, 3: 38-51.

Vedder, H. (2001). 'The New Community Guidelines on State Aid for Environmental Protection – Integrating Environment and Competition?'. European Community Law Review, 9: 365-379.

Zhang, Z.X. (1998). 'Greenhouse Gas Emission Trading and the World Trading System'. *Journal of World Trade*, 32 (5): 228.

Chapter 9

JÜRGEN LEFEVERE[1]

The EU Greenhouse Gas Emission Allowance Trading Scheme

1 Introduction

The Kyoto Protocol will require the European Union to reduce its aggregate emissions of a 'basket' of 6 greenhouse gases[2] (GHGs) by 8% over the period 2008–2012 compared to its 1990 emissions. Although in the year 2000 community-wide greenhouse gas emissions stabilized in relation to 1990 emissions, recent inventories have shown a rise in emissions to 2.1% above the Kyoto target by the end of 2001. Member States' projections furthermore suggest that existing policies and measures will not be sufficient to continue the EU-wide reductions of total EU greenhouse gas emissions. Instead, progress made so far will be outweighed by further increases. The 'business-as-usual' scenario, with existing measures, suggests that in 2010

[1] At the time of writing Jürgen Lefevere was Programme Director of the Climate Change Programme of the Foundation for International Environmental Law and Development (FIELD), London. He has been involved in a number of studies for the European Commission's Environment Directorate General that have laid the foundations for the Green Paper on Emissions Trading (March 2000), the proposal for a Directive establishing a scheme for greenhouse gas emission allowance trading within the Community (EATD Directive) (October 2001) and the proposal for a Directive to link the EATD directive with the project-based mechanisms (July 2003). This chapter is an updated version of a contribution to the 2001 Yearbook of European Environmental Law.

[2] Carbon dioxide (CO_2), methane (CH_4), nitrous oxide (N_2O), hydrofluorocarbons (HFCs), perfluorocarbons (PFCs) and sulphur hexafluoride (SF_6).

MICHAEL BOTHE AND ECKARD REHBINDER (EDS.), Climate Change Policy, 259–308.

EC-wide emissions will have decreased by only 0.5%, which leaves a significant gap of 7.5% to be achieved through new measures[3].

In 1998 the need to reinvigorate the debate on the development and adoption of effective policies and measures to reduce the EU's GHG emissions led the Commission to focus on the introduction of an innovative instrument to tackle the EU's greenhouse gas emissions: emissions trading. Since the start of the discussions on EU-wide emissions trading, the instrument has rapidly gained support across a broad range of stakeholders within the European Union, although each for their own reasons. Together with the Commission's eagerness to establish a trading regime that could serve as the 'flagship' of the Community's strategy to implement the Kyoto Protocol this led to the adoption in October 2001 of the Proposal for a Directive establishing a scheme for greenhouse gas emission allowance trading within the Community[4]. After an unusually short decision making process the final text of the Directive was adopted by the European Parliament and the Council in July 2003[5].

This contribution will discuss the background to the Emission Allowance Trading Directive (EATD), as well as some of its key design issues. This contribution will first explain the background to the development of the EATD proposal. It will continue with a more in-depth discussion of a number of key aspects of the EATD. It will end with an overview of the Commission proposal for linking the EATD with the Kyoto Protocol's project-based mechanisms, which the Commission issued in July 2003[6].

---

[3] Report from the Commission under Council Decision 93/389/EEC as amended by Decision 99/296/EC for a monitoring mechanism of Community greenhouse gas emissions, COM(2003)735 of 28 November 2002.

[4] Commission's proposal for a Directive establishing a scheme for greenhouse gas emission allowance trading within the Community and amending Council Directive 96/61/EC, COM(2001)581, 23 Oct. 2001.

[5] Directive 2003/87/EC of 13 October 2003 establishing a scheme for greenhouse gas emission allowance trading within the Community and amending Council Directive 96/61/EC, [2003] OJ L 275/32.

[6] Commission Proposal for a Directive of the European Parliament and of the Council amending the Directive establishing a scheme for greenhouse gas emission allowance trading within the Community, in respect of the Kyoto Protocol's project mechanisms, COM(2003)403 of 23 July 2003. Available through: http://europa.eu.int/comm/environment/climat/home_en.htm

2 The Development of Emission Allowance Trading in the EU

This section will give a brief background to the development of the EATD in the EU as well as of some of the instruments that preceded or influenced the Directive[7].

A first indication of the Community's interest in the use of market-based mechanisms can be found in its 5th environmental action programme[8]. In this programme, under the heading 'broadening the range of instruments', the Community proposed to use a broader mix of instruments, which would include 'Market-based instruments, designed to sensitize both producers and consumers towards responsible use of natural resources, avoidance of pollution and waste by internalising of external environmental costs (through the application of economic and fiscal incentives and disincentives, civil liability, etc.) and geared towards 'getting the prices right' so that environmentally-friendly goods and services are not at a market disadvantage vis-à-vis polluting or wasteful competitors'. Emission allowance trading was, however, not specifically mentioned.

The specific interest in emission allowance trading developed quickly after the adoption of the Kyoto Protocol in December 1997. A first reference to an 'EC-wide approach to emissions trading' can be found in the Commission's Communication on the EU's post-Kyoto strategy from 3 June 1998[9]. In this Communication the Commission asked the Council to:

- endorse the introduction of the flexible mechanisms in a step-by-step and coordinated way within the Community;
- and endorse the objective of the gradual inclusion of private entities over time, and that, as national use of the flexible mechanisms will have to respect the Community law, it would be desirable to have a Community framework to safeguard the internal market.

In the same Communication the Commission suggested that the 'Community could set up its own internal trading regime by 2005 as an expression of its determination to promote the achievement of targets in a

---

7 See for an evolution of the thinking on emissions trading in the EU also Zapfel and Vainio (2002).

8 Resolution of the Council and the Representatives of the Governments of the Member States, meeting within the Council of 1 February 1993 on a Community programme of policy and action in relation to the environment and sustainable development – A European Community programme of policy and action in relation to the environment and sustainable development, [1993] OJ C 138/1.

9 Communication from the Commission to the Council and the European Parliament, Climate Change – towards an EU Post-Kyoto Strategy, COM(1998)353, 3 June 1998.

cost-effective way'. In its Communication 'Preparing for the Implementation of the Kyoto Protocol' of 19 May 1999, the Commission expanded this idea and suggested that the Commission should adopt a Green Paper on EU greenhouse gas emissions trading and organize a wide consultation on the basis of this issue in 2000[10]. In preparation for this Green Paper, a study was commissioned to a group of consultants in autumn 1998. The results of this study[11] provided the foundation for the Commission's Green Paper on greenhouse gas emissions trading within the European Union, which was released in little more than three months after the finalization of the study. On 8 March 2000 the Commission published the Green Paper on Greenhouse Gas Emissions Trading within the European Union (the Green Paper)[12], together with a Communication on EU Policies and Measures to Reduce Greenhouse Gas Emissions: Towards a European Climate Change Programme (ECCP)[13].

Within less than two years the Commission moved from a first interest in the instrument to the adoption of a Green Paper on the issue. The quick development of the Commission's interest in trading and the very short time span within which the Green Paper was produced are remarkable. It is interesting to note that shortly after the adoption of the Kyoto Protocol most of the staff of the Commission's Climate Change unit, including the head of unit, were replaced. Key members of the new staff previously worked in the economic instruments unit and had participated in the preparation and negotiation of the Commission's CO_2 taxation proposals. This may explain their eagerness to embrace this new market-based mechanism and make it a key feature of the Community strategy to implement the Kyoto Protocol.

The aim of the Green Paper was to start a discussion within the European Union on the merits and design of greenhouse gas emissions trading system, and the relation between such a system and existing and future policies and measures. The ECCP started a stakeholder dialogue to prepare further common and co-ordinated policies and measures. It created a number of Working Groups, each of these providing a platform for stakeholder discussions on a specific issue. The ECCP and the Green Paper were mutually supportive: Working Group I of the ECCP addressed the Flexible Mechanisms under the Kyoto Protocol, focussing on developing the Community's GHG emissions trading regime. The ECCP set out the larger policy context and created the platform for discussion, while the Green Paper focused on the further development of one specific instrument.

[10] Communication from the Commission to the Council and the European Parliament, Preparing for Implementation of the Kyoto Protocol, COM(1999)230, 19 May 1999.

[11] The results of this study are published on the internet at: http://www.field.org.uk/climate_4.php.

[12] COM 2000(87).

[13] COM 2000(88).

2.1 The Green Paper on Emissions Trading

The Green Paper makes a strong case for the introduction of GHG emissions trading in the European Union. It points out the economic benefits and the benefits for the internal market of applying emissions trading at the EU level rather than only at the national level. It explores the design options for a regime with a degree of harmonization at the EU level enough to reap those benefits, but with sufficient freedom for the Member States to adapt the specifics of the regime to their national circumstances and preferences.

The Green Paper advocates a step-by-step approach. Starting in January 2005, a Community regime could initially confine itself to large fixed point sources of carbon dioxide. The Green Paper sets out a list of possible industry sectors to include in the emissions trading regime, which covers in total 45.1% of all Community CO_2 emissions. Although it points out the advantages of having an identical coverage of sources across the Community, it recognizes the need for flexibility. It proposes a number of alternative approaches for doing so, including variations where Member States choose to either opt-in or opt-out certain sectors from an EU list of participating sectors and installations.

The Green Paper also addresses the central issue of allocation of allowances, between sectors and for individual sources. It sets out the various options available, including auctioning and different approaches to free allocation, such as grandfathering, and discusses the case of new market entrants. It also addresses the question whether the allocation and choice of allocation methods should be left to the Member States or whether some degree of harmonization is appropriate.

Both the EU and its Member States already have a body of environmental legislation in place, including measures addressing climate change. The Green Paper discusses the relation between a possible EU trading system and existing and planned policies and measures. It specifically addresses the relation with the IPPC Directive[14], negotiated agreements, energy taxation and points at potential links with the Monitoring Mechanism[15].

Robust compliance and enforcement are the basis for any GHG emissions trading regime. Recognizing this, the Green Paper describes the compliance and enforcement at both the Member State level (vis-à-vis companies) and the EU level (vis-à-vis the Member States). It asks whether elements of the

---

[14] Council Dir. 96/61/EC Concerning Integrated Pollution Prevention and Control, [1996] OJ L257/26.

[15] Decision 93/389/EEC for a monitoring mechanism of Community CO_2 and other greenhouse gas emissions, as amended by Council Decision 1999/296/EC [1999] OJ L 117/35.

national compliance and enforcement regime need to be harmonized at the Community level to avoid gaming by its participants.

The Green Paper laid the foundations for the development of an EU-wide GHG emissions trading regime. Rather than proposing a specific regime, the Green Paper's aim clearly was to stimulate a focused debate. It set out the available design options, and some of their implications. To stimulate the discussion within the EU, it contained a list of specific questions addressed to stakeholders and interested parties and invited interested parties to give their reactions and opinions to the Green Paper by 15 September 2000.

The Green Paper on Emissions Trading was discussed at the Environment Council Meeting on 22 June 2000[16]. In its conclusions the Environment Council welcomed the Green Paper and encouraged the Commission to take the process forward, recognizing that emissions trading could play an important role in the Community's strategy to reduce its GHG emissions. Parliament discussed the Green Paper on 26 October 2000. In its resolution in reaction to the Green Paper Parliament supported the introduction of an EU-wide trading regime by 2005. Parliament also urged that the trading regime should be based on quantified greenhouse gas abatement targets, set in advance, per country and per sector[17]. The consultation process launched by the Green Paper triggered an unexpectedly large amount of reactions. In total 90 reactions were submitted by a wide range of interested parties, virtually all of which supported the development of an EU-wide trading regime[18].

2.2 European Climate Change Programme, Working Group I

In parallel to the discussion of the Green Paper in the Council and the European Parliament, the Commission organised the stakeholder meetings under the European Climate Change Programme, Working Group I on Flexible Mechanisms (WG I). In total ten meetings were held between July 2000 and May 2001. These meetings served not only as a platform for different stakeholders to exchange their views on the development of the trading regime, but also as a capacity building exercise, where the exchange of opinions between the various participants helped clarify different concepts and approaches and generated new ideas.

The final report of WG I was presented at the ECCP conference, which was held in July 2001. Even though the stakeholder group participants came

[16] Council Conclusions 2278th meeting, Environment, Luxembourg, 22 June 2000.

[17] European Parliament resolution of 26 October 2000, [2001] OJ C197/219 & 400.

[18] All submissions as well as a summary of the comments made are published on the Internet at: http://europa.eu.int/comm/environment/docum/0087_en.htm. See also Krämer (2001) who discusses the Green Paper and some of the reactions to it.

from very different backgrounds and represented a broad range of interests, the degree of consensus on a number of key design issues was encouraging. The group was unanimous on the need for emissions trading to be introduced as soon as practicable. Differences of opinion did, however, remain on key design issues, such as the role of relative targets in the regime and the degree of harmonization of emissions trading at the EU level[19].

2.3 The Adoption of the Proposal

At the end of January 2001 the Commission started with the drafting of the Emissions Trading (ET) Proposal, helped by a small group of consultants[20]. The US rejection of the Kyoto Protocol in March 2001 caused the Commission to significantly advance its agenda for bringing out a proposal. Environment Commissioner Wallström decided that there was a need for the Community to give a clear signal to the outside world that it was still taking the Kyoto Protocol seriously and was preparing for its implementation. The Commissioner instructed the Climate Change Unit to prepare a draft proposal to be adopted by the Commission before the start of the international climate change meeting in Bonn on 16 July 2001. A draft proposal was sent to other Commission services for informal consultation in mid-May 2001, and went into inter-service consultation at the end of May 2001.

The widely leaked draft, only a few weeks before the ECCP conference, caused uproar among a number of industry lobby groups. Industry had not expected the proposal to be presented so soon, and had hoped on another round of stakeholder consultations before a final proposal would be presented[21]. The short deadline for the adoption of the ET Proposal caused a panic reaction with a small number of key industry representatives, who heavily pressured the Commission to delay the adoption of the proposal. Faced with the heavy industry lobbying and growing opposition within the Commission, Wallström decided on 28 June 2001, while senior international negotiators where discussing the progress of the negotiations on the

---

[19] The terms of reference, background documents, meeting reports, interim and final report are published on the internet at: http://europa.eu.int/comm/environment/climat/eccp.htm.

[20] FIELD (2001).

[21] See, for instance, UNICE (2001a), which states that ‘the Commission had made clear its intention to propose a Community emissions trading scheme towards the end of this year, with further consultation of stakeholders already planned. We regret that the current proposal pre-empts those sound intentions, without taking into account outcomes of the European Climate Change Programme, where major issues were raised of how an EU framework should be linked to separate Member State approaches, and how individual companies should become involved’.

implementation of the Kyoto Protocol at a high-level meeting in The Hague, to postpone the proposal.

The success of the international climate change negotiations in Bonn in July 2001, which led to the adoption of the Bonn Agreements[22], reinvigorated the Commission's determination to push for the adoption of the ET Proposal before the end of 2001. After an extra round of consultations on 4 September 2001 (with Industry and NGOs) and on 10 September 2001 (with Member States, EEA and accession countries), the Commission published on 23 October its proposal for a Directive establishing a scheme for greenhouse gas emission allowance trading within the Community[23], together with a proposal for a Council Decision to ratify the Kyoto Protocol[24] and a Communication on the implementation of the first phase of the ECCP[25]. The adoption was timed to occur just before the start of the final act in the international negotiations on the completion of the package of measures necessary to implement the Kyoto Protocol in Marrakech on 29 October 2001.

2.4 The Adoption of the Directive

The Directive was negotiated and adopted by the Council of the European Union and the European Parliament under the so-called 'co-decision procedure'[26]. This procedure, set out in Article 251 of the EC Treaty, provides for qualified majority voting in the Council, composed of relevant Ministers of the Member States, allowing Member States to be outvoted. Parliament has the right to propose amendments and can veto the adoption of the entire proposal if it feels that its amendments have not sufficiently been reflected in the final text. The adoption of the proposal usually goes in two so-called 'readings'. Parliament can propose amendments in each of these readings. The Council adopts a common position after the first reading, which provides the first political compromise between the Member States, taking into account the Parliament's amendments, and is usually the basis for the final text of the piece of legislation. Parliament can propose new amendments after the adoption of the common position, in second reading. If

---

[22] Decision 5/CP.6, The Bonn Agreements on the implementation of the Buenos Aires Plan of Action, FCCC/CP/2001/5, 36–49.

[23] N. 4 above.

[24] Council Decision 2002/358/EC concerning the conclusion, on behalf of the European Community, of the Kyoto Protocol to the United Nations Framework Convention on Climate Change and the joint fulfilment of commitments thereunder.

[25] Communication from the Commission to the Council and European Parliament on the implementation of the first phase of the ECCP, COM (2001)580, 23 Oct. 2001.

[26] An elaborate description of the co-decision procedure and the roles of the various EU institutions therein can be found in Craig and de Búrca (2002).

no agreement is reached between Parliament and the Council on these amendments, a conciliation committee is convened, which is to seek agreement between the two institutions.

Within little over a year after the proposal's adoption by the Commission, the Council agreed its Common Position on 9 December 2002. Only six months after that, twenty months after the original Commission Proposal, the final text of the Directive was agreed in second reading between the European Parliament and the Council on 23 June 2003. The compromise was approved by the European Parliament on 2 July 2003 and the Council on 22 July 2003 and formally entered into force on the date of its publication, 25 October 2003.

The unusually fast adoption of a Directive on such a complex topic can be explained by the high priority given to this Proposal by the Commission, combined with the political pressure to adopt a meaningful measure at Community level to prepare for the implementation of the Kyoto Protocol and almost unprecedented widespread support for the proposal from key sectors of industry and NGOs. Although the key elements of the original Commission Proposal have remained intact, the adoption process did not go without difficulty, and the final proposal shows the scars of the negotiations that took place in Parliament and Council. The key issues in these negotiations included the binding nature of the trading regime, its coverage of sectors and gases, the link with the Kyoto Protocol flexible mechanisms, the allocation of allowances (method and total quantity) and the issue of penalties.

The most intensive discussions on these various issues took place before the adoption of the Council's common position on the Proposal on 9 December 2002. Parliament proposed in total 73 amendments[27], and the discussions in Council had to be postponed due to negotiations within Germany on its position on the Directive. Unlike virtually all other major pieces of environmental legislation that have been recently adopted by the European Community under the co-decision procedure, the EATD did not go into conciliation. Uniquely, Parliament was lobbied by a broad range of stakeholders, including industry, environmental NGOs, the European Commission and most Member States, to avoid proposing too many amendments and accept the common position as it was adopted by the Council in December 2002. Had the Directive gone into conciliation, its adoption would have been delayed at least until the end of 2003, which would not only have jeopardized the start of the trading regime by 1 January 2005, but could even have undermined the adoption of the Directive

[27] See European Parliament, 1st reading report on the proposal for a European Parliament and Council Directive establishing a scheme for greenhouse gas emission allowance trading within the Community and amending Council Directive 96/61/EC, 13 September 2002, Rapporteur: Jorge Moreira da Silva, PE 232.374.

altogether. The unique compromise between the Council and Parliament on 23 June 2003 avoided the conciliation procedure and allowed the Directive's timetable to remain unchanged. As part of this compromise, the Council did, however, have to make a number of concessions to the European Parliament, relating to the coverage of sectors and gases, the method and amount of allocation and the link with the Kyoto Protocol's project-based mechanisms.

The key issues under negotiation during the adoption of the Directive are elaborated in the next section, describing the contents of the Directive.

3 Key Elements of the EATD

This section provides an overview and background of the key aspects of the EATD as it was finally agreed upon in June 2003. When analyzing the EATD it should not be overlooked that this is a 'Directive'. The EATD merely provides the backbone of the European trading regime. For it to become effective it will need to be implemented in the domestic legislation of the Member States, each of which has a certain amount of freedom to choose the methods of implementation, as long as the results prescribed by the Directive are actually achieved. By choosing the Directive as the backbone of the European trading regime, there will not actually be one single trading regime, but a series of 15, and, from 1 May 2004, 25 harmonized and connected domestic trading regimes, all of which may be slightly different in specific details of their design.

The following part of this contribution will give an overview of the key elements of the EATD, starting with a discussion of the timing and nature of the regime, followed by a discussion of its coverage, the instruments 'permit' and 'allowance', the allocation of allowances, enforcement and the relation with the IPPC Directive. The role of the project-based mechanisms will be discussed in the next section.

3.1 Compulsory Participation from 1 January 2005 Onwards

The EATD lays down the framework for a Community-wide compulsory greenhouse gas emission allowance trading scheme in all EU Member States from 1 January 2005 onwards. Although the original Commission Proposal put forward a compulsory regime from 1 January 2005 onwards, it was not until shortly before the adoption of the Common position that it became clear that participation in the regime would indeed be compulsory. The discussions on the binding nature of the regime focused mostly on the period from 1 January 2005 to 31 December 2007. While the Commission and the environmental NGOs[28] favoured a binding regime, a number of industry groups argued that EU emissions allowance trading regime should be voluntary, at least during the pre-Kyoto period[29]. Experience with voluntary trading regimes in for instance Canada[30] and the United Kingdom[31] has, however, shown that these regimes do not bring the large-scale reductions needed to meet the Kyoto Protocol targets, or that the voluntary setting of targets leads to targets that are easily achievable and, thus, in turn cause an oversupply of allowances on the market. The Commission had in addition argued in its original proposal that a pre-Kyoto trading regime would benefit the Community in allowing it to gain important experience with trading.

The final compromise on the EATD keeps the trading regime binding from 1 January 2005 onwards, thus including the pre-Kyoto period. To allow for this, a number of concessions had to be made. The first concession is laid down in a new Article 27 of the EATD. This provision, inserted at the insistence of in particular the United Kingdom, allows Member States to request permission from the Commission to exclude certain installations during the period up to 31 December 2007, subject to a range of conditions. A further concession is laid down in the new Article 29 of the EATD. This provision allows Member States to apply to the Commission for permission to issue additional, non-transferable, allowances to certain installations in limited cases of *force majeure*.

But perhaps the most interesting addition can be found in the form of a new Article 28 on 'pooling'. This article is the result of strong lobbying by German industry and ensuing pressure from the German government during the negotiations. German industry has since the beginning of the discussions on trading strongly opposed a mandatory regime[32]. The reason for this is that in Germany agreements were negotiated between industry and government to reduce greenhouse gas emissions from key industry sectors, with the agreement that no further requirements would be imposed upon the sectors included in those agreements. Industry feared that mandatory participation in an emissions trading regime could not only lead to a revision in the domestic legal framework and the administrative burdens attached to that, but also to reduction obligations going beyond those already agreed in the climate

28 Climate Action Network Europe (2001).

29 See, for instance, the UNICE (2001b), which states that 'Most believe that at least this initial phase should be on a voluntary basis, since a prime principle of emissions trading should be to offer motivation and clear market signals to companies'. Stronger opposition to mandatory trading came from the German industry association BDI, which opposed a mandatory trading regime. See for instance BDI (2002).

30 See, for instance, Haites and Hussain (2000).

31 The evaluation report of the first year of the UK emissions trading scheme can be found at: http://www.defra.gov.uk/environment/climatechange/trading/index.htm.

32 See the BDI (2002), n. 29 above.

change agreements. Under pressure from the BDI, the main German industry association, Germany proposed during the negotiations leading up to the Common Position that the possibility be created for sectors of industry, or even all industry within a country, to pool their allowances together and allow for a single trustee to manage the allowances held by the pool. The original idea behind this was that Germany would make participation in this pool mandatory for all sectors covered by the trading regime and the pool would be managed by the German government as the trustee. Under this scenario there would therefore be no emissions trading within Germany, but the German government would hold all allowances and do the buying and selling to other Member States. During the discussions on the pool proposal, the proposed text was however significantly weakened. The compromise text as set out in the EATD no longer allows for mandatory pools and it requires that in case a trustee fails to comply with the penalties under the trading regime, each operator of an installation remains responsible in respect of emissions from its own installation. The current provisions on pooling have been weakened to the extent that it is now unlikely that this possibility will be used in practice.

Interestingly, the EATD is designed to continue after 2012. From 2008 onwards the Directive works with 5-year periods, for each of which Member States are required to draw up a national allocation plan. The second 'commitment period' under the EATD overlaps with the first commitment period of the Kyoto Protocol. After 2012 a new five-year commitment period starts, regardless of the outcome of the forthcoming negotiations on the 2nd commitment period of the Kyoto Protocol, although the Directive could be amended to be brought in line with any future international agreements if necessary.

3.2 Coverage of Sectors, Gases and Member States

The EATD contains in its Annex II a list of all greenhouse gases, but it initially only applies to CO_2 emissions from activities set out in its Annex I. The sources listed in Annex I are a subset of the installations listed in Annex I of the IPPC Directive, limited to installations that emit large quantities of CO_2. One addition is that combustion installations from more than 20 MW, rather than the 50 MW threshold set in the Annex of the IPPC Directive, are covered by the EATD. The EATD is thus expected to cover approximately 46% of the estimated EU CO_2 emissions in 2010. Initial Commission estimates were that 4000 to 5000 installations would be covered[33]. More recent estimates have, however, shown that, in particular in light of the

[33] These figures are from the Explanatory Memorandum to the Proposal, n. 4 above, 10.

planned accession of 10 new Member States in May 2004, up to 15000 installation may be covered by the regime, of which around 2000 installations in the UK alone[34].

In the original Commission Proposal expansion of the list of activities and the gases emitted by those activities was only possible through amendment of the Directive through the normal co-decision procedure. Such amendment could be proposed by the Commission on the basis of Article 26 in the original proposal, now renumbered to Article 30. Paragraph 1 of that Article states that 'the Commission may make a proposal ... by 31 December 2004' to expand the sectors and gases covered by the regime. Paragraph 2 of that same Article requests the Commission to draw up a report on the application of the Directive, 'accompanied by proposals as appropriate' by 30 June 2006, which is to include the coverage of the regime. The Commission had in the explanatory memorandum to the original proposal defended this approach on the basis of the need to further reduce monitoring uncertainties for other gases and to limit the amount of sources that would initially enter the regime. This approach was also supported by the environmental NGOs[35]. Industry however lobbied heavily for an expansion of the trading regime to include all six Kyoto gases[36]. The European Parliament also urged the Commission to include other gases at the earliest possible stage, although it recognized that this should not go at the cost of jeopardizing the simplicity of the regime and expansion to other gases and sources should only be possible if the quality of monitoring and measurement of these emissions could be ensured. Parliament therefore requested that the Commission develop the methodologies for monitoring these emissions to be finalized in time for the Directive's revision in 2004[37].

The result of the discussions in Parliament and Council is the addition of a number of new provisions. A first relevant provision, already mentioned above, was included more as a result of the discussions on the compulsory nature of the trading regime in the pre-Kyoto period. A new Article 27 was included in the EATD, which allows for the temporary exclusion of installations up to 31 December 2007, under strict conditions. Member States must apply for such temporary exclusion with the European Commission. In the Common Position a temporary exclusion could be given for 'certain installations and activities', giving the impression that entire categories could be opted out. Due to pressure from the European Parliament during the 2nd reading of the Directive, the temporary exclusion in the final text of the EATD is limited to installations. On applying for a temporary exclusion the Member State has to list and publish each installation for

[34] Estimate based on personal Communication with the Commission and ENDS (2003).
[35] See n. 28 above.
[36] See n. 29 above.
[37] See n. 27 above, 89.

which it seeks exclusion. A temporary exclusion is only given if the installations will limit their emissions as much as they would have under the trading regime, are subject to equivalent monitoring, reporting and verification requirements and equivalent penalties.

The EATD, however, also allows expansion of the coverage of the activities and gases beyond those included in Annex I. As a result of the negotiations on coverage a new Article 24 on 'procedures for unilateral inclusion of additional activities and gases' was included in the final text of the EATD. This provision allows Member States, with the approval of the Commission, to expand the coverage of the regime to other activities and gases from 2008 onwards and from 2005 onwards to activities below the thresholds in Annex I. Expansion of coverage under this provision is approved by the Commission following the 'comitology' procedure[38] referred to in the Directive's Article 23(2), 'taking into account all relevant criteria, in particular effects on the internal market, potential distortions of competition, the environmental integrity of the scheme and reliability of the planned monitoring and reporting system'. Article 24(3) requires the Commission, on request by a Member State, to adopt monitoring and reporting guidelines for these activities 'if the monitoring and reporting can be carried out with sufficient accuracy'.

In addition to these new provisions, preambular paragraph 14 was amended to point out the possibility to extend the coverage of the regime to 'emissions of other greenhouse gases than carbon dioxide *inter alia* from aluminium and chemicals activities' in the final compromise between the Council and Parliament. Similarly, Article 30(a) was amended to explicitly mention chemicals, aluminium and transport as sectors to be included in the 2006 review. These amendments were included at the insistence of the European Parliament, which had argued for the inclusion of the aluminium and chemicals sectors in the EU trading regime in both first and second reading[39].

---

[38] 'Comitology' refers to a legislative procedure where the adoption of the legislation has been delegated to the Commission, assisted by a committee composed of Member State representatives. The EATD uses the so-called regulatory procedure. Under this procedure the Commission can only adopt implementing measures if it obtains the approval by qualified majority of the Member States meeting within the committee. In the absence of such support, the proposed measure is referred back to the Council which takes a decision by qualified majority. However, if the Council does not take a decision within three months, the Commission finally adopts the implementing measure provided that the Council does not object by a qualified majority. See also EIPA (2000).

[39] See n. 27 above and European Parliament, 2nd reading report on the Council common position for adopting a European Parliament and Council Directive establishing a scheme for greenhouse gas emission allowance trading within the Community and amending Council Directive 96/61/EC, 12 June 2003, Rapporteur: Jorge Moreira da Silva, PE 328.778.

The original Commission proposal briefly states that it would not only apply to the current 15 EU Member States, but also to new Member States[40]. None of the accession agreements closed between the EU and the 10 new accession states includes transitional measures for the EATD. The proposal will therefore by default become applicable to those new Member States at their accession on 1 May 2004, as part of the so-called *acquis communautaire*, encompassing the total body of Community law, including legislation and related instruments such as judgements of the European Court of Justice, that are in place in the Community. Interestingly, a number of the new accession states have negotiated transitional measures for the closely related IPPC Directive[41]. Although the absence of such extended deadlines for the new accession states means that significant effort will have to be put into the timely implementation of the EATD, the Directive can also be an important vehicle for these states to exploit their generous allocations under the Kyoto Protocol and use their tremendous potential for further emission reductions. At the time of writing it was still unclear whether the Directive would also apply to the European Economic Area (EEA) States (Norway, Iceland and Liechtenstein) and whether Switzerland is interested in participating in the framework of its bilateral agreements with the EU. On 1 January 2005 there will therefore be at least 25, and potentially even 29, states participating in the trading regime.

3.3 Permits and Allowances

The EATD distinguishes between a 'Greenhouse Gas Emissions Permit' and 'Allowances'. Article 4 of the Directive requires Member States to ensure that from 1 January 2005 no installation undertakes any activity listed in Annex I of the Directive unless it holds a greenhouse gas emissions permit. Article 5 of the Directive sets out the information that needs to be included in the permit application, which includes a description of the installation, the raw and auxiliary materials used that are likely to lead to greenhouse gas emissions, the sources of greenhouse gas emissions and the planned monitoring and reporting measures. Article 6 of the Directive contains the conditions for and contents of the greenhouse gas emissions permit. It requires that permits are only issued if the operator is capable of monitoring and reporting emissions. It also specifies that the permit includes the monitoring requirements, specifying monitoring methodology and

---

40 N. 4 above, 3.

41 The application of the IPPC Directive to existing installations has for instance been postponed to 2010 for Latvia and Poland and to 2011 for Slovenia, compared to 2007 for existing Member States. For a summary of the various transitional measures see: http://www.europa.eu.int/comm/enlargement/negotiations/chapters/chap22/index.htm.

frequency, as well as the reporting requirements, and, importantly, the obligation to surrender allowances equal to the total emissions of the installation in each calendar year. As the procedure for granting the permit is based on that of the IPPC permit, Article 8 of the EATD requires the permitting procedure to be coordinated with the granting of the IPPC permit and allows Member States to fully integrate the two permitting processes[42].

Article 3(a) EATD defines 'allowance' to mean 'an allowance to emit one tonne of carbon dioxide equivalent during a specified period, which shall be valid only for the purposes of meeting the requirements of this Directive and shall be transferable in accordance with the provisions of this Directive'. Allowances are initially allocated to the operators of installations covered by the Directive, but can be transferred between natural and legal persons within the Community and between persons within the Community and persons in third countries if a bilateral agreement under Article 25 EATD exists with those countries. According to Article 13 EATD, allowances are only valid to offset emissions during the period for which they are issued. Four months after the end of the pre-Kyoto period, the allowances issued for this first period are automatically cancelled. Member States *may*, but are not obliged to, issue new allowances for the period starting in 2008 to replace the allowances that were cancelled. The reason for this 'may' is that a large surplus in allowances from the pre-Kyoto period could jeopardize a Member States' compliance with its Kyoto targets if these can be used to offset emissions during the first Kyoto Protocol commitment period. These allowances would after all not be valid Kyoto Protocol units, and therefore any emissions offset by pre-2008 EU allowances that take place within the Kyoto Protocol's first commitment period would need to be offset by reductions elsewhere. The 'may' thus gives the flexibility to Member States to disallow the banking of allowances into the first Kyoto commitment period. The situation is different for the surplus allowances at the end of 2012. As the Kyoto Protocol also allows for the banking of most of its tradable units, Article 13(3) EATD requires that Member States *shall* issue allowances to replace those cancelled in May 2012.

Permits and allowances should therefore be clearly distinguished under the EATD. The permits are the prerequisite framework for installations to participate in the trading regime, as they set out the conditions for the monitoring and reporting of their emissions. Permits are however not the units that are traded, those are the allowances.

With the introduction of the term 'allowances', the EU has introduced its own tradable unit. An important question is what the relation between the EU allowances and the units created under the Kyoto Protocol is. This

[42] For a more elaborate discussion of the relation between the EATD and the IPPC Directive, see below.

question is addressed below in the section on the links to the project-based mechanisms.

3.4 Allocation of Allowances

The allocation of allowances is one of the core parts of the Directive. The ATD neither lays down targets for Member States nor contains detailed allocation criteria. Instead, it requires Member States to decide upon their own targets for the sectors covered by the EATD and their own criteria for the allocation of the allowances to individual installations. The lack of targets in the Commission Proposal was criticized by the environmental NGOs, who would have liked more certainty on a robust environmental outcome of the operation of the trading regime[43]. The call for targets from the environmental NGOs was supported by the European Parliament. In first reading of the Proposal, Parliament put forward a number of amendments by means of which a ceiling was to be placed on the emission allowances to be allocated by each Member State, representing 50% of the emissions forecast annually for each Member State on a linear curve converging with the Kyoto commitments[44]. In the negotiations on the second reading, agreement between the Council and Parliament was reached to include two new sentences in paragraph 1 of Annex III of the EATD, which point out that 'the total quantity of allowances to be allocated shall not be more than is likely to be needed for the strict application of the criteria of this Annex' and that 'prior to 2008, the quantity shall be consistent with a path towards achieving or over-achieving each Member State's target' under the Kyoto Protocol.

It is difficult to see how the EATD could have been adopted within such a short period of time if targets would have been included. At the time of the drafting of the Directive there was very little information available on the number of installations that would be covered by the regime, let alone their emissions and their potential for emission reductions. The composition of the sectors and the distribution in the types of installations furthermore differs too much between the various Member States to define EU-wide targets. An EU-wide debate on targets for specific industry sectors would have been much more complicated than the discussions on the burden sharing agreement, as it would have exposed the participating governments to much more intensive lobbying from these industry sectors, which could have made the adoption of the Directive altogether impossible. Although certainly not perfect, the approach in the text of the EATD, set out below, provides the

---

43 See n. 28 above.
44 See n. 27 above.

advantage that it sets boundaries for the national target setting, while avoiding protracted discussions on specific targets at the EU level.

The Member State targets and criteria for allocation must be laid down in an 'Allocation Plan', which is submitted to other Member State and can be rejected by the Commission. The EATD requires Member States to allocate at least 95% of the allowances free of charge for 2005–2007 and at least 90% for the period 2008–2012. It also contains a number of requirements for the Allocation Plan in its Annex III.

As already stated above, the issue of allocation turned out to be one of the most hotly debated during the discussions on the adoption of the EATD[45]. In choosing the Directive's approach to allocation a number questions had to be addressed.

A first question relates to the division of competence between the Member States and the Community. To what extent is it desirable or necessary for the Community to prescribe not only allocation methods, but also the exact allocation to individual sources at Community level, or what control can the Community exercise over this allocation by Member States? To provide an answer to this question it is interesting to take a closer look at which aspects of allocation have a Community dimension under current Community law.

The initial allocation of the Community and the Member States' total greenhouse gas emissions during the first commitment period was agreed upon in the Kyoto Protocol, and set at an 8% reduction for the EU as a whole and for each Member State individually. The EU Bubble agreement, discussed in a separate chapter in this book, subsequently re-distributed this reduction target among the Member States. During the stakeholder discussions in the ECCP working group I, German industry stated on several occasions that it did not want Member States or the Commission to directly translate the burden sharing target into targets for the sectors under the ET regime. It clearly feared that Germany's 21% reduction target would be translated into a similar reduction target for each individual source[46]. While this is highly unlikely, it does reflect the fear within a number of Member States that their targets under the burden sharing agreement may directly reflect upon the competitiveness of their industry under the ET regime[47]. After all, a number of Member States are allowed large increases in their emissions, whereas others must make significant reductions. As seen above,

---

45 See also the analysis by Krämer (2001), n. 18 above, 27–30, written before the Commission issued the ET Proposal.

46 See also the BDI (2002) position paper on the original Commission Proposal, n. 29 above.

47 In the UK a discussion started on whether the UK government should take its bubble target of –12.5% or its self-proclaimed target of –20% by 2010 as the starting point for the allocation under the EATD. See recent coverage of the UK's allocation discussion in ENDS (2003), n. 34 above.

the burden sharing targets do not necessarily reflect a Member State's marginal reduction costs but are to an important extent based on a political redistribution of the Community's overall target.

Related to the Bubble distribution is the 'Dutch situation'. The Netherlands has indicated that it will meet half of its reduction effort through the use of the Kyoto Protocol's flexible mechanisms. Since May 2000, the Dutch government has started two major investment programmes, the ERUPT programme and the CERUPT programme. Using these programmes the government, through public tendering procedures, is purchasing CDM and JI credits. The government has set aside a total of 450 million US $ for buying these credits, and is looking into the use of other instruments as well. It aims to buy credits representing between 20–25 million tonnes of CO_2 emissions annually between 2008 and 2012[48]. The question is to what extent this large-scale purchasing of credits will have an impact on the allocation of allowances to installations covered under the ET proposal. A Member State could choose to use these credits in order to allocate larger amounts of allowances to its industry sectors, thus exempting these sectors from significantly reducing their emissions under the Community trading regime. The question is, however, how the distribution of these 'extra' credits by a Member State government is different from the allocation of targets between trading and non-trading sectors to comply with a Member State's burden sharing target. But this does not take away the fear within other Member States that the Dutch government could be indirectly subsidizing its industry participating in the trading regime.

This leads to the question on the relation between the allocation of allowances and the Community's state aid rules[49]. The application of the Community's state aid rules to the various aspects of emissions trading is not entirely clear. The issue was not addressed in the recent Community Guidelines on State aid for environmental protection[50]. Requests for more clarity on the relation between Community competition law and emissions trading were made on various occasions throughout the stakeholder discussions in ECCP Working Group 1[51]. The Commission, however,

---

48 More information on these programmes can be found on the Internet at: http://www.carboncredits.nl.

49 See for a more elaborate discussion on state aid and the EATD: König, Braun and Pfromm (2003).

50 Community guidelines on State aid for environmental protection, (2001) OJ C 37/3. See for a discussion of these guidelines Fernández Armenteros (2001) and Vedder (2001).

51 See, for instance, the remark by C. Boyd during the first meeting of ECCP Working Group I on 4 July 2000, stressing the need to have input from DG competition in the discussions and asking the Commission to ensure that officials from this DG attend the WG I meetings. This remark is reflected in the summary record of that meeting, published on the Internet at: http://europa.eu.int/comm/environment/climat/eccp.htm. The Commission's

decided that it would be too early to draft guidelines for the application of state aid rules under the ET Proposal. Instead, it preferred to gather more experience with the application of the existing rules.

Recent Commission Decisions on the state aid implications of the UK and Danish domestic trading regimes and domestic energy taxation in combination with the state aid guidelines do, however, provide some guidance on the application of those rules. The Commission has found that the free allocation of allowances to entities constitutes state aid, which brings a host of interesting but also complicated legal issues on state aid into the allocation debate[52]. The Commission has reached a similar conclusion for the UK's payment of the incentive money[53]. The Commission has, however, found the state aid aspects of both the Danish and the UK trading regimes to be admissible under 87(3)(c) of the EC Treaty. The main reason for allowing the UK incentive money was that it allows companies to go beyond existing Community standards and provides a net environmental benefit. The main reason for allowing the free allocation in the UK scheme was that this is matched by further reduction of the source's emissions[54]. The Commission similarly allowed the free allocation under the Danish trading scheme because of the further emission reductions sought by the trading regime. The Commission did, however, require Denmark to change its regime to ensure that new entrants on the market will receive allowances 'based on criteria that are objective and non-discriminatory in relation to those applied to incumbent producers'[55].

Although initial internal Commission proposals for the EATD said very little about the allocation methodology[56], it became clear during the preparation of the proposal that carefully balanced guidance on allocation was needed to address a number of the issues above[57].

---

cont.

Competition DG was in fact regularly invited by DG Environment officials to attend the ECCP meetings, but never turned up.

[52] See Commission decisions on State aid No N 416/2001 – United Kingdom Emission Trading Scheme; N 653/99 – Denmark, CO_2 quotas and State aid; and N 123/2000 – United Kingdom, Climate Change Levy for the free allocation of allowances for companies entering into Climate Change Agreements. Various authors have warned that clarity on the state aid aspects of free allocation is urgently required if a multitude of state aid cases is to be avoided from undermining the allocation process and thereby the functioning of the trading regime. See n. 49 above.

[53] Commission decision on State aid No N 416/2001 – United Kingdom Emission Trading Scheme.

[54] Ibid.

[55] Commission decision on State aid No N 653/99 – Denmark, CO_2 quotas and State aid.

[56] N. 20 above.

[57] UNICE (2001a), in its comments to the proposal that was leaked in June 2001, for instance stated that 'guidance to member States in Annex III, on criteria for allocation plans,

The allocation guidance in the EATD consists of a number of elements. A first element is the virtually complete harmonization of the method of allocation. The Commission's original Proposal stated that for the period between 2005 and 2008 allowances were to be allocated free of charge and required the Commission to specify a harmonised method of allocation for 2008–2012, using the comitology procedure. By doing so, the Commission addressed the fear existing with many industry groups that the ET Proposal or individual Member States in the implementation of the proposal would opt for auctioning as the allocation method, an allocation method that is strongly opposed by industry[58]. Free of charge allocation of valuable emission allowances, however, significantly complicates the allocation debate. This is not only because it sets governments the challenging tasks of finding allocation criteria that are acceptable to the sectors involved, but also because it brings in a host of state aid questions that could undermine the allocation process through judicial challenges of allocation decisions[59]. Environmental NGOs strongly favoured auctioning as an allocation methodology, as it rewards companies that have taken early action, thus following the 'polluter pays principle' and because it allows the recycling of the revenues of the auction for example to the development of renewable energy sources[60]. Parliament equally argued strongly for compulsory auctioning of at least part of the allowances[61]. As a result of Parliament's insistence that at least part of the allowances should be auctioned, the final text of Article 10 EATD now requires that for the three-year period beginning 1 January 2005 Member States shall allocate at least 95% of the allowances free of charge, and from 2008 onwards at least 90% of the allowances is allocated free of charge. At the insistence of the European Parliament, Article 30(c) was also amended in second reading to include the consideration of auctioning as the method of allocation after 2012 in the Commission's review of the Directive in 2006.

The second element of the guidance on allocation given in the EATD is the 'national allocation plan'. Article 9 EATD requires Member States to submit this plan by 31 March 2004 for the period starting on 1 January 2005 and at least 18 months before the start of each subsequent period. In this

---

cont.
is not nearly clear enough to avoid a danger of single market distortions being caused'. See n. 21 above.

[58] See in particular UNICE (2003a).

[59] N. 52 above.

[60] N. 28 above.

[61] In first reading Parliament proposed that from 2005–2007 70% be allocated free of charge and the remaining 30% were to be allocated by means of auction. For the second period covered by the scheme Parliament proposed that all allowances were to be allocated by means of an auction. See n. 27 above, 52.

allocation plan the Member State has to indicate the total quantity of allowances that it intends to allocate for that period and how it proposes to allocate them. Member States are required to base the plan on objective and transparent criteria, and the Proposal lists a number of those criteria in its Annex III (see Box 1).

The EATD also sets out a procedure for the assessment of the national allocation plans. Member States have to submit these plans to the Commission and to other Member States. These plans are subsequently discussed in the Committee set up under the Proposal, which will be the same Committee as used under the Community's Monitoring Mechanism[62]. The Commission may reject the Member State's allocation plan within three months after its submission on the basis of its incompatibility with Annex III of the Proposal or the Community's prescribed allocation method, set out in Article 10 of the EATD. Member States are only allowed to allocate allowances to the installations once proposed amendments to the national allocation plan have been accepted by the Commission.

With the deadline for the submission of the Member State's allocation plans rapidly approaching, the issue of allocation is increasingly providing food for discussions. In February 2003 the Commission issued a non-paper for discussion in the Monitoring Mechanism Committee, addressing various process-related allocation issues[63]. Article 9(1) EATD requires the Commission to develop by 31 December 2003 at the latest guidance on the implementation of the allocation criteria listed in Annex III.

Allocation issues that remain open are for instance whether Member States will be required to 'fix' their allocation at the beginning of each commitment period, or whether this allocation can be changed. The DG Environment's non-paper on Allocation clearly states that 'before the trading period commences, the issue of allocation concerning the period is closed', which seems in line with Article 9(3) EATD. This may, however, bring problems in the case of the Dutch scenario, where the government of a Member State decides to buy a large amount of credits through the Protocol's flexible mechanisms. If the Dutch government wishes to use these credits to relieve the reductions required from the sectors covered under the EATD, it will need to make prior estimates of the amount of units that it is likely to be able to buy, as it is not allowed to allocate these units in addition to allocations foreseen in the national allocation plan[64].

---

[62] N. 15 above.

[63] European Commission (2003a).

[64] The recently issued Commission Proposal to link the EATD with the Kyoto Protocol's project-based mechanisms (discussed below) could provide some answers. The government could, for instance, decide to convert Kyoto Units that it buys into allowances and transfer these to installations covered by the regime. Depending on how much and whether installations would have to pay for this state aid issues are again likely to come up.

Box 1: Annex III of the EATD:

Criteria for National Allocation Plans Referred to in Article 9, 22 and 30

The total quantity of allowances to be allocated for the relevant period shall be consistent with the Member State's obligation to limit its emissions pursuant to Decision 2002/358/EC and the Kyoto Protocol, taking into account, on the one hand, the proportion of overall emissions that these allowances represent in comparison with emissions from sources not covered by this Directive and, on the other hand, national energy policies, and should be consistent with the national climate change programme. The total quantity of allowances to be allocated shall not be more than is likely to be needed for the strict application of the criteria of this Annex. Prior to 2008, the quantity shall be consistent with a path towards achieving or over-achieving each Member State's target under Decision 2002/358/EC and the Kyoto Protocol.

The total quantity of allowances to be allocated shall be consistent with assessments of actual and projected progress towards fulfilling the Member States' contributions to the Community's commitments made pursuant to Decision 93/389/EEC.

Quantities of allowances to be allocated shall be consistent with the potential, including the technological potential, of activities covered by this scheme to reduce emissions. Member States may base their distribution of allowances on average emissions of greenhouse gases by product in each activity and achievable progress in each activity.

The plan shall be consistent with other Community legislative and policy instruments. Account should be taken of unavoidable increases in emissions resulting from new legislative requirements.

The plan shall not discriminate between companies or sectors in such a way as to unduly favour certain undertakings or activities in accordance with the requirements of the Treaty, in particular Articles 87 and 88 thereof.

The plan shall contain information on the manner in which new entrants will be able to begin participating in the Community scheme in the Member State concerned.

The plan may accommodate early action and shall contain information on the manner in which early action is taken into account. Benchmarks derived from reference documents concerning the best available technologies may be employed by Member States in developing their National Allocation Plans, and these benchmarks can incorporate an element of accommodating early action.

The plan shall contain information on the manner in which clean technology, including energy efficient technologies, are taken into account.

The plan shall include provisions for comments to be expressed by the public, and contain information on the arrangements by which due account will be taken of these comments before a decision on the allocation of allowances is taken.

The plan shall contain a list of the installations covered by this Directive with the quantities of allowances intended to be allocated to each.

The plan may contain information on the manner in which the existence of competition from countries or entities outside the Union will be taken into account.

It is also unclear how the Commission is going to apply its allocation check. The original Commission Proposal required that the national allocation plan set out the total quantity of allowances that a Member State intends to allocate and how it proposes to allocate them. While this language was retained in the EATD, a new paragraph 10 was added to Annex III, which requires that the plan 'shall contain a list of installations covered by this Directive with the quantities of allowances intended to be allocated to each'. It is unclear whether the Commission's allocation check will be limited to how Member States propose to allocate, or whether it will apply additional state aid checks on how many allowances are allocated to individual installations. Although the emphasis of the allocation check is likely to be on the first, the Commission is unlikely to be willing to exclude the possibility to scrutinize allocation on an installation by installation basis in order to test the compatibility of that particular allocation with the Community's state aid rules. The allocation check will also bring interesting capacity issues for the Commission. 15 Member States will have to submit their allocation plans by 31 March 2004, and another 10 Accession States will follow in May 2004. This means that the Commission will have to check at least 25 allocation plans, between 1 April and 31 July 2004, a Herculean task. It is therefore not unlikely that the Commission's checks may not be as in-depth as they could be, given more time, and that it will allow itself to be assisted in providing its comments by the information it obtains from stakeholders and other Member States.

Apart from the issues set out above, the EATD is also likely to raise important questions in relation to the implementation of the EATD's allocation provisions at the national level. Most Member States have elaborate administrative appeal procedures in place, which allow entities and stakeholders to challenge specific types of government decisions. Should Member States also allow appeal against allocation decisions, or is excluding such an appeal even possible? If an appeal is allowed, challenges to the allocation decisions cannot be prevented. This may cause very long delays in the allocation process which in turn may delay the start of the regime or bring the Member State in violation with the requirements of the Directive. Challenges may be prevented through a wide acceptability of the allocation methods, but these may in turn take a long time to negotiate. Constitutional law could prevent Member States from excluding the allocation decision from appeal. To solve this, Member States may choose to institute special fast-track appeal procedures.

The allocation provisions in the EATD are a brave attempt to create an allocation framework which addresses the various issues raised above. Perhaps the biggest advantage of the EATD's approach is that it has managed to bring the main components of the allocation debate to the Member State level. By doing this it has avoided lengthy allocation discussions at the EU level and has set time limits for the debate within the Member States, thereby limiting the impact of the allocation discussion on the entry into force of the trading regime. With the EATD being the first instrument of its kind and the scale of its application, there are, however, still a number of important allocation issues that will need to be resolved in the implementation of the EATD.

3.5 Enforcement

The EATD pushes the prescription of enforcement measures to the limits of what Member States usually find acceptable in an environmental Directive. It can, however, be argued that the enforcement issue is still insufficiently addressed, and that the package of measures currently on the table to address the implementation of the Kyoto Protocol and the EU trading regime leaves a number of important gaps.

Four categories of compliance issues can be identified in the ET proposal. The first three of these relate to the behaviour of the installations and their operators:

- Compliance with the monitoring, reporting and verification requirements;
- Compliance with the requirement to surrender allowances for each tonne of emissions controlled under the regime;
- Compliance with other, more general requirements set out in national legislation implementing the ET proposal including the duty to undertake transfers in accordance with these provisions and the prohibition to commit fraud.

The fourth relates to the behaviour of the Member State and concerns:

- The correct implementation of the Directive.

In relation to the first compliance issue, Article 15 EATD requires Member States to 'ensure that an operator whose report has not been verified as satisfactory in accordance with the criteria set out in Annex V by 31 March each year for emissions during the preceding year cannot make further transfers of allowances until a report from that operator has been verified as satisfactory'. This sanction bans an operator from selling his allowances until the installation's emissions have been verified in accordance with the requirements of the Directive. The provision does not preclude the operator from buying allowances.

It is now generally accepted that Directives can include a provision which requires that 'Member States shall determine the sanctions applicable to breaches of the national provisions' adopted pursuant to the Directive. Since the European Court of Justice's judgement on a case involving the enforcement of Community Funds in Greece in 1989[65], it is also general

[65] ECJ 21 September 1989, Case 68/88, Commission v Greece (community funds).

practice to include language in Community Directives requiring that 'sanctions determined must be effective, proportionate and dissuasive'. More recent Community Directives have also included language requiring Member States to 'take all necessary measures' for the implementation or application of the sanctions as well as the requirement for Member States to notify their provisions on sanctions, as well as any amendments to them, by a certain date. Article 16(1) follows this trend by requiring Member States to lay down the rules on penalties applicable to infringements of the national legislation implementing the Directive and ensuring that these rules are implemented. It also requires that the penalties must be effective, proportionate and dissuasive and Member States must notify these provisions to the Commission by 31 December 2003. This standard provision serves as a 'catch-all' for different types of non-compliance with the national legislation implementing the EATD, as required to address the 3rd compliance issue above.

This standard requirement is, however, not sufficiently strong to address non-compliance with the requirement to surrender allowances for each tonne of emissions controlled under the regime. This is where the EATD pushes the boundaries of what Member States have in the past found acceptable in EC environmental legislation. It requires Member States to impose a minimum financial penalty on operators that do not hold sufficient allowances to cover the emissions from installations under their control. Experience in the US has shown that a high penalty for excess emissions is important to ensure compliance with the trading regime, and through that its success[66]. Although Community environmental law has in the past never prescribed specific enforcement measures, it is important that all Member States apply a minimum-level of penalties. Doing otherwise would result in companies shifting their allowances away from the Member State in which the compliance penalty was low, allowing their operators in that Member State to be in non-compliance and pay the low penalty, rather than buying more expensive allowances. This could in turn have serious consequences for that Member State's compliance with its Kyoto targets.

The EATD's unique penalty regime is set out in Article 16 paragraphs (2), (3) and (4). For the period between 2005 and 2008 it sets a minimum penalty rate of 40 Euro for each tonne that an installation is in non-compliance with its obligation to cover its emissions by sufficient allowances. This minimum penalty rate is increased to 100 Euro per tonne from 2008 onwards. On top of the requirement to pay the penalty, operators of installations have to compensate for their excess emissions in the following compliance period. The Directive also contains the requirement for Member States to publish the names of the operators who are in breach

[66] Schwarze and Zapfel (2000), 288–289.

of their obligation to surrender allowances equal to the total emissions from their installation.

The final version of Article 16 differs on a number of points from the original Commission proposal, but in view of the innovative nature of these provisions these differences are marginal. The proposal provided for a penalty of 50 Euro or twice the average market price during the preceding year, from 2005 to 2007, and 100 Euro or twice the average market price of the preceding year, from 2008 onwards. Industry and a number of Member States objected to the link between the penalty and the average market price, which they argued would cause too much uncertainty on the risks of non-compliance[67]. A number of Member States felt that a penalty of 50 Euro during the trial period from 2005 to 2007 was too high. During the negotiations on the Common Position the link to the market price was removed and the penalty from 2005 to 2007 was lowered to 40 Euro. The requirement for the operator to surrender sufficient allowances to compensate for the excess emissions, however, remains. The original proposal furthermore contained the requirement for Member States to publish the names of operators who are in non-compliance with their obligations in general, which was limited to cases of non-compliance with the obligation to surrender sufficient allowances to compensate for excess emissions in the final version of the Directive.

The fourth compliance issue, the correct implementation of the Directive by the Member States, is where the EATD may yet show its greatest weakness. For the operation of the EU trading regime it is vital that the Directive is fully implemented and on time. Experience with Member State implementation of Community environmental Directives has, however, shown that the track-record of most Member States continues to be deplorable[68]. The Commission's current enforcement tools, set out in Articles 226 and 228 of the EC Treaty, have proven insufficient and especially too time-consuming to provide for the effective enforcement tool necessary to ensure the correct operation of a Community-wide greenhouse gas emission allowance trading regime[69]. The final version of the EATD

[67] See UNICE (2002).

[68] The latest Commission Annual Survey on the implementation and enforcement of Community environmental law starts with the conclusion that: 'The last five years have seen a growing difficulty in the timely and correct implementation as well as proper practical application of EC environmental legislation. This is reflected in the number of complaints received and infringement cases opened by the Commission every year. As in the earlier years, in 2002 the environment sector covered over one third of all infringement cases investigated by the Commission. The Commission brought 65 cases against Member States before the Court of Justice and issued 137 reasoned opinions on the basis of Article 226 of the EC Treaty'. See European Commission (2003b).

[69] See also Krämer (2001), n. 18 above, 38–40.

leaves the various implementation deadlines that were included in the Commission's proposal untouched. This means that Member States have to implement the EATD into domestic legislation by 31 December 2003 at the latest. The usual time given to Member States for the implementation of Community environmental legislation is two years. The time for the implementation of the EATD is likely to be little over 3 months after its formal adoption. In this time Member States have to adopt the domestic legislation needed to make the EATD operational, a deadline which is unlikely to be met by most Member States, especially if this legislation has to be approved by their national parliament. This situation is further complicated by the fact that all Accession States will be required to have implemented the Directive on the moment of their accession on 1 May 2004.

One option which the Commission could have chosen was to require Member States to impose caps on installations in their territory, but make the participation of installations in a Member State conditional on the Member State's compliance with a set of eligibility criteria. This approach is similar to the approach chosen in the rules elaborated under the Kyoto Protocol. The eligibility of installations in a Member State to participate in the trading regime would be determined through a periodic assessment of whether each Member State is fulfilling its obligations under the Directive and not undermining the environmental integrity or reducing the efficiency of the Community scheme. The advantage of this approach is that it would have separated the requirement to cap emissions from the operation of the trading regime. This approach not only maintains the integrity of the regime's objective, the cap, but also gives industry a clear incentive to pressure government to ensure that the necessary rules for the operation of the trading regime are in place – the trading component is after all what makes the emission caps palatable for industry.

Now the EATD is adopted, its implementation is likely to become a major issue. Implementation will pose challenges for the Member States and the Commission and will be decisive for the regime's success.

3.6 Relation with the IPPC Directive

Defining the relation between the EATD and the IPPC Directive was one of the Commission's key challenges in its development of the ET proposal. An underlying strategic issue which had to be resolved was whether emission trading was to be introduced as a separate proposal or as an amendment to the IPPC Directive. Although the latter could, for reasons of legislative consistency, have been preferable, this was not achievable in practice. Introducing the trading proposal by amending the IPPC Directive could have undermined the implementation process of the IPPC Directive. Introducing emissions trading into this Directive would have required a significant amendment of a Directive of which a considerable part has not yet entered into force[70]. By doing so, it could have intervened in the implementation process that is currently ongoing in most Member States. Instead, the Commission chose to design a parallel instrument which very closely follows the approach and language of the IPPC Directive.

Because of this strategic choice, the Commission had to deal in its ET proposal with the relation between the ET proposal and the IPPC Directive. Although the ET proposal addresses this relation in a number of places, industry and Member States raised questions on the Commission's approach during the negotiations on the Directive. In January 2002, in reaction to these questions, the Commission's Environment Directorate General released a non-paper on the synergies between its ET proposal and the IPPC Directive[71]. The Commission's non-paper clarifies a number of these questions, specifically in relation to the overlaps in the use of terminology, coverage, and the permitting procedures, as well as on the relation between the energy efficiency requirements and the emission limits in the IPPC Directive and in the ET proposal.

The close link between the two instruments is most relevant in relation to three issues: the overlap in coverage of installations, the overlap in coverage of greenhouse gas emissions and the relation between the permitting procedures.

The first issue, the overlap in coverage of installations, has already been discussed above. Almost all sectors covered by the EATD are also covered by the IPPC Directive. The only exception relates to combustion installations, where the ET proposal lowers the threshold for inclusion in the regime from a rated thermal input of 50 MW, the threshold used in Annex I of the IPPC Directive, to 20 MW. The overlap between the sectors covered by both instruments allows the EATD to address the largest industrial sources of CO_2 emissions and creates additional flexibility for reducing greenhouse gas emissions which, as the Commission argues, were already covered by the IPPC Directive.

The second overlap between the EATD and the IPPC Directive relates to the extent to which greenhouse gas emissions are already covered under the IPPC Directive. The IPPC Directive addresses greenhouse gas emissions in a number of places. Article 3(d) imposes a general obligation upon Member States to ensure that energy in the installations covered by the Directive is used efficiently. Article 6 requires that the application for an IPPC permit includes a description of 'the raw and auxiliary materials, other substances

[70] Member States only have to apply the national legislation implementing the IPPC Directive to existing installations from October 2007 onwards.

[71] European Commission (2002).

and energy used in or generated by the installation' as well as 'the nature and quantities of foreseeable emissions from the installation into each medium'. Article 9(3) requires that the IPPC permit includes 'emission limit values for pollutants, in particular, those listed in Annex III'. The IPPC Directive defines in Article 2(2) 'pollution' as the 'direct or indirect introduction as a result of human activity, of substances, vibrations, heat or noise into the air, water or land which may be harmful to human health or the quality of the environment, result in damage to material property, or impair or interfere with amenities and other legitimate uses of the environment'. In view of the contribution of greenhouse gases to global warming it can be argued that the impacts of global warming fall within the definition of pollution in the IPPC Directive. While Annex III of the IPPC Directive does not specifically list any of the greenhouse gases, it does include 'volatile organic compounds', which covers CH_4, 'oxides of nitrogen', which covers N_2O and 'fluoride compounds' which covers HFCs, PFCs and SF_6. Pollutants not included in the indicative list should also be subject to emission standards if they are 'likely to be emitted from the installation concerned in significant quantities'. It can thus be argued that the IPPC Directive requires that significant emissions of greenhouse gases must be covered by an emission limit value or by 'equivalent parameters or technical measures'. Support for this argument can furthermore be found in the Directive's definition of 'best available techniques' (BAT). The IPPC Directive requires that Member States use BAT to determine the emission limit values and the equivalent parameters and technical measures. Annex IV of the IPPC Directive lists among the considerations to be taken into account in determining BAT 'the consumption and nature of raw materials ... used in the process and their energy efficiency'. Energy efficiency criteria are indeed included in a number of the BAT reference documents (BREFs)[72]. The final, important, argument that the IPPC Directive covers greenhouse gas emissions is that all six gases are included as 'pollutants' in the Commission's decision on the European Pollutant Emission Register (EPER), elaborated under to Article 15(3) of the IPPC Directive[73]. In its non-paper the Commission confirms this interpretation of the IPPC Directive[74].

The Commission could have opted to allow emissions trading to take place within the limits of the IPPC Directive. Installations could then have traded beyond the minimum emission limit values set in their IPPC permit. This is in fact the approach that is chosen explicitly by the Dutch NO_x

[72] BREF documents adopted so far can be found at: http://eippcb.jrc.es/pages/FActivities.htm.

[73] Commission Decision 2000/479/EC on the implementation of a European pollutant emission register (EPER) according to Article 15 of Council Directive 96/61/EC concerning integrated pollution prevention and control (IPPC), [2000] OJ L192/36.

[74] European Commission (2002), n. 71 above.

trading regime[75] and implicitly by the UK greenhouse gas trading regime. One of the reasons that the Commission approved the UK's incentive payments to the sectors participating in the regime was that it requires those sources to achieve a level of protection higher than the Community standards, including those based on the IPPC Directive[76]. Different application of the discretion given to national authorities in the IPPC Directive could, however, have disadvantaged certain sectors or installations. This would in particular be the case for situations in which Member State authorities would make full use of the possibilities offered by the IPPC Directive to apply strict emission limit values for greenhouse gas emissions. Perhaps more importantly, trading within the IPPC framework could have significantly reduced the scope of trading regime, which could have defied the objective of introducing the instrument in the first place[77].

The EATD therefore excludes emission limit values for greenhouse gas emissions covered under its Annex I from the scope of the IPPC permit. Article 26 EATD amends Article 9(3) of the IPPC Directive by adding the subparagraphs set out in Box 2 below.

This provision guarantees that only the emission limit values for direct emission of greenhouse gases are removed from the scope of the IPPC Directive, unless local environmental quality standards require that a minimum standard is set. By doing so it only intervenes to the minimum extent possible in the scope of the IPPC permit, thus leaving its integrated approach as much as possible intact. It also guarantees that emissions from other substances that are not covered under the EATD remain covered by the IPPC regime.

In its original proposal, the Commission had opted to explicitly retain the requirement in the IPPC Directive to set targets related to energy efficiency for installations covered under the ET Proposal. This requirement, included in Article 2(2) of the ET Proposal, set compliance with energy efficiency requirements as a clear baseline below which trading cannot take place. In its non-paper the Commission explained that it did not expect this to be problematic, and that both the UK and Danish trading regime apply the same minimum baseline[78]. In addition it can be argued that BREF documents that do contain energy efficiency standards mostly do not set those standards far beyond general technology levels. Combined with the significant margin of interpretation afforded to national authorities in the application of BAT, this

75 See FIELD and IEEP (2002).

76 Commission decision on State aid No N 416/2001 – United Kingdom Emission Trading Scheme. Since the IPPC Directive does not come into force for new installation until October 2007 there is significantly more leeway.

77 For a discussion on the use of emissions trading within the boundaries of the IPPC Directive see n. 75 above.

78 European Commission (2002), n. 71 above.

requirement does indeed seem to set a sensible minimum level, which would prevent emissions trading to allow installations to operate in a very energy-inefficient manner but would not unreasonably restrict the possibilities for trading. Opposition by the Member States to maintaining the energy efficiency requirement as a baseline for trading, however, led to the requirement being dropped in the final version of the EATD. Article 2(2) EATD now only reads 'this Directive shall apply without prejudice to any requirements pursuant' to the IPPC Directive. The amendment to the IPPC Directive set out in Box 2 now explicitly allows Member States not to impose requirements relating to energy efficiency.

Box 2: Article 26 EATD:

Amendment of the IPPC Directive (96/61/EC), Article 9(3)

In Article 9(3) of Directive 96/61/EC the following subparagraphs shall be added:

"Where emissions of a greenhouse gas from an installation are specified in Annex I to Directive 2003/ /EC of the European Parliament and of the Council of ... establishing a scheme for greenhouse gas emission allowance trading within the Community and amending Council Directive 96/61/EC in relation to an activity carried out in that installation, the permit shall not include an emission limit value for direct emissions of that gas unless it is necessary to ensure that no significant local pollution is caused.

For activities listed in Annex I to Directive 2003/ /EC, Member States may choose not to impose requirements relating to energy efficiency in respect of combustion units or other units emitting carbon dioxide on the site.

Where necessary, the competent authorities shall amend the permit as appropriate.

The three preceding subparagraphs shall not apply to installations temporarily excluded from the scheme for greenhouse gas emission allowance trading within the Community in accordance with Article 27 of Directive 2003/ /EC.

The third area, the relation between the two permitting procedures, is explicitly addressed in Article 8 of the EATD. This Article states that:

> Member States shall take the necessary measures to ensure that, where installations carry out activities that are included in Annex I to Directive 96/61/EC, the conditions of, and procedure for, the issue of a greenhouse gas emissions permit are coordinated with those for the permit provided for in that Directive. The requirements of Articles 5, 6 and 7 of this Directive may be integrated into the procedures provided for in Directive 96/61/EC.

In its non-paper the Commission stressed that this allows Member States to combine the permitting procedures for the ET permit with the IPPC permit. Although the Commission expects that Member States will make use of this, it stresses that this is not obligatory[79]. In view of the similarity and complementarity of the two permitting procedures, it is even likely that Member States will implement the EATD in the same legal framework they use to implement the IPPC Directive. By doing so, the two procedures could be fully integrated and any disadvantages created by the Commission's opting for a separate proposal rather than the amendment of the IPPC Directive would be fully removed. It is even imaginable that Member States choose to merge the ET permit and the IPPC permit into a single permit. This is not precluded by the EATD and would indeed further promote the integration between the two regimes.

3.7 Other Issues

The EATD furthermore lays down the basic requirements for the monitoring and verification of greenhouse gas emissions by the installations covered under its Annex I. It also sets out the public participation requirements and basic requirements for a Member States' registry to track the trade in allowances, as well as various reporting requirements for Member States and the Commission. The proposal allows the linking of the EU trading regime with regimes in non-Member States on the basis of bilateral agreements between the Community and these other States. It furthermore delegates a wide range of tasks to a Committee under a comitology procedure. These tasks include the elaboration of more detailed monitoring and reporting guidelines, a regulation on the standardization of the national registries, criteria for mutual recognition of allowances from non-Member State regimes, the consideration of national allocation plans, the revision of the allocation criteria in Annex III and establishing harmonized allocation method to be used from 1 January 2008 onwards.

4 The EATD and the Project-based Mechanisms

The Commission has been criticised for not including an explicit link to the project-based mechanisms (JI and CDM) in its original ET proposal. The original proposal was designed to operate independently from the Kyoto Protocol's flexible mechanisms. The reason for this was that when the Commission issued the proposal, the negotiations on the Marrakech

---

[79] Ibid.

Accords, spelling out the rules for the functioning of the flexible mechanisms, had not been finalized. These rules were only adopted in Marrakech in November 2002, a month after the Commission had issued its proposal[80]. The Community used the ET proposal, together with the proposal for a ratification Decision and the ECCP communication[81], as an important political signal before the commencement of the negotiations in Marrakech to show the world that the European Union was serious about its intentions to ratify and implement the Kyoto Protocol. Waiting to issue the proposal for the EATD until after the conclusion of the Marrakech accords would have significantly weakened this important signal.

But perhaps the most important reason for not including a direct link in the EATD was that the EATD was in first instance a domestic implementation measure. It was felt that creating a direct link in the original proposal could complicate and delay the adoption of the EATD because of the differing views between stakeholders on the desirability of linking, but also because of the uncertainty on the future of the Kyoto Protocol at the time of the elaboration of the proposal. The environmental NGOs, from the start of the emissions trading debate, strongly objected to any links with the Kyoto Protocol's flexible mechanisms[82]. Industry on the other hand has traditionally been a strong proponent of a direct link with the Protocol's mechanisms[83]. The European Parliament has throughout the negotiations maintained that any link with the Kyoto mechanisms would only be acceptable after 2008, as long as the projects do not include carbon sinks or sources of energy that use nuclear power. Although the Commission did find a link with the Kyoto flexible mechanisms to be desirable, such link was to be 'subject to the satisfactory resolution of outstanding issues regarding their environmental integrity'. The Commission therefore preferred to discuss the link with the Kyoto mechanisms in the context of the elaboration and adoption of a separate legislative instrument[84]. This proposal (the linking Directive) was published by the Commission on 23 July 2003[85].

This section will give a background to the linking Directive. It will first discuss the linking provisions in the EATD and then provide a more in-depth analysis of its contents.

[80] These texts are part of the Marrakech Accords, adopted in November 2002, and can be found in FCCC/CP/2001/13/Add.2.

[81] N. 24 and n. 4 above, as well as the Communication from the Commission on the Implementation of the First Phase of the European Climate Change Programme, COM(2001) 580, 23 Oct. 2001.

[82] See n. 28 above. More recently CAN Europe (2003a).

[83] See n. 29 above. More recently UNICE (2003b).

[84] Explanatory memorandum, n. 4 above, 17.

[85] See n. 6 above.

4.1 Linking Provisions in the EATD

Although the original Commission Proposal did not include the flexible mechanisms, it did refer to linking the EU scheme to these mechanisms in a number of places. The first reference was included in the review clause in the proposal's Article 26 (Article 30 in final text), which included 'the use of credits from project mechanisms' as one of the issues to be considered in the review. The second reference was included in the proposal's Article 24 (Article 25 in the final text) which allowed the Community to conclude agreements with third countries to link with their trading regimes.

With both industry and a number of Member States strongly advocating a direct link with the Kyoto mechanisms, a number of concessions had to be made to show a stronger commitment to use the mechanisms in the European trading regime. As a result of the discussions two new paragraphs 17 and 18 were added to the preamble. These paragraphs stress the advantages of linking the Community scheme with schemes in third countries and the importance of the project-based mechanisms CDM and JI in increasing the cost-effectiveness of the Community scheme. Importantly, the Council also changed the language of the linking provision (Article 24 in the Proposal, Article 25 in the final text) from 'the Community may conclude agreements with third countries' into 'agreements should be concluded with third countries'. The Council also added a third new paragraph to the review provision (previously Article 26, now Article 30), stating that:

> Linking the project-based mechanisms, including Joint Implementation (JI) and the Clean Development Mechanisms (CDM), with the Community scheme is desirable and important to achieve the goal of both reducing global greenhouse gas emissions and increasing the cost-effective functioning of the Community scheme. Therefore, the emission credits from the project-based mechanisms will be recognised for their use in this scheme subject to provisions adopted by the European Parliament and the Council on a proposal from the Commission, which should apply in parallel with the Community scheme in 2005.

In second reading the European Parliament insisted that a new sentence be added to this paragraph, stating that 'the use of the mechanisms shall be supplemental to domestic action, in accordance with the relevant provisions of the Kyoto Protocol and the Marrakech Accords'. Parliament also succeeded in significantly altering the language of preambular paragraph 18[86]. With the Parliament's amendments the question on whether credits

[86] In the Council's common position this paragraph read: 'The recognition of credits from project-based mechanism for fulfilling obligations under this Directive as from 2005 will increase the cost-effectiveness of achieving reductions of global greenhouse gas emissions and will be provided for by a Directive for linking project-based mechanisms including Joint Implementation (JI) and the Clean Development Mechanism (CDM) within the Community Scheme'. The final text of this paragraph, after the second reading compromise, now reads:

from the project-based mechanisms may come into the Community scheme from 2005 or from 2008 onwards was left to be decided in the Directive on the project mechanisms.

4.2 CDM and JI in the EU Trading Regime: Opportunities and Threats

Table 1 gives a short overview of the different types of flexible mechanism and the units created by those mechanisms under the Kyoto Protocol.

Table 1: Mechanisms and types of credits under the Kyoto Protocol

Mechanism	Unit of Trade
International Emissions Trading (IET) (Article 17 Kyoto Protocol)	Assigned Amount Units (AAUs) Emission Reduction Units (ERUs)* Certified Emission Reductions (CERs)* Removal Units (RMUs)*
Joint Implementation (JI) (Article 6 Kyoto Protocol)	Emission Reduction Units (ERUs)
Clean Development Mechanism (CDM) (Article 12 Kyoto Protocol)	Certified Emission Reductions (CERs)*

*CERs are created through a CDM project; RMUs are created as a result of eligible Article 3.3 and 3.4 KP forestry activities. ERUs are created by converting AAUs or RMUs through a JI project. CERs, ERUs and RMUs are fully fungible and can be further traded under Article 17, where they keep their original name. The name of CDM credits resulting from eligible forestry activities has not yet been decided upon and is part of the ongoing negotiations on the rules for forestry CDM projects, which are to be concluded by COP-9 in December 2003.

Before discussing the threats and opportunities for allowing these units to be used to comply with the obligations under the EU trading regime, it should first be pointed that an *indirect* link between the EATD and these project-based mechanisms already exists. As already discussed above, the

---

cont.
'Project-based mechanisms including Joint Implementation (JI) and the Clean Development Mechanism (CDM) are important to achieve the goals of both reducing global greenhouse gas emissions and increasing the cost-effective functioning of the Community scheme. In accordance with the relevant provisions of the Kyoto Protocol and Marrakech Accords, the use of the mechanisms should be supplemental to domestic action and will thus constitute a significant element of the effort made.'

EATD does not determine the size of the cap for the sectors covered by the proposal. The size of the cap is to be determined by the Member States, depending on their GHG reduction commitments and the allocation of the responsibility to achieve these commitments across the different sources and sinks in their territory, including the sources covered by the EATD. During 2008–2012 Member States can use CERs, ERUs, RMUs and AAUs to meet their commitments under the Protocol and the EU 'Bubble' agreement. The Member State could buy such units[87] and provide larger allocations to entities in the trading programme, subject to the allocation rules and the Member State and Community review of proposed allocations for 2008–2012. This will affect the stringency of the cap, in particular the impact of the cap on the amount of domestic reductions that will be achieved under the EATD.

A proposal for a Directive on project-based mechanisms is to envisage a *direct* link between project-based mechanisms and the EATD. The link would allow participating entities in domestic emissions trading programmes under the EATD to use credits generated by the project-based mechanisms to offset reduction obligations under the EATD.

There are a number of reasons for allowing entities to directly use project-based mechanisms for compliance with their targets under the EATD. The most frequently used reason is that the inclusion of project-based mechanisms can reduce the compliance costs for sectors covered under the EATD by broadening the range of opportunities to reduce emissions elsewhere at lower costs. A second reason is that the use of project-based mechanisms can engage sources and sinks not covered by the emissions trading Directive in implementing cost-effective reduction options. Including project-based mechanisms can bring a net benefit to project developers who are not under an obligation to limit their emissions pursuant to the EATD, by giving them a financial incentive to reduce their emissions if those emissions can be reduced at a lower cost than the market price for allowances under the EATD. The use of the project-based mechanisms within the EU could thus be seen as a piece-meal and ad-hoc way to extend of the coverage of the trading regime to gases and sectors not covered under the EATD, starting where the coverage of the EATD ends. Including CDM and JI in the EATD will also provide an important boost to the use of these instruments. Before the Commission proposed the linking Directive there was only very little interest from companies in the use of CDM and JI. The reason for this was that these companies had little incentive to use these mechanisms since they were not subject to limitations

[87] For instance through initiatives such at the Dutch ERUPT and CERUPT programmes, see above. For more information on the ERUPT and CERUPT programmes: http://www.carboncredits.nl

on their emissions for which they could use CDM and JI credits to contribute to their compliance with such limitations. The combination of emission caps under the EATD with the possibility to use CDM and JI will create an important incentive for companies to invest in these instruments and thus kick-start the CDM and JI market.

There are, however, also a number of important arguments that caution against creating a direct link. A first argument is that the EATD is a domestic implementation measure and should in the first place be used to reduce GHG emissions *inside* the EU, rather than embarking on a path which will make the EU's ability to achieve its emission reduction target dependent on buying sufficient credits from elsewhere. Linked to this argument are fears that opening the EU trading regime for CDM and JI credits may flood the EU market with these credits and thus avoid the need for intra-EU reductions. While the Kyoto Protocol allows private entity trading, it only does so under the responsibility of Parties and to the extent that Parties wish to allow this. The EU trading regime would be the first and hitherto only regime that would allow the use of these credits at the domestic level, and indeed creates an incentive for doing so by imposing a cap on the sources included in the regime. Since the regime only covers a relatively small number of sources (recent estimates are around 15,000 installations within the EU[88]) and will initially concentrate on CO_2 only, exposing this regime to the full global CDM and JI market could, however, be disproportionate and have grave impacts on its functioning. A further argument is that using JI to expand the coverage of the EU trading regime within the EU does not actually lead to reductions, as the credits earned by those investments will be used to avoid reductions in the sectors covered by the regime. Instead this broadening of coverage can provide an impetus for sectors to avoid other policies and measures limiting their emissions or avoid being brought directly into the EU trading regime. An important argument used by environmental NGOs against 'linking' are the doubts that some of the CDM and JI projects that are currently in the pipeline do not provide sustainable development benefits or represent real reductions compared to what would have happened otherwise (i.e. these projects are not 'additional'). Using doubtful credits inside the EU trading regime would undermine its environmental credibility.

The European Commission has attempted to balance these various threats and opportunities in its proposed linking Directive. The proposal has however drawn sharp criticism, not only from the environmental NGOs[89], but also from industry[90]. Environmental NGOs, however, find themselves

[88] See n. 34 above.
[89] See the CAN Europe (2003a), n. 82 above as well as CAN Europe (2003b).
[90] See the UNICE (2003b), n. 83 above.

isolated in their opposition to any link with the project-based mechanisms. With a proposal for a linking Directive already on the table and not only the Commission, but also the European Parliament, the Council and industry in favour of establishing a link with the project-based mechanisms, the question is no longer *whether* such link will be created but *how* such link will be created.

A remark should be made about the widely heard criticism, in particular from industry and a number of Member States, that the proposed linking Directive may be restricting access to the Kyoto Mechanisms. It can be argued that this criticism is unfounded; the linking Directive will do exactly the opposite. As already discussed above, the proposal for the linking Directive, once in force, will for the first time allows legal entities to use the project-based mechanisms to comply with their obligations. Rather than limiting the market, this will create a whole new market for CDM and JI credits and provide a true incentive to the use of those mechanisms, even if the EU decides, for whatever reasons, not to allow all Kyoto units to be used under the EATD.

4.3 The Proposal for a Linking Directive

Soon after the adoption of the EATD proposal and the success of the Marrakech negotiation in November 2001, the Commission placed the elaboration of a proposal for a Directive to linking the project-based mechanisms to the EATD high on its agenda. In January 2002 the Commission revived ECCP Working Group I. The first meeting of this new working Group I took place on 27 February 2002, which was followed by three further meetings[91]. These meetings were used to exchange ideas among stakeholders on a number of issues related to the possibility of linking project-based mechanisms with the trading regime set up under its ET proposal. Important issues which were discussed included the timing of the inclusion of the various project-based mechanisms, in particular whether any of these mechanisms can be linked before 2008, as well as the question on which types of projects can be linked to the trading regime. In relation to the latter, environmental NGOs argued that should any projects be linked at all, which they oppose, these should be limited to specific types of projects that conform to high environmental standards[92]. In particular projects using carbon sinks or nuclear energy should be excluded. Industry continued to take the view that the Kyoto Mechanisms should be incorporated 'as is' in the EATD. The conclusions of the group, adopted at its 4th meeting in

[91] The terms of reference, agenda, minutes and conclusions of these meetings are published on the internet at: http://europa.eu.int/comm/environment/climat/ji_cdm.htm.
[92] See, *inter alia*, CAN Europe (2001), n. 28 above.

September 2002, called for 'the early adoption of legislation regarding the recognition of project credits' 'as a matter of particular priority' and stated that 'the Commission should aim to make its proposal for a Directive linking JI/CDM credits with the EU emissions trading scheme early in 2003' and that 'the Council and the European Parliament should aim at adopting this legislation so as to allow its implementation as from the commencement date of the EU emissions trading scheme'[93].

The Commission had planned the adoption of its proposal early in the first half of 2003, but the proposal was delayed in the adoption process within the Commission. One of the reasons for the delay was the disagreement between DG Environment and other Commission services on the need to limit the influx from credits from project-based mechanisms into the EU trading regime. This disagreement will be further discussed below.

The Commission finally issued its proposal on 23 July 2003[94]. The text of the Directive itself is little over 5 pages and takes the form of an amendment to the EATD Directive, thus fully integrating the use of the CDM and JI into the EATD Directive. Although the proposal is very short, the wide spectrum of views on whether and how CDM and JI credits can be used under the EU trading regime, has already caused lively discussions in Council and Parliament. The following paragraphs will provide a background to the five key elements of the proposal in the form of questions that are playing a key role in the ongoing negotiations:

- How should a link be established?
- Which projects should a link be established with?
- When should a link be established?
- Should there be any quantitative limits on a link?
- How can double counting be avoided?

4.4 How to Link

There are a number of important differences between the operation of the EU trading regime and the functioning of the Kyoto Mechanisms. The most salient differences are that units traded under the EU trading regime are fully 'fungible' (all units are fully interchangeable and thus have the same value)

---

[93] Environmental NGOs, in a letter to the Commission, disagreed with the representative nature of these conclusions, objecting in particular to this paragraph. ECCP Working Group on JI/CDM, Conclusions, 15 November 2002, available at: http://europa.eu.int/comm/environment/climat/jicdm/jicdm_final_conclusions.pdf.

[94] N. 6 above.

and that Member States are required to accept all EU allowances for compliance, irrespective of the company that the allowance was originally allocated to and the country it was originally allocated by. All EU allowances are furthermore treated the same under the EATD's banking rules. Under the Kyoto Protocol trading regime there are differences between the various credits (CERs, ERUs, AAUs and RMUs) with regard to their generation, banking and use. Parties under the Kyoto Protocol are furthermore free to choose whether they wish to accept credits from another Party or project.

The question thus arises how the differences between these regimes will be reconciled. The features for linking the two regimes are built into the 'entry point' of the Kyoto units into the EU trading regime in the linking proposal. The linking proposal provides for a new Article 11(bis) to be included in the EATD. This article allows for the conversion of CERs and ERUs from CDM and JI projects for use in the Community scheme. This conversion is done by a Member State, who *may* issue one new allowance in exchange for one CER or ERU. The use of the word 'may' signifies that Member States continue to keep the freedom to impose other criteria for the conversion of Kyoto credits into allowances. By converting the Kyoto units into EU allowances, the EU appears to maintain the full fungibility of its units. The advantage of this is that it will give business more certainty about the possibilities to use, and therefore the value of, these units.

The linking proposal, however, leaves many questions unanswered. It is for instance not clear what happens when the newly created allowance is traded between installations in different EU Member States. Preambular paragraph 9 of the EATD states that the 'transfer of allowances to another Member State will involve corresponding adjustments of assigned amount units under the Kyoto Protocol'. But it is does not say whether there is also a corresponding transfer of ERUs or CERs under the Protocol. If this would be the case, there could be implications resulting from the application of the Kyoto Protocol's rules on the banking, generation, transfer and use of those credits, which could in turn have consequences for the fungibility of the EU allowances. On the other hand this preambular paragraph could also be seen to say that once an ERU or CER has been converted into an EU allowance, any subsequent inter-Member State trade will be 'shadowed' by an AAU, rather than the original ERU or CER that the allowance was converted from. The proposal also does not mention what happens once an EU allowance that originated from an ERU or CER is sold outside the States participating in the trading regime. The proposal leaves these questions to be resolved in the forthcoming Registries Regulation, which is to be released and adopted by the Commission on the basis of Article 19 EATD in the first half of 2004.

4.5 What Projects to Link With?

A key question is whether the linking proposal should allow all project credits generated under the Kyoto Protocol to be included in the EATD, or whether these should be limited to a subset of these projects. It should be noted that this question would arise even in the absence of an EATD as it would be for the EU and Member States to decide whether to give domestic actors access to international project-based mechanisms to offset domestic obligations, and if so, on what terms.

During the negotiations on the rules on the project-based mechanisms, the EU put forward a number of proposals to guarantee the environmental integrity of the mechanisms. These proposals included a definition of 'supplementarity', excluding sinks credits in the CDM, prohibiting nuclear energy projects, as well as opposing Article 3.4 forestry credits in the first commitment period[95]. To achieve international agreement and ensure the survival of the Kyoto Protocol, the EU compromised significantly on a number of its positions. It has not been decided whether this means that the EU has accepted this compromise also for its own application of the Kyoto rules.

There are a number of 'environmental integrity' arguments for Member States and stakeholders to advocate a selective inclusion of project-based mechanisms. These arguments include questions on whether specific types of projects constitute real emission reductions, and whether those can be measured or verified; the need to pursue real reductions rather than temporary storage; and the environmental and social impact of specific project types.

There are a number of approaches to selectively allow the use of JI and CDM credits for compliance with the obligations under the EU trading. A first possibility could be to limit the types of instruments to link with. The EU could decide not to link with CDM or JI. A number of EU members have hitherto had a policy of complying with their targets under the EU Bubble on the basis of domestic action alone, which in principle amounts to a decision not to use the flexible mechanisms. In view of the Council's common position requesting a link to be established, this type of limitation is, however, unlikely. A second possibility is to limit the types of projects. Such restrictions could take the form of a positive list or a negative list. A positive list specifies the projects from which credits can be used. Thus a positive list could specify project types (e.g., renewable energy), project sizes (e.g., less than 100 ktCO_2) and other characteristics of projects that generate acceptable

[95] Various EU position papers for the international negotiations can be found at: http://europa.eu.int/comm/environment/climat/cop.htm.

credits[96]. A negative list accepts all project-based credits except those specifically excluded. A negative list might for example specify that credits generated by nuclear, forest management projects and large dams cannot be used for compliance with obligations under the EATD. A further possibility could be to limit any link to credits resulting from reductions in the emission of specific types of gases. The proposal for an EATD only covers CO_2. This excludes the creation of allowances to offset the emissions of other gases under the EU trading regime.

Concerns related to the true level of 'additionality' of a project could be addressed through a more specific definition of this concept in the linking Directive. Such more specific definition would, however, be difficult to apply in practice as it would mean that the underlying project of every credit surrendered for compliance with emission limitations under the EATD would need to be checked on compliance with the EU additionality test. Concerns related to the environmental and social impact of projects could be addressed through the application of existing EU and regional rules on environmental impacts, access to information and access to justice[97]. A key question here is whether the application of those rules should be limited to projects within the EU or whether these rules, or the concepts underlying them, should also be applied to projects outside the EU.

Effective enforcement of any type of limited access can be complicated or even impossible. Restrictions would need to be implemented on an EU-wide level to avoid any such restrictions to be made effective and avoid their circumvention by benefiting from the different domestic linking arrangements of another Member State. The possibility to circumvent EU-wide restrictions would also pose itself in case the EU trading regime was to be linked to trading regimes outside the EU.

The proposed linking Directive tackles these issues through a combination of methods. As already seen above, the 'gateway' in the proposal states that Member States 'may' convert CDM and JI credits into EU allowances, thus allowing Member States to define further criteria for conversion. The proposal furthermore contains a 'negative list' of types of projects that are not admissible. This list contains nuclear facilities and land use, land-use change and forestry projects ('sinks'). The proposal, however, qualifies the exclusion of nuclear projects to projects that are excluded 'in accordance with the Kyoto Protocol and subsequent decisions adopted thereunder'. This language was included in the proposal as certain Member

[96] During the international negotiations the EU in fact advocated such a positive list.

[97] These include Council Directive 85/337/EEC of 27 June 1985 on the assessment of the effects of certain public and private projects on the environment, Directive 2003/4/EC of the European Parliament and of the Council of 28 January 2003 on public access to environmental information and Council Directive 96/61/EC of 24 September 1996 concerning integrated pollution prevention and control.

States and Commission services did not want to see a categorical and indefinite exclusion of nuclear projects. The current formulation brings any uncertainties associated with the Kyoto exclusion of nuclear projects into the EU trading regime[98] and allows nuclear projects to be included should such decision be taken in the international negotiations. The exclusion of sinks projects could be seen as being in line with the EU's strong opposition to the inclusion of sinks projects in the CDM during the international negotiations. It would, however, be technically difficult to include sinks projects at this point, as the international rules for the inclusion of sinks in the CDM have not yet been agreed upon, leaving important questions such as the name of sinks credits unanswered (see Table 1 above). The question on the inclusion of sinks is therefore likely to come up again in the next revision of the EATD under its Article 30. Earlier unofficial versions of the Commission's proposal also excluded 'hydro-electric power production incompatible with the criteria and guidelines of the Wold Commission on Dams in its year 2000 Final Report'. This exclusion was removed in the final proposal. The proposal, however, adds a new paragraph 2(k) to review the provision set out in Article 30 of the EATD stating that the review shall also consider 'the impact of project mechanisms on host countries, particularly on their development objectives, including whether JI and CDM large hydroelectric power production projects have been established which have negative environmental and social impacts'.

The proposal furthermore contains a number of guarantees in a new Article 11(ter) that is to be included in the EATD. Paragraph 1 of this new provision requires that Accession countries take into account the *acquis communautaire*[99] in the establishment of baselines for JI and CDM projects. This provision was included to avoid Accession countries using CDM and JI projects to bring their infrastructure in line with EU requirements. As these countries are obliged to do so under their Accession Treaty, providing ERUs or CERs in return for such projects would clearly not be 'additional' under the rules of the Kyoto Protocol. Paragraph 4 of the new Article 11(ter) EATD requires Member States participating in CDM and JI projects outside the Community to take into account the environmental and social impacts of these projects. It also requires Member States to 'ensure that these projects are developed and implemented in such a manner to contribute to sustainable

[98] The preambular paragraphs of the Marrakech decisions on JI and CDM include a statement that 'Parties included in Annex I to the Convention are to refrain from using emission reduction units generated from nuclear facilities to meet their commitments..', therefore not excluding nuclear energy CDM projects, but excluding the use of the resulting credits to meet the Kyoto commitments. See FCCC/CP/2001/13/Add.2, p. 5 and p. 20.

[99] The *acquis communautaire* encompasses the total body of Community law, including legislation and related instruments such as judgements of the European Court of Justice that are in place in the Community.

development, and to the specific development needs and objectives of the host countries'. Paragraph 5 furthermore requires Member States to ensure that project activities result in:

- real, measurable and long term benefits related to the mitigation of climate change;
- reductions in emissions that are additional to any that would occur in the absence of the proposed project activity; and
- the transfer of environmentally safe and sound technology and know-how.

While each of these requirements are already included in the Kyoto Protocol and the Marrakech accords[100], the proposal clarifies the obligation to respect them. Perhaps the most interesting part of this the new Article 11(ter) EATD is Paragraph 6. This paragraph allows the Commission to adopt further guidance on all the prior paragraphs under the 'comitology' procedure set out in Article 23(2) of the EATD[101].

4.6 When to Link

The EU emissions allowance trading regime is scheduled to start on 1 January 2005. The Kyoto Protocol's first commitment period starts on 1 January 2008 and ends on 21 December 2012. Credits resulting from JI and CDM currently only have value for developed country Parties to comply with their quantified emission limitation and reduction obligations under the Protocol. JI credits can only be issued for emission reductions and carbon sequestration projects from 2008 onwards. CDM credits can, however, be issued for emission reductions or carbon sequestration projects starting from 2000.

If a link is created between the EATD and the project-based mechanisms, the question is from when the installations covered under the EATD will be allowed to use JI and CDM credits. JI credits (ERUs) cannot be used pre-2008, as they will not yet exist. Although a JI-like structure could be created pre-2008, such structure could be cumbersome to design and implement for only a 3-year period, and its links with actual JI crediting under the Kyoto Protocol would not necessarily be guaranteed.

CDM credits (CERs) could be used pre-2008. Although under the Kyoto Protocol these can only be used to fulfil commitments from 2008 onwards,

---

[100] These requirements are included in Article 12(5) of the Kyoto Protocol and in para. 43 and the preamble of the CDM text, FCCC/CP/2001/13/Add.2, pages 36 and 20.

[101] See n. 38 above.

their recognition for compliance purposes pre-2008 would help the early start of the CDM by generating the interest of the private sector in these credits for the 2005–2008 period. As already mentioned above, the European Parliament has strongly opposed this option. The reason for this is that the use of CERs pre-2008 will allow installations to avoid making reductions inside the EU pre-2008, thus making achieving the EU's Kyoto target during the Protocol's first commitment period even harder. From a Member State perspective it may be more interesting to wait until the first commitment period, as this will reduce intra-EU emissions pre-2008 and ensure an increased availability of CERs during the Protocol's first commitment period, since CERs for reductions pre-2008 will not come onto the EU market until 2008. Industry has, however, already indicated that it wishes to see CDM credits included in the 2005–2007 period of the EU trading regime[102].

Allowing CDM credits to enter the EU trading regime would also raise some interesting questions. What would for instance happen to the CERs once they have been surrendered by an installation to compensate for emissions during the 2005–2008 period? Logically they should be cancelled by the Member State authority, as they have already been used to offset greenhouse gas emissions. But since Member States have no emission limitation objective themselves pre-2008, strictly there is no need for doing so, and Member States could decide to hold on to these CERs and use them to offset emissions during the first commitment period, although by doing so they would use one CER to offset two tonnes of GHG emissions. Allowing CDM credits to come into the EU trading regime pre-2008 could also have an impact on the decisions of Member States to allow banking from the pre-2008 to the post-2008 period.

In line with the European Parliament's position and thereby avoiding the various complications above, the linking proposal limits the inclusion of CDM and JI credits in the EU trading regime to the post-2008 periods.

4.7 How Much to Link

As already stated above, the discussion on the amount of CDM and JI credits coming into the EU trading regime was one of the reasons for the delay in the adoption of the linking proposal by the Commission.

Fears have been expressed that an unlimited linking of the EU trading regime with the Kyoto Protocol's project-based mechanisms could have a significant impact on the price of the allowances traded in the EU trading regime. A large influx of credits could also undermine the incentive for

[102] N. 83 above.

companies to reduce their emissions within the EU. This would undermine the EU's leadership in showing that it can achieve important greenhouse gas reductions domestically. It could also violate the (loosely worded) supplementarity obligations in the Kyoto Protocol and the Marrakech Accords.

One way to address the fear for a large influx of credits is through a quantitative limitation on the amount of credits that are introduced into the EU trading regime. In the negotiations on the Kyoto Protocol and the subsequent Marrakech Accords, the EU tried to introduce a limit on the use of credits traded under the Kyoto mechanisms as a means to implement the supplementarity principle[103]. Although no specific cap was introduced to comply with the supplementarity requirement, the Marrakech Accords did introduce caps for a number of other issues. The Decision on the modalities for the accounting of assigned amounts under Article 7(4) of the Kyoto Protocol introduces a cap on the amount of sinks credits for other projects than afforestation and reforestation projects that each Annex I country can issue under Article 3(4) of the Kyoto Protocol[104]. It also prohibits the banking of RMUs and limits the banking of CERs and ERUs into the following commitment period[105].

The draft linking proposal that went into inter-service consultation included a paragraph stating that 'Member States may convert CERs and ERUs from project activities for use in the Community-scheme up to 6% of the total quantity of allowances allocated by the Member State' for each of the EATD's compliance periods. Industry, as well as a number of Member States and other stakeholders, strongly lobbied other Commission DGs to oppose the inclusion of this paragraph[106]. As a result of this opposition the proposed cap was replaced by a provision that triggers a review at 6%, on the basis of which a cap may be proposed through a comitology process of 'for example' 8%.

4.8 Double Counting and JI in Accession Countries

The current text of the EATD omits to address the issue of 'double counting'. 'Double counting' refers to a situation where CERs or ERUs are issued as a result of reductions that also lead to a reduction in emissions from an installation covered by the EATD. Double counting could, for instance, occur if an installation decides to stop its on-site power generation, and instead buys its power from an external source. By doing so it would

[103] See also the discussion on international emission trading in chapter 4 in this book.
[104] FCCC/CP/2001/13/Add.2, p. 63, para. 28.
[105] Ibid., p. 61, para. 16.
[106] See, for instance, UNICE (2003b), n. 83 above.

reduce its own GHG emissions and could free up allowances. The operator could decide to buy electricity generated by renewable energy. The operator of the renewable electricity power plant could, however, also attempt to obtain credits for the installation of renewable electricity capacity. By freeing up allowances through moving power-generation off-site and by giving credits for the generation of renewable electricity that replaces this, the reduction in GHG emissions would be credited twice. The issue of double counting is not limited to a situation under which the EATD and the project-based mechanisms are directly linked. It can also occur without a link between the two regimes, although in case of such a link the double counting issue could be brought directly into the EU trading regime if the CDM or JI credits are subsequently converted into EU allowances.

The linking proposal includes a provision that will prevent double counting from occurring. The linking proposal will amend the EATD to include a new Article 11(ter)(2). This provision will require Member States to ensure that no ERUs or CERs are *issued* for reductions of anthropogenic emissions of greenhouse gases or removals by sinks resulting from project activities that reduce or claim to reduce greenhouse gas emissions from installations covered under the EATD. By prohibiting the *issuance* of CERs and ERUs rather than their *exchange* for allowances, this provision also prevents that any units generated through such projects are traded and used for compliance outside the scope of the EATD. This provision prohibits JI and CDM to be used for installations covered under the EATD.

Prohibiting the issuance of ERUs and CERs for emissions for installations covered under the EATD could, however, pose a problem for specific JI projects that are currently being implemented in Accession countries. The moment those countries become an EU Member State and are subject to the EATD, these projects would no longer be able to issue ERUs and will thus cease to exist as JI projects. While JI projects could in principle be converted into emission trades under the EATD, such conversion could create legal and contractual difficulties for projects that are already in the pipeline.

The linking proposal solves this issue in a new Article 11 (ter). This provision will allow Member States to issue ERUs for JI projects that are approved before 31 December 2004, or the date of the Member State accession, but no allowances may be allocated for emission reductions resulting from the JI project. It is, however, likely that the number of projects wishing to use such an exemption will only be very small. Project developers may instead opt to move into the EATD and sell the reductions achieved through their JI project under the EU trading regime. A key question in such a situation would, however, be the relation between the number of allowances allocated to an installation and the baseline under the JI project.

5 Conclusion

This article has given a background to a range of issues which have played and will play an important role in the design, adoption and implementation of the EATD and the new linking proposal. Although the EATD itself has now been adopted, the elaborating of the 'flanking' legislation, including the linking proposal, the registries regulation, but also the monitoring guidelines, will continue to pose a number of challenges. The adoption of the EATD has initiated the process of implementation in the Member States, a process which will be vital in determining the instrument's success.

With the adoption of the EATD the EU has started a unique and challenging experiment with a unique and new regulatory tool. Unexpectedly, Europe is now taking the lead, both domestically and at the EU level, on the development and implementation of this, of which the idea originated in the United States. The developments in Europe are anxiously watched by US industry and academics, many of whom feel that they are losing out on a unique opportunity to participate in this learning experience and shape the future of emissions trading globally.

The carbon constrained economy is a fact – at some point even the US will have to face this. Europe, by taking the initiative on implementing this new instrument is not only demonstrating its determination to take climate change seriously and adopt instruments to tackle it now, it is also giving itself a major advantage by exposing itself to a valuable learning exercise, which will give it a head-start over other countries.

References

BDI (2002). *Statement of the German Business on the Proposal for a Directive Establishing a Framework for Greenhouse Gas Emissions Trading within the European Community*. 21 January 2002. http://www.bdi-online.de.

CAN Europe (2003a). *Commission shoots its own emission trading system full of holes*. 23 July 2003. http://www.climnet.org.

CAN Europe (2003b). *Letter to the Commissioner for the Environment Margot Wallstrom*. 28 February 2003. http://www.climnet.org.

Climate Action Network Europe (2001). *Emissions trading in the EU: let's see some targets*! 20 December 2001. http://www.climnet.org.

Craig, P. and de Búrca, G. (2002). *EU Law – Text, Cases and Materials*. Third Edition. Oxford: Oxford University Press.

European Institute of Public Administration (EIPA) (2000). *Governance by Committee, the Role of Committees in European Policy-Making and Policy Implementation*. Maastricht: EIPA. http://www.eipa.nl/Topics/Comitology/comitology.htm.

ENDS (2003). 'UK opens European emissions trading debate'. *Environment Daily 1494*.

European Commission (2002). *Non-Paper on Synergies between the EC Emissions Trading Proposal and the IPPC Directive.* http://europa.eu.int/comm/environment/climat/non-paper_ippc_and_et.pdf.

European Commission (2003a). 'The EU Emissions Trading Scheme: How to develop a National Allocation Plan'. Non-Paper for the 2nd meeting of Working Group 3 of the Monitoring Mechanism Committee, DG Environment. http://europa.eu.int/comm/environment/climat/emission_plans.htm.

European Commission (2003b). 'Commission's Fourth Annual Survey on the implementation and enforcement of Community environmental law'. *Commission Staff Working Paper*: SEC(2003) 804. http://europa.eu.int/comm/environment/law/4th_en.pdf.

Fernández Armenteros, M. (2001). 'Overview of the Community Guidelines on State Aid for Protecting the Environment'. *Environmental Law Network International Review*, 1:36–42.

FIELD and IEEP (2002). *Assessment of the relation between Emissions Trading and EU Legislation, in particular the IPPC Directive.* http://www.field.org.uk/PDF/FINALReport31Oct.pdf.

FIELD (2001). *Study on the Legal/Policy Framework needed for Establishment of a Community Greenhouse Gas Emissions Trading Scheme.* http://www.field.org.uk/PDF/ETreport.PDF.

Haites, E. and Hussain, T. (2000). 'The Changing Climate for Emissions Trading in Canada'. *Review of European Community and International Environmental Law*, 3:264–275.

König, C., Braun, J.D. and Pfromm, R. (2003). 'Beihilferechtliche Probleme des EG-Emissionsrechtehandels'. *Zeitschrift für Wettbewerbsrecht*, 2:152–186.

Krämer, L. (2001). 'Grundlagen aus europäischer sicht, Rechtsfragen betreffend den Emissionshandel mit treibhausgasen der Europäischen Gemeinschaft', in Rengeling, H.W. (ed.), *Klimaschutz durch Emissionshandel*. Köln: Carl Heymanns Verlag, 30.

Pfromm, R. (2003). 'Die entgeltfreie Allokation von Emissionszertifikaten – eine wettbewerbsrechtliche Sackgasse?'. *Zeitschrift für Wettbewerbsrecht*, 6:537–542.

Schwarze, R. and Zapfel, P. (2000). 'Sulfur Allowance Trading and the Regional Clean Air Incentives Market: A Comparative Design Analysis of Two Major Cap-and-Trade Permit Programs?'. *Environmental and Resource Economics*, 17:279–298.

UNICE (2001a). Letter from 25 June 2001 to James Currie, Director General of DG Environment of the European Commission. http://www.unice.org.

UNICE (2001b). Letter from 10 December 2001 to the Belgian Council Presidency.

UNICE (2002). UNICE Comments on the Proposal for a Framework for EU Emissions Trading, of 25 February 2002.

UNICE (2003). Letter to Caroline Jackson, Chairman of the European Parliament's Committee on the Environment, Public Health and Consumer Policy of 2 June 2003.

UNICE (2003b). Preliminary comments on linking the Kyoto Protocol mechanisms (JI and CDM) with the EU emission trading scheme.

Vedder, H. (2001). 'The New Community Guidelines On State Aid For Environmental Protection – Integrating Environment And Competition'. *European Competition Law Review* 9: 365–373.

Zapfel, P. and Vainio, M. (2002). 'Pathways to European Greenhouse Gas Emissions Trading History and Misconceptions'. Milan: Fondazione Eni Enrico Mattei (FEEM), Working Paper 85. http://www.feem.it/web/activ/_activ.html.

Chapter 10

DIETRICH BROCKHAGEN

Inhomogeneous Allocation and Distortions of Competition in the Case of Emissions Trading in the EU

1 Introduction

Greenhouse gas emissions allowance trading under the Kyoto Protocol and the Marrakech Accords allows for inhomogeneous allocation. Inhomogeneous allocation means that different national regulators allocate emissions allowances with different methods and/or stringencies to their domestic industries, where the latter compete in the same international product market. A practical example is the draft EU directive on emissions allowance trading for European energy intensive industries, where discretion over the allocation stringency is left to the national regulators of the Member States. We will show in this article by means of theoretical and empirical analysis that differences in allocation may lead to distortions in competition. We find that at a permit price of Eur 20/ton of CO_2 in the steel industry, national allocations that differ more than about 40 percentage points in terms of free allowances given per ton of steel product would probably lead to market exits of firms with stringent allocations.

We will first discuss the economic theory and identify potentially distorting effects (Sections 1–5). Second, we will develop empirical indicators, to evaluate the likelihood of these effects (Section 6). Finally, we will apply our indicators to four energy intensive industries in the EU (steel, cement, refineries, electricity generation) (Section 7).

Sections 1–5 review the economic theory. In Section 1, we discuss the abatement cost functions and homogeneous and inhomogeneous allocations for output optimising firms. Section 2 discusses the combination of different allocation methods and stringencies in international trading systems and its effects on output optimising firms. Section 3 reviews the product price as a part

MICHAEL BOTHE AND ECKARD REHBINDER (ED.), Climate Change Policy, 309–350.

of a firm's business strategy. Section 4 looks into the effects of changes in financial positions due to different allocations. Section 5 outlines potential distorting effects.

1.1 Definition of Distortions of Competition

First, we have to define what we mean by distortions of competition. We see distortions of competition as economic inefficiency with regard to the resource allocation among competing firms (cf. Van der Laan and Nentjes, 2001). For practical reasons, we propose to measure distortions of the optimal resource allocation as the changes in market share among competing firms that are brought about by inhomogeneous allocation. Thus, we take the initial situation without emissions trading as undistorted (whether there are pre-distorting factors such as taxes or not) and analyse whether inhomogeneous allocation brings about changes in market shares.

1.2 Abatement Costs

We analyse, using neoclassic economic theory, how output-optimising firms might change their output and thus their market shares due to allowance allocation. We consider first the situation of a single firm that comes under an emissions trading system and has to constrain its emissions. For reducing its emissions, the firm has the choice to reduce either its specific emissions (emissions per unit of product output) or the output itself, or both. Marginal abatement costs will in general depend on both specific emissions and the output of the firm. Figure 1 shows two examples for the marginal abatement cost spheres of a firm. In both panels the abatement costs increase as specific emissions are lowered. However, in panel (a) we encounter the case where variable costs of abatement prevail and where production facilities of peak output show higher abatement costs than those of average output (for instance, in the case of parallel usage of several cement kilns). Panel (b) shows an example where variable abatement costs are zero (such as in the case of the upgrade of an electricity generator). The intersection of the marginal abatement cost sphere with a given level of an allowance price is in general a curve in the plane of a given allowance price. We have projected this curve into a plane on the bottom of the two panels.

We assume that the marginal abatement costs spheres of competing firms in the same industrial sector are all equal. This will not be the case in reality. However, the differences in marginal abatement costs cannot be counted as relevant for distortions of competition, since all firms have the same choice with regard to technologies. Technologies can be bought on the market and are thus

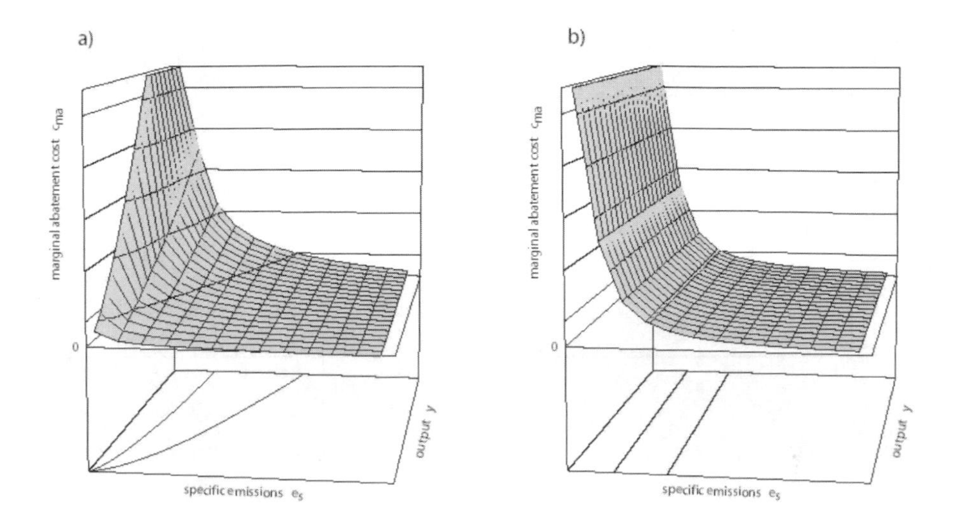

Figure 1. Marginal abatement costs as function of specific emissions and output

part of the investment decisions that are taken by firms. The only exception may be in the case of electricity generation, where natural endowments of a country and energy policies may influence the choice of electricity generation technology. This case will be discussed below. The assumption of equal marginal abatement cost spheres within an industrial sector implies that emissions trading within that industrial sector may only be of interest in the case of inhomogeneous allocation. Under homogeneous allocation, firms within one industrial sector would abate specific emissions to the same level and would not engage in allowance trading among themselves. Emissions trading would, however, remain efficient among different industry branches with different abatement costs.

1.3 Finding New Equilibriums for Emissions and Output under Emissions Trading

We first consider the case of a firm that becomes subject to emissions trading in an international system with homogeneous allocation (Figure 2). As outlined above, emissions trading is not an interesting option for the firm as long as the system is confined to a single industrial sector.

The firm will thus reduce its total emissions to the level e_t, imposed by the regulator, marked by the full curve in the figure (every point in the figure represents a specific amount of total emissions e_t). The firm is initially located in

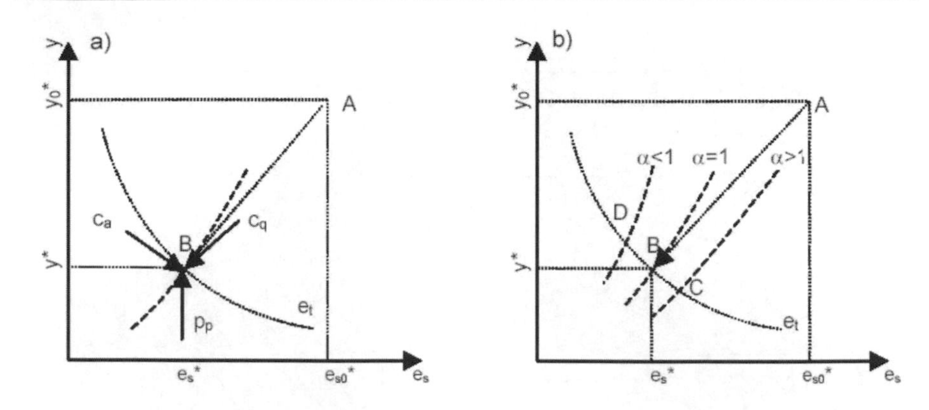

Figure 2. Finding new optima for specific emissions and output under emissions trading

point "A", with regard to its optimal output y_0* and specific emissions e_{s0}* and subsequently comes under the trading regime. The firm will abate emissions until marginal abatement costs equal the allowance price. We take as an example for the marginal abatement costs the case of Figure 1, panel (a). Hence, at a given allowance price, the firm will find its new optima with regard to specific emissions e_s* and output y* on the curve shown in the bottom plane of Figure 1 (a), that is repeated as the broken curve in Figure 2, panel (a). Hence, the firm will find its new equilibrium in point "B" at the intersection of the two curves. The corresponding driving forces for the firm on its way to the new equilibrium are also shown in Figure 2, panel (a). Abatement costs c_a are lowest to the bottom-right of the chart (see Figure 1, panel (a)). Therefore, they steer the firm this way. However, allowance costs c_q for the firm are lowest with lowest emissions and steer the firm thus towards the bottom-left of the chart. In this case of homogeneous allocation, allowance costs would be the opportunity costs of allocated allowances that cannot be sold, since they are needed to cover the firm's emissions. Finally, the firm has an interest to keep the output at a level closest to the initial situation, thus steering the equilibrium towards the top of the chart. These business-related forces are marked "p_p" in the figure. In the final equilibrium "B", both output and specific emissions will be reduced as compared to the situation in the absence of emissions trading.

The exact location of the new equilibrium depends on the marginal abatement cost sphere. Figure 2 (b) shows two further examples for final equilibriums (Points "C" and "D"), where marginal abatement costs increase over- or under-proportionally with the output (marked $\alpha > 1$ or $\alpha < 1$ in the figure, respectively).

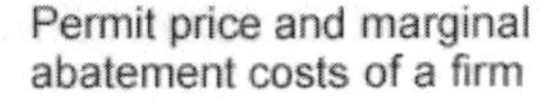

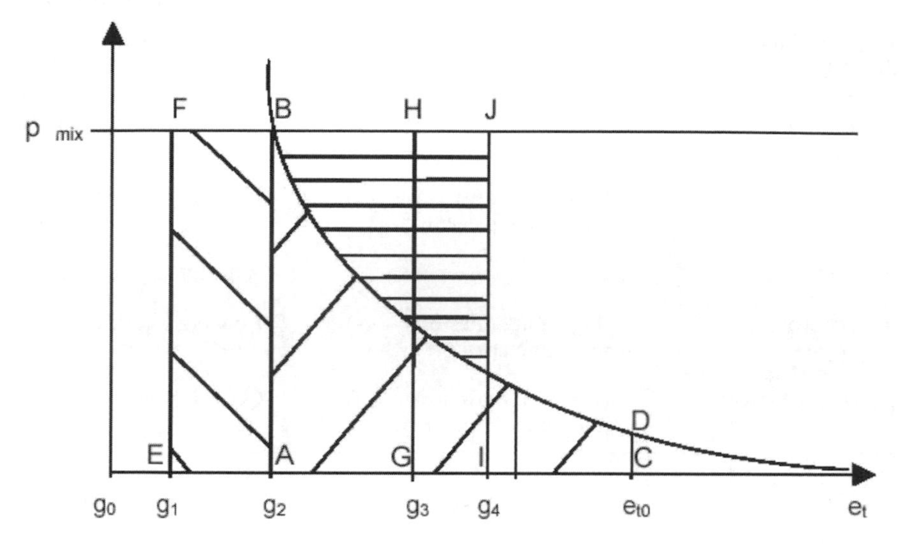

Figure 3. Total control costs of a firm in an international emissions trading regime

1.4 Cases of Inhomogeneous Allocation

Next, we consider cases of inhomogeneous allocation.

An international emissions trading system comprises national regulators from several countries. A positive allowance price p_{mix} emerges as soon as the sum of all national allocations is less than the amount of allowances needed by the regulated firms. The allowance price emerges only as a function of the marginal abatement costs of the firms and the total number of allowances allocated by national regulators. The allowance price does not, in general, depend on whether the allowances are given for free or at a price to the firms (it may, however, be influenced by different modes of auctioning). However, the amount of free allowances received will have a direct influence on the total costs of the firm. In an international trading system, the allocation in one country will in general differ from the average allocation in the whole system. In Figure 3, where a firm becomes subject to five alternative cases of allocation with different stringencies, the allowance price p_{mix} in the international system remains unchanged.

In the first case, the regulator sets an allocation to the firm with allowances covering an amount g_2 of initial total emissions e_{t0}. Total control costs for the firm

Table 1. Total costs under emissions trading and convention on allocation stringencies

Convention on allocation (Figure III)	Total costs from emissions trading	Aggregated cost impact
g_0, auction	Optimal total emissions * allowance price + c_a	Maximum cost increase
g_1, stringent	(Optimal emissions – g_1) * allowance price + c_a	High cost increase
g_2, average	c_a	Average cost increase
g_3, generous	(Optimal emissions – g_3) * allowance price + c_a	Low cost increase
g_4, over generous	(Optimal emissions – g_4) * allowance price + c_a	Cost decrease

are then represented by the area ABCD. We call this type of allocation "average allocation".

In the second case, the firm gets a more stringent allocation, receiving allowances equal to emissions g_1. Total control costs are presented by the area ABEF+ABCD, since the firm would abate emissions to the level of g_2 and buy allowances for the remaining emissions to cover. We call this type of allocation "stringent allocation".

Even in the third case, where the firm receives allowances equal to emissions of g_3, it would abate to the level of g_2 and sell the remaining allowances, represented by the area ABGH. Since ABGH < ABCD, this type of allocation still means net costs to the firm. We call this type of allocation "generous allocation".

Consider now the case of the allocation g_4. Since ABIJ exceeds ABCD, this case implies an overall cost decrease to the firm. We call this case "over-generous allocation".

If, finally, the firm receives an allocation g_0 (no allowances for free) we have the case equivalent to an auction, since all allowances would have to be acquired at the allowance market. Obviously, for the firm, this implies the highest cost increase of all five cases discussed here.

Table 1 summarizes allocation stringencies and the associated costs under emissions trading. Note that according to our previous discussion, we do not consider different abatement costs c_a as relevant for distortions of competition. Hence, in Table 1, we assume that c_a is equal for all competing firms. We only take the differences of total allowance costs or benefits as relevant for distortions of competition. They are directly proportional to the amount g_{0-4} of free allowances.

Note that in an international system with several national regulators, there will, in general, always be stringent and generous allocations, as long as not all regulators allocate exactly with the same stringency. If the allocation in the international system is homogeneous, the allowance costs of all firms are zero, due to the assumed same abatement costs. The maximum difference in allocation occurs, if one industry receives an allocation corresponding to g0 and another industry an over-generous allocation. Hence, the biggest cost increase resulting from a national allocation is a measure for the inhomogeneity of the national allocations in the international trading system.

2 Output Optimisation and Inhomogeneous Allocation

Now we consider different cases, where in an international emissions trading system two identical groups of output optimising firms, (a) and (b), become subject to different allocations of their national regulators. First, we consider cases where the allocation method is identical, but where allocation stringency differs.

2.1 Allocation Based on Historic Emissions (Grandfathering)

In this case, the allocation of the national regulators corresponds to free allowances covering the emissions of the firms in a past baseline year, minus a reduction target for the target year. The two targets are assumed to differ for the two groups of firms. The resulting situation under emissions trading is depicted in Figure 4. In isolated emissions trading systems, the two groups, (a) and (b), starting from the common origin "A" without emissions trading, would find their new equilibriums in points C and D, respectively, which are associated with different allowance prices. Due to the higher stringency of their allocation, the firms of group (b) would face a higher allowance price p_b in their trading system and vice versa for group (a). If the two trading systems are combined, the firms in group (b) can evade high abatement costs by purchasing emissions allowances from the firms in group (a). Hence, a common allowance price p_{mix} would emerge, in between p_a and p_b. Trade continues, until eventually all firms find their new equilibrium in point B. In the figure, the shaded areas mark the volume of the emission allowances that are traded between the two groups of firms. Hence, after emissions trading, the financial positions of the firms in group (a) would have improved and those of the firms in group (b) would have deteriorated.

We can now ask, whether this outcome means distortions of competition in the sense of our definition. The answer is no, since all firms change their output in

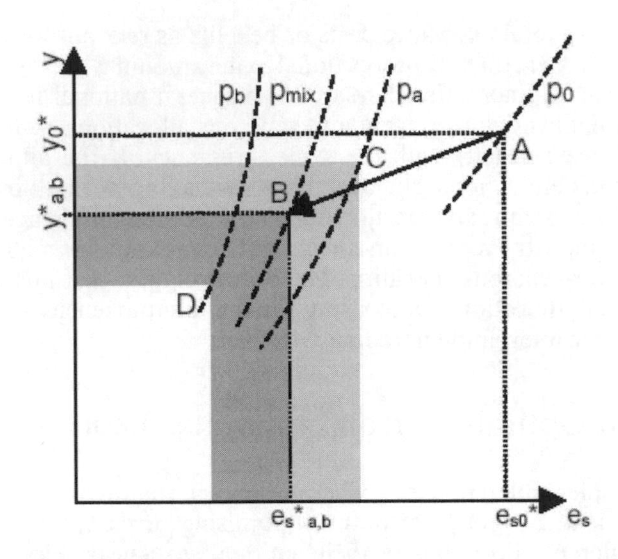

Figure 4. Grandfathering with different stringencies to two competing groups of firms

the same way (from y_0* to $y^*_{a,b}$ in Figure 4), independent of their allocation. (The fact that financial positions have altered and the possible consequences of these alterations are examined below.) Note that this result also represents the extreme case, where one group of firms gets all emission allowances needed free of charge, whereas the other group has to purchase all allowances at an auction. We refer below to this case of allocation based on historic emissions as the "reference case".

2.2 Allocation Based on Output

In this case, firms get their individual allocation as a function of their output. In practice, the national regulator sets aside a distinct amount of allowances for an industrial sector. At the end of the commitment period, these allowances are distributed to the firms in the sector. A firm contributing a specific share of the total product output of the sector receives the same share from the total allowances issued by the regulator.

Figure 5 sets out this case of allocation. Panel (a) shows the comparison of two isolated systems with allocation based on historic emissions (reference case) and output-based allocation. Both systems have the same stringency in terms of total

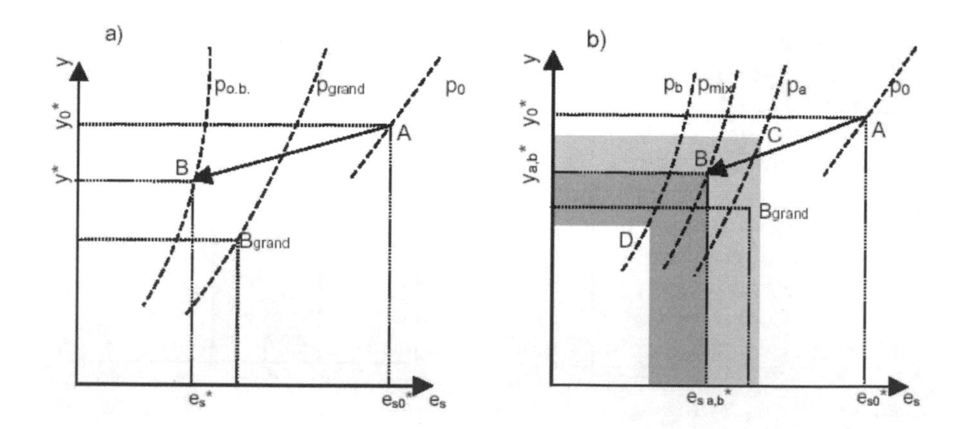

Figure 5. Output based allocation. (a) Comparison with grandfathering, (b) allocation with different stringencies to two competing groups of firms

allowances provided by the regulator. The firms in such isolated systems would find their new optima in points B_{grand} and B for the reference case and output-based allocation, respectively. Firms perceive an incentive to increase their output under output-based allocation, since this increases their allocation as well. However, since all firms tend to increase their output, but the amount of allowances remains fixed, the allowance price increases as compared to the reference case. Thus, under output-based allocation, output is increased, specific emissions are reduced and overall compliance costs are increased as compared to the reference case, rendering this allocation form less efficient than the reference case from a macroeconomic point of view.

Panel (b) of Figure 5 shows the case of the linkage of two emissions trading systems with output-based allocation of different stringency. We assume that both groups of firms, (a) and (b), receive their allocation output based, however, the two regulators set aside different amounts of allowances. In isolated trading systems, the groups (a) and (b) would find their new optima in points C and D, respectively, at the different allowance prices p_a and p_b. In a joint system, the firms in group (b) can evade high abatement costs by purchasing emission allowances from the firms in group (a), leading to a common allowance price p_{mix}. Since the resulting allowance price is in between p_a and p_b, trade continues until the amounts of allowable emissions are the same for all firms. Hence, all firms will eventually find their new optima in point B, which is marked by higher output and more reduced specific emissions as point B_{grand}, standing for the combined system in the reference case.

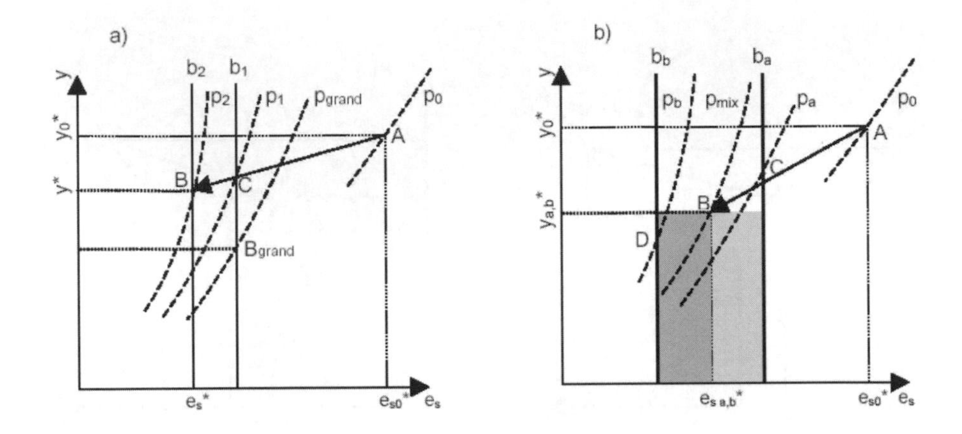

Figure 6. Benchmark based allocation. (a) Comparison with grandfathering, (b) allocation with different stringencies to two competing groups of firms

The evaluation of this case with regard to distortions of competition is the same as for the reference case. Only the system efficiency is deteriorated in this case, due to the perceived incentives towards higher output.

2.3 Allocation Based on Emissions Benchmarks

In this case, emission allowances are allocated output based also according to a fixed benchmark b_{fix} for specific emissions set by the regulator. An individual firm receives allowances corresponding to the product of output times the benchmark. Hence, the total allocation to an industry is not fixed, but depends on the industry output. Such emission benchmarks are part of the allocation formula in the US SO_2 allowance system and they have, for example, been proposed for the industry sector in the Netherlands (Mulder 2002). On top of the free allowances, the regulator only sells additional allowances at a prohibitive price. Consequently, firms that miss the emissions benchmark and search for allowances on the market provide incentives for other firms to over-achieve the benchmark. If we assume all competing firms have the same marginal abatement cost, every firm will thus reduce its specific emissions to the benchmark level.

Figure 6 sets out this case of allocation. In panel (a) we compare isolated systems of benchmark-based allocation and reference case allocation. The benchmark b_1 results in the same optimal specific emissions (point C) as in the reference case (B_{grand}). However, in order to achieve the same amount of total emissions as in the reference case, the benchmark has to be set to the level b_2, leading to the higher allowance price p_2. Panel (b) shows the effects of combing two emissions trading systems with benchmarks of different stringencies. In isolated systems, the groups (a) and (b) would find their new optima in points C and D, respectively. When linking the two systems, the firms in group (a) would have an incentive to reduce their emissions even further than the benchmark b_a and sell the remaining allowances to the firms of group (b), since this allows them to evade high abatement costs. As in the case of output-based allocation, emissions trading continues until the amount of total emissions is equalized between the firms (point B), with a net financial flow from the firms in group (a) to those in group (b). Again, the resulting allowance price p_{mix} is higher than that of the reference case, signifying an efficiency loss from the system point of view.

The evaluation of benchmark-based allocation with regard to distortions of competition is the same as for the reference case. Only the system efficiency is deteriorated in this case, since a given total emissions abatement is achieved at a higher allowance price and thus higher total abatement costs.

2.4 Combining Different Methods of Allocation

So far, we have analysed isolated systems and combined systems with the same allocation method, but different stringencies. Now we analyse the combination of two emissions trading systems with different allocation methods. As for stringency, we have seen that for all allocation methods the firms subject to more stringent allocation purchase emission allowances from the firms subject to less stringent allocation. Hence, it is now sufficient to look into the case, where different allocation methods of the same stringency are combined. Combining systems of different allocation methods and different stringencies will then deliver the same results, only superimposed by the stringency effects already discussed. We will first analyse the effects of linking different allocation methodologies and then embark on their interpretation with regard to distortions of competition.

First, we consider the combination of two trading regimes: one with grandfathering (a-firms), the other with output-based allocation (b-firms). This case is illustrated in Figure 7. Both isolated systems limit total emissions to the same amount. As we have discussed above, the output-based system leads to a higher allowance price (point C) as compared to the reference case (point B). Hence, when the two systems are linked, the b-firms feel incentives to acquire allowances from the a-firms. Consequently, the a-firms benefit from a financial flow from the b-firms for the traded allowances, but also have to reduce their total emissions further and vice versa. A medium allowance price with $p_{grand} < p_{mix} < p_{output\ based}$ emerges, which drives the a-firms to lower output and lower specific emissions (point E) as compared to the b-firms (point D). Due to the elevated allowance price as compared to the reference case, the mixed system will be less cost efficient

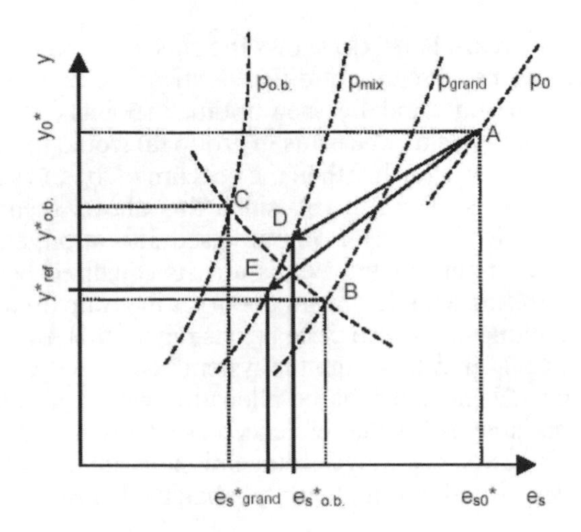

Figure 7. Linking two emissions trading systems with allocations of the same stringency, output based and grandfathering

than the isolated reference case, but more cost efficient than the isolated output based system.

Second, consider the case of combining grandfathering with benchmark-based allocations of the same stringency. As we have seen before, output-based and benchmark-based allocations have the same effects on allowance price, specific emissions and output, if they are set equally stringent in terms of the resulting total emissions. Hence, under the premise of the same stringency, the combination of benchmark allocation with grandfathering leads to the same results as when linking output-based and grandfathering, as discussed above.

Finally, consider the case of combining a system with benchmark-based allocation and a system with output-based allocation of the same stringency. As we have seen above, the isolated systems would be identical with regard to the changes of output and emissions. Therefore, we do not expect any change of the behaviour of the firms, when combing the two systems.

In conclusion, there are only the two cases of combining grandfathering with output or benchmark-based allocation, where we expect different output and emissions adaptation behaviour of the a- and b-firms, respectively.

2.5 Conclusion on Output Optimisation

We can now summarize our findings for the case of output optimising firms coming under an emissions trading system. In isolated systems and compared to the case of grandfathering, both output and benchmark-based allocation result in less reduced output and more reduced specific emissions in order to achieve the same total reductions, leading to a higher permit price. When combining systems with the same allocation method of different stringencies, firms with more stringent allocation buy allowances from firms with more generous allocation, thus facing a financial disadvantage. However, this would not imply distortions of competition in the sense of our definition, since all firms reduce output to the same extent. Of course, one can wonder whether and how firms benefiting financially from allowance trades might use this financial advantage against their competitors. This will be looked into below.

However, there would be a direct effect on outputs, when a grandfathering system is combined with a benchmark or output-based allocation system. For instance, firms with output-based allocation would reduce their output less than the other firms and hence have a greater market share. However, since the total allocation is assumed as being the same to both groups of firms, this advantage is entirely offset financially by the losses incurred for the acquisition of allowances. In fact, under all combinations of different trading systems, an advantage in market share can always be bought by each firm at the expense of expenditures for allowances and the creation of an equal financial disadvantage. Hence, we do not count this type of effect as distortions of competition.

3 Price Setting in Oligopolies

In this section, we expand the basis of output optimisation and also consider price setting by the firms. Our focus is on oligopolies, since most industries and particularly energy intensive industries are characterized by this form of competitive environment. A number of price-setting theories for oligopolies have been developed, but none of them being without challenges. In the field of distortions of competition, two theories may be of importance: the theory of predatory pricing; and the theory of administered prices.

3.1 Predatory Pricing

Predatory pricing represents a deliberate strategy of a firm to get rid of its competitors (see Milgrom and Roberts 1982). In the context of our research, the likelihood for predatory pricing to happen depends on two factors:

- The competitive environment, including the necessary dominant position of the predator, the financial situation (deep purse) and the necessary structural cost advantages e.g. due to technology advantages and economies of scale.
- The differences in allocation of free allowances faced by the competing firms.

We will see below that in none of the four EU industrial sectors analysed, the necessary first conditions are met. Hence, we can conclude that predatory pricing is a non-issue in the field of distortions induced by inhomogeneous allocation.

3.2 Administered Prices

We have noted before on a general level that allocation directly influences total costs of a firm. We will now look in detail at the potential effects. Hall and Hitch (1939) show by means of interviews that firms mostly fix prices at the average costs of the product plus a certain profit margin, determined by the undertaking. Firms that try to obtain excessively high profit margins are penalised by the market. Firms that suffer from a factor cost increase, e.g. due to emissions trading, increase their prices. However, since demand is generally elastic, firms cannot put through the entire cost burden to the customers.

In the following section, we assume that an oligopoly starts out without environmental policy at a product price of p_0. Second, it becomes subject to an international emissions trading system with a positive allowance price. In this system, every firm receives an allowance allocation from its national regulator. In the new final equilibrium under emissions trading, we assume that there will be the new product price p_1. For the way from p_0 to p_1, we allow for pricing strategies of competing firms. However, we assume an oligopoly without a dominant firm with structural cost advantages. In other words, we exclude a dumping scenario, where the product price is lowered by firms that each receive a generous allocation.

In Figure 8 we have depicted heuristically the situation of four competing firms in an oligopoly. For the purposes of this study, initial profits of the firms differ. This may be, for instance, due to differences in management or in factor costs, such as labour. The vertical axis carries average costs of production and emissions control. The emissions control costs comprise abatement costs and allowance costs. The total number of emission allowances to the oligopoly is the same in both panels of Figure 8. Individual allocation does not determine the increase of marginal costs of an individual firm, but the common allowance price does. Thus, marginal cost increases will be the same for all firms. Furthermore, as discussed above, we ignore differences in abatement cost functions. Hence, all

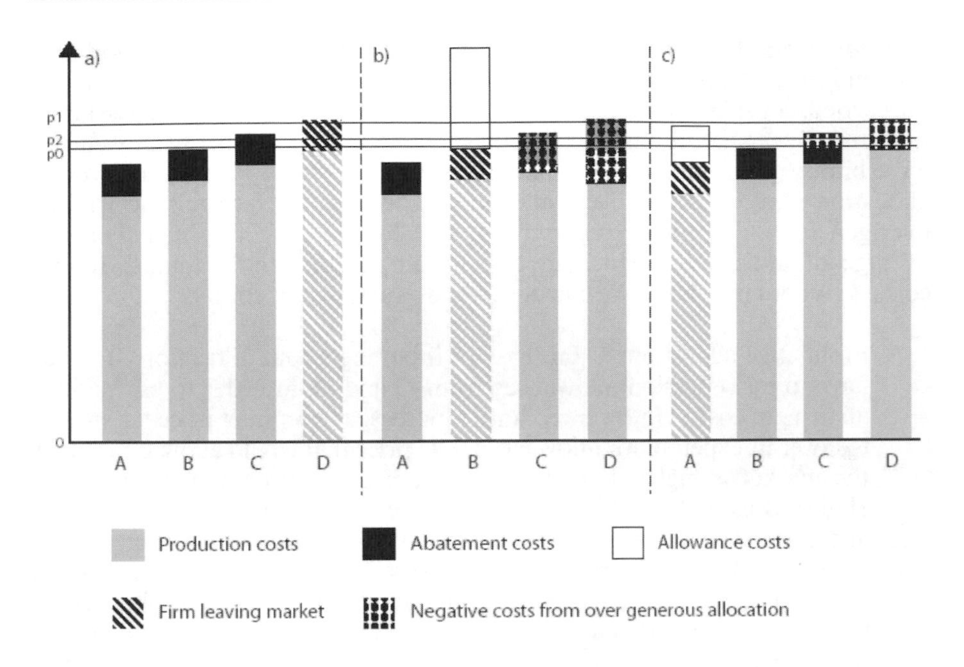

Figure 8. (a) Average allocation, (b) prohibitive allocation and extended lifetime, (c) passive dumping and extended lifetime in an international emissions trading system

firms can be considered to abate emissions down to the level marked g_2 in Figure 3. Hence, abatement costs are considered the same for all firms and cost differences stem from differences in allocation only.

First, we consider the case of homogeneous allocation. Figure 8, panel (a) illustrates the case of uniform allocation from the four national regulators. The average costs of all firms will uniformly increase by the abatement costs. The allowance costs will be zero for all firms, since the homogeneous allocation and equal abatement cost functions do not require trades for firms shifting to their new optima. Hence, all firms have an interest to increase the product price. However, as noted above, the feasible price increase to the level p_1 will be below the cost increase. As drawn, the new product price p_1 does not allow firm D to stay in the market.

We can ask, whether firm D closing its doors represents distortions of competition in the sense of our definition. The answer is no, since there is no inhomogeneous allocation from the regulators. Panel (a) shows market forces in a situation, where emissions have become an additional factor of competition between firms. D's performance is too low in order to survive this sort of "environmental Darwinism". This is in line with a functioning market and does not imply distortions of competition.

Second, consider the case of inhomogeneous allocation. If targets are set non-uniformly by regulators, things may change. In Figure 8 panel (b), we have chosen an arbitrary, distribution of national allocations. Firm A receives an average allocation, B receives a stringent allocation, and C and D receive generous and over-generous allocations, respectively. The top parts of the bars represent average allowance costs, which are directly proportional to the total allowance costs, as set out in Table 1. We can observe several different effects.

1. Prohibitive allocation: in this case of inhomogeneous allocation, B would have to buy emission allowances from C and D, in order to escape high abatement costs. However, B cannot increase the product price far enough to cover its expenses for allowances. The price that would allow B to stay in the market is higher than the maximum demand compatible price p_1. Hence, B cannot impose this price on the market against its competitors and has to cease business. The fact that B has to quit the market is independent from the price that is realised within the range of possible prices (from p_0 to p_1). Hence, one can assume that all firms would strive to set prices at p_1. In the new equilibrium, the three firms remaining would take over a part of B's market share that is compatible with the new product price. According to our definition, B leaving the market implies distortions of competition. We will see later that the proportions as drawn in Figure 8 may be realistic in some energy intensive industries. We call this type of distortion "prohibitive allocation", since B's allocation is so stringent that B is in the same situation as if it was obliged to acquire an essential production factor at prohibitive prices.

2. Extended lifetime: another effect shown in panel (b) of Figure 8 is that D stays in the market, while its marginal costs exceed the product price. However, due to the over-generous allocation, D's average costs are below the product price, rendering D profitable by means of the sold allowances. D would have the choice to cease production, which would be profit maximising, or continue production at a level, where the variable production costs are lowest. This could be an option in order to stay in the market for a transitional period of the firm or the whole industry, when an upwards trend of the firm's cost situation or the overall economic situation can be expected. Hence, over-generous allocation may constitute an obstacle to competitive markets: the exit of firms that have become unprofitable in a more competitive environment. According to our definition, D staying in the market implies distortions of competition. We will call this effect "extended lifetime".

3. Passive dumping: if differences in allocation are smaller than discussed above for prohibitive allocation, there may nevertheless be a distorting effect. In Figure 8, the total volume of emission allowances for the oligopoly remains in all panels the same. However, in panel (c) we have chosen a less extreme distribution of allocations than the one shown in panel (b). Firm A receives a stringent allocation, B receives an average one, and C and D receive generous allocations, respectively.

   If the firms increased the price to the level of p_1, all firms would stay in the market. However, this would imply that all firms have to reduce output, since the price p_1 is only compatible with a lower demand relative to p_0. However, the firms B, C and D could decide to only increase the price to the level p_2. This would be too low for A to stay in the market, even though its marginal costs would be below this price level. However, the necessary acquisition of emissions allowances would render A unprofitable. B, C and D would take over a part of A's market share which is compatible with the new price level p_2.

   Note that his behaviour of B, C and D cannot be explained without considering strategies of firms B, C and D. In the short run, B, C and D reduce their profit margin and render themselves willingly in a non-optimal situation. The optimum price level would be the price p_1. However, this price level would mean that all firms stay in the market but have to reduce their output. This is an incentive for B, C and D to keep the price low. In the longer run, the three remaining firms may raise the price to p_1, however, as drawn in Figure 8, panel (c), they would have got rid of competitor A. This effect is only possible, when the competitors B, C and D have sufficient capacities to take over the market share of A. Furthermore, this effect is contingent on the assumption of imperfect mobile capital. However, under this assumption, A may not re-enter the market, once B, C and D have waited long enough before raising the product price. The duration of this period depends on the cost structure of the industry concerned. However, we can suppose that an industry with high profitability cannot deter the re-entry of firms in the long run. Hence, the effect depicted in Figure 8, panel (c) can be considered as a dynamic effect, which may hold for a shorter or longer period, depending on industry specific circumstances.

   As in the above case of extreme inhomogeneous allocation, this effect constitutes a distortion of competition, according to our definition. This form of conduct resembles the case of predatory pricing, the difference being that in this case the "predators" remain inactive, keeping prices lower than optimal for a certain period of time. We call this effect "passive dumping", since the outcome is the same as in the case of dumping (driving out competitors), but the way of achieving this is a passive one.

4 The Impact of Financial Positions

In this section, we include more characteristics of real firms than output and price variation. The broadest approach is to investigate potential effects of altered financial positions of competing firms. As we have seen above, the financial position of firms is in general altered by inhomogeneous allocation. We will now discuss how firms may use beneficial allocation in order to increase their market shares outside price or output variation. Formally, we distinguish between fair and anticompetitive practice of non-price-output competition.

4.1 Anticompetitive Practices of Non-price Output Competition

Two groups of anti-competitive practices can be classified. These are horizontal and vertical restraints on competition (Brennan, 2000):

- Vertical restraints: this includes practices such as exclusive dealing, geographic market restrictions, refusal to deal/sell, resale price maintenance and tied selling.
- Horizontal constraints: this includes specific practices such as cartels, collusion, conspiracy, mergers, predatory pricing, price discrimination and price-fixing agreements.

Practices such as exclusive dealing or the formation of cartels do not or only weakly depend on the financial position of a firm. They are based on strategic decisions, taken by a few persons and do not require any significant investment in the human forces or technical infrastructure of a firm. Hence, we conclude that in this field, the financial position of a firm does not matter for distortions of competition.

Mergers and acquisitions can be considered here also. It could be argued that mergers and acquisitions may be influenced by the financial positions of firms, altered by national allocation. Indeed, it may not be ruled out that free allowances could influence take-overs, mergers or acquisitions. However, this would not distort competition by itself. An acquisition always takes place through a market procedure and does not itself therefore distort competition. Companies have a certain market value and can always be bought, provided that their shareholders agree. The market value of a firm may be altered by allocation. However, this will not happen directly, but only as a consequence of the changed activities of the firm, which may be induced by differences in allocation. Hence, if we want to analyse the impact of allocation on acquisition and mergers, we have to examine the induced change of activities rather than the acquisition itself. A change of activities (e.g. output) is what we have already discussed so far.

4.2 Fair Practices of Non-price Output Competition

As for fair competition other than price and output competition, we consider the following activities:

- Publicity (commercials, sponsoring, image (booklets, web pages) etc.)
- Research and development (R&D)
- Internal management of the firm, pertaining e.g. to bonus schemes for employees, the working environment (buildings, office equipment), training courses, etc.

For all of these practices, there is empirical evidence (cf. Bienaymé, 1998). The difference to price and output competition is not a clear-cut one. The costs of the above activities are included in the fixed or variable costs of a firm, covered by price or output optimisation as discussed above. However, there is a fundamental difference. In energy intensive industries, the marginal costs of production are changed directly through emissions trading and the associated carbon costs. However, costs of publicity, and R&D are not influenced by carbon constraints. In any case, a firm that decides to spend a certain portfolio in the above categories will have to finance the expenditures from its income. It could borrow them also, but in the long run all activities have to be financed from income, otherwise the firm is unprofitable. It is here, where allocation in emissions trading makes a difference. The differences between getting additional capital for free (allowances) and borrowing it at interest rates may become decisive for the decision of a firm to maintain or start one of the above activities. Koutstaal (1997) notes that the smaller the amount of equity of the firm, the higher will be the interest rate charged by the bank and the less worthwhile the project appears to a firm. The financial advantage of free allowances over the normal debt contract is evident when the borrowing firm has to repay its debt at the end of a certain period, which is shorter than the payback period of the investment. However, even in the case of a credit facility, where the borrower only has to pay back the interest in each period, this does not dissuade the financial advantage of free allowances since in the long run the interest accumulates. This is why we argue that the financial advantage of a firm benefiting from free allowances constitutes a principal difference between competitors with regard to the engagement into potential activities. The question is rather, whether the financial difference from allocation will be quantitatively significant.

Table 2. Synthesis of distorting effects of inhomogeneous national allowance allocation in an international emissions trading system

Distorting effect	Description	Conditions
Prohibitive allocation	A firm subject to an excessively stringent allocation cannot increase prices to the level sufficient for survival and has to leave the market.	Excessively stringent allocation to a firm as compared to the average allocation.
Passive dumping	Firms with average or generous allocation increase product prices to a small extent. Competing firms with stringent allocation have to increase prices more and must leave the market.	Sufficient cost differences through stringent allocation and pricing strategies of competitors.
Predatory pricing	Deep purse of predator is established or further deepened due to generous allocation and allows for predation.	Dominant position of predator, structural technological advantage, sufficient cost advantage through allocation.
Extended lifetime	Firms stay in the market that otherwise would have to leave the market. Allocation turns out as hidden subsidy.	Generous or over generous allocation.
Investment advantage	Firms receiving generous allocation have a cash-flow advantage for non-price/output related activities such as R&D and publicity.	Non-price/output related practices being a relevant means of competition for the sector in question

5 Synthesis of Potential Distortions of Competition under Emissions Trading

Table 2 summarizes the potential distortions of competition, found in the above analysis.

6 Developing Empirical Criteria for the Likelihood of Distortions

In this section, we develop indicators that can be applied empirically in order to evaluate the likelihood of distortions of competition identified in Section 5 above.

6.1 Two-tier Approach

So far, we have discussed distortions on the general level, where firms do not belong to a certain industry and only compete in an abstract common market with few characteristic properties. However, it is clear that the likelihood for distortions will in reality also depend on the competitive environment. The established practice of ruling on alleged cases of predatory pricing consists of a two-tier approach (cf. OECD, 1989): in the first tier, competition authorities examine the market in question and determine whether it is susceptible to successful predation. Only where this is the case, the second tier applies, in which price-cost relations are investigated. For our purposes, this two-tier approach would look as follows:

- The first tier verifies whether the five potential effects are possible in the given competitive environment of the relevant market.
- If the first tier is positive, the second tier applies, using different indicators based on the allocations of national regulators.

6.2 First Tier: The Competitive Environment

As outlined above, the objective of the first tier is to screen out irrelevant cases. This would occur where the geographical market is of national or sub-national rather than international size. One could also investigate how fiercely competition is exerted among the different national competitors. The following 15 criteria characterize the means and degree of international competition among firms in an industry.

- criteria 1–4: general criteria on the macro level
- criteria 5-6: specific criteria on investments and predatory pricing
- criteria 7–9: market dynamics on the firm level
- criteria 10–15: geographical demarcation of the relevant market

  We discuss below these criteria and specify how they can be put into practice.

1. Privatisation: if competing firms are mainly state owned, competition will be low, since any losses incurred by the firms are normally born by the state. For empirical evaluation, we estimate the part of the market that is state owned, using various data sources.

2. Subsidies: subsidies impede competition and are thus controlled by competition authorities. The more subsidies are given, the less relevant will be differences in national allowance allocation. For empirical evaluation, we use official sources from the EU State Aid Board.

3. Mark-up index: the mark-up index is the classic indicator used to measure how close a market is to perfect competition. The empirical determination of the mark-up index is difficult. For the manufacturing industry, only one study was available (OECD 1996), presenting mark-up indicators for OECD countries.

4. Market saturation: stagnant or only moderately growing demand is not likely to fuel competition. We characterise the market as either expanding or saturated. This will be done empirically by comparing output, demand and production capacities.

5. Expenditures in R&D: these expenditures are linked to the distorting effect of "investment advantage", depending on the financial position of a firm. We only use expenditures for R&D for the analysis, since, for energy intensive industries, no data was available for other fields such as publicity. We estimate, whether typical R&D related expenditures are high or low, based on an empirical demarcation (see below). If they are high, we relate them in the second tier to the relative cost advantage from generous allocation. If they are low, we consider that in this industry R&D is not of importance. Consequently, even an enhanced financial position due to generous allocation would not cause harm. A report of the enterprise and innovation scoreboard of the European Commission on the performance of European Industry classifies percentages below 2% of production value as being low and vice versa (COM Performance, 2001). We use the same threshold.

6. Past occurrence of predation: if in the past, predatory pricing occurred in the industry, this is an indication for the structural vulnerability through this effect. If, however, there were no such events, it would not necessarily follow that they could not occur in the future. For the evaluation of this criterion, we use the database "European Competition law", which allows comprehensive inquiries over all primary sources of EU legislation, going back to the very beginnings of the European Community (Kluwer, 2002).

7. Exits and new entrants: high dynamics in terms of market entries are an indication of strong competition. However, the opposite is not always true. Low dynamics could be an indication of weak competition, but could also be due to poor profitability of the industry. The latter will be captured by one

of the quantitative indicators established in the second tier. Hence, we have to see these two variables together. Empirically, we measure the share of annual exits and entries as compared to the existing market population.

8. Concentration: one would expect that the less competitive a market is, the more concentrated it is. This criterion is directly relevant to predatory pricing, since successful predation requires the predator to be already dominant. However, high concentrations need not to be critical per se. A trend towards concentration may also be driven by rising competitive pressure in a liberalised market. Hence, we have to see concentration in the light of the historical development of an industry. Empirically, we measure concentration rates in terms of output of the biggest firms in the industry and compare them to established critical thresholds. The German competition law has established critical concentration rates for the largest one, three and five companies in Germany. Apart from this, European legislation on merger control envisages a critical concentration on the European scale, if the biggest firm has a market share of more than 50%. This is assumed to be a necessary condition for predatory pricing. For market shares of the biggest undertaking below 25%, distortions of competition are considered not being an issue (EU Council 1989).

9. Internationalisation: this is a newly derived criterion, unique to the case of emissions trading and allocation. If internationally operating groups own production sites for the same product in different countries, these sites will become subject to the allocations of their respective national regulators under emissions trading. Irrespective of these national allocations being equal or being totally different, they will always sum up to the total allocation received by the international group. The group is then in a position to internally balance out any differences from the national allocations, e.g. by internally redistributing the received allowances among the different national production sites. The lack of homogeneity of a given distribution of national targets tends to be lowered through international groups. Empirically, we provide an overview of the industrial structures, using data from industry associations and official sources. This allows for an evaluation of the extent to which the industry is composed of international groups.

10. Trade intensity: high trade intensity across borders indicates strong competition between producers in different countries. However, if product prices are the same in all countries and transport costs are low, even zero trade will not preclude fierce competition. Hence, these criteria of transport

costs and price differences have to be viewed together (see below). We choose the ratio of total EU internal cross border imports to total EU production. Imports were measured in volume or mass units.

11. Transport costs and economic trade distances: high transport costs tend to lower international competition since they limit economic distances of trade and thus constrain the market. Empirically, transport costs are put into perspective with product prices and unit profit margins. This enables estimates of the economic distances of the product to be transported by different transport means.

12. Deviation of product prices: high price variances between different countries indicate low competition, since apparently the competitors do not exert price pressure. This is, however, contingent on a homogeneous product that cannot be differentiated. Empirically, in the absence of a statistical sufficient database, we replace the standard deviation by a simple measure, which is half of the difference of the extreme to average prices.

13. Price transparency: price transparency is known as a basic requirement for competitive markets. If the market is not transparent, we can thus take this as an indication of lowered competition. Empirically, we use information on how readily available price information is for consumers.

14. Product differentiation and customer services: product differentiation and customer services are classic means for firms to mitigate price competition. If products cannot be differentiated at all, price competition will be high. Empirically, we qualitatively distinguish means of product differentiation within different product classes.

15. The downstream market and customers choice: a competitive market is characterised not only by competing producers, but also by the free choice of consumers. Empirically, this concerns in general the wholesale and retail level. Analysing qualitative information on the industry in question, we ask whether customers can switch to a producer abroad, if national producer increase prices. This may require market intermediates that offer supply alternatives.

The evaluation form for these criteria can be found in Table 3.

6.3 Second Tier: Quantitative Indicators I and II

If the first tier is evaluated positively or if doubts remain, the second tier applies, using quantitative indicators. In the following section, we derive and discuss such indicators for the potential distortions identified in Section 5.

6.3.1 Prohibitive Allocation and Indicator I

Looking at Figure 8, illustrating the effects of prohibitive allocation and passive dumping, we note that the following factors influence the situation: the initial profitability π; abatement costs c_a; allowance costs c_q; and the difference between initial and final product price ($p_1 - p_0 = \Delta p$). Looking at Figure 8 and put into plain language, we can identify two necessary conditions for the effect of prohibitive allocation to happen:

- First, the firm in question must be profitable enough to survive the cost increase c_a under emissions trading. Otherwise, the firm would have left the market regardless of its allocation. This would constitute "environmental Darwinism" rather than distortions of competition.
- Second, the burden of the allocation must be strong enough to drive the firm out of the market.
- We can combine both conditions into one equation [I understand, that (1) is not an equation, but an inequation since it uses "<" rather than "=". But I can live with "equation" also, if "inequation" is not common in economic mathematics]:

$$0 < \pi - (c_a - \Delta p) < c_q \qquad (1)$$

The first part of (1) describes the first condition. Under homogeneous allocation, the new profits under emissions trading will be those noted in the middle part of (1), where from the initial profit that part of costs is deducted, which cannot be put through to the customers. However, if the allowance costs c_q due to allocation exceed the new profits under emissions trading, the firm becomes unprofitable and suffers from prohibitive allocation. This is noted in the second part of (1). However, due to empirical data constraints, we do not know the exact size of Δp in (1) (this would require demand and supply functions for different outputs and product classes). Therefore, we apply approximations. Furthermore, all variables in equation (1) bear empirical margins of uncertainty. An analysis of these margins shows that together, the relative error margin may amount to around +-50%. Therefore, we formulate the Indicator I for prohibitive allocation as follows:

$$\text{Indicator } I_a: \quad \pi > c_a \qquad (2)$$

Indicator I_b: $c_q/(\pi - c_a) < 0.5$ (low likelihood),
$0.5 < c_q/\pi < 1.5$ (medium likelihood),
$c_q/\pi > 1.5$ (high likelihood). (3)

In this formulation, (2) expresses the first condition on the initial profitability. The latter should be in the order of magnitude of the abatement costs c_a as discussed above. Outcome (3) expresses the second condition for prohibitive allocation to happen. It differentiates the allocation burden according to the error margins that we have to cope with empirically (50%). If (3) is smaller than 0.5, we can be almost sure that the allocation burden is too small to provoke prohibitive allocation. In the same sense we can use (3) with the ranges from 0.5–1.5 and above 1.5.

Indicator I can be applied if national allocation plans are known. If national allocation plans are not yet known, we can nevertheless apply Indicator I in a modified form. As we have discussed above, the larger c_q is in the country with the most stringent allocation, the more inhomogeneous the allocation is altogether in the international trading system. Hence, we can apply a sensitivity test by assuming the most inhomogeneous allocation possible. This would be the case, if we replace c_q in (3) with the allowance costs of the most stringent allocation possible, corresponding to a zero allocation (see Table 1). If in this case the cost increase is inferior to the profit margin (i.e. $I_b < 0.5$ in (3)), the industry considered is not vulnerable to prohibitive allocation, since it would remain profitable in this extreme case even if it cannot increase the product price at all. In other words, if $I_b < 0.5$ for all firms in all countries, the sector is too profitable for prohibitive allocation to happen at all.

We can also use the Indicator I for an *ex ante* guess regarding tolerable margins of differences between national allocations, before prohibitive allocation becomes an issue. Therefore, we assume *ex ante* the worst case, where the least profitable industry is hit by the most stringent allocation. At the maximum of tolerable, non-distorting diversity of national allocations, for this industry the allowance costs will then equal the initial profits. Provided that this industry is still profitable enough to pass Indicator I_a, this allocation sets the benchmark for the tolerable stringency of national allocation, before distortions arise. The critical allowance costs can be converted into CO_2 emissions at a given allowance price and be expressed as tolerable CO_2 allocation per unit of product output.

6.3.2 Extended Lifetime and Indicator I

Since the effect of extended lifetime is inverse to the effect of prohibitive allocation, we can use the same analysis as for prohibitive allocation; however, with the opposite direction. The only difference is that we have to reverse the negative sign of the allowance costs p_q in order to take into account that generous allocation involves negative allowance costs. We note the Indicator I for extended lifetime:

Indicator I_a: $\pi < c_a$ (4)

Indicator I_b: $|c_q| / (\pi\text{-ca}) < 0.5$ (low),
$0.5 < |c_q| / \pi < 1.5$ (medium),
$|c_q| / \pi > 1.5$ (high). (5)

Note that this effect of extended lifetime may occur with an allocation, which is only slightly more generous than average allocation. The problem here is that this allocation itself does not look critical, since it may still require an absolute reduction of the emissions of the firm, both in absolute terms and per unit of product output. However, it would still in fact constitute distortions of competition. This may become a political challenge, when the allocation plans are evaluated by the European Commission.

6.3.3 Passive Dumping and Indicator I

As for passive dumping, we note that the conditions are also similar to those for prohibitive allocation. The first condition for the effect to happen remains completely unchanged, since both effects require a firm to be initially profitable enough to survive with increased costs c_a under emissions trading. As for the second condition, the burden of the allocation must be strong enough to drive the firm out of the market even at the lower product price p_2, as shown in Figure 8. Hence, the equation (see above) (1) remains unchanged, with the only difference that Δp_p now denotes the price difference between p_2 and p_0. Hence, passive dumping will have the same negative proxy as prohibitive allocation, but may occur already in less stringent allocations. Whether passive dumping will actually occur once the conditions are met depends on the pricing strategies of firms, their market shares and the distribution of their allocations. This is the crucial difference to prohibitive allocation, which will always happen, once the conditions are met. Furthermore, since Δp is the smallest term in (1), we cannot explore it further with the above approximations. For these reasons, we evaluate passive dumping also with Indicator I for prohibitive allocation, knowing that in fact it may be always more likely than prohibitive allocation.

6.3.4 Predatory Pricing and Indicator I

When it comes to predatory pricing, the meaning of Indicator I is less straightforward. Strictly speaking, Indicator I has no significance on the potential for predatory pricing. If there is a firm with a dominant market position benefiting from generous allocation, the relative financial advantage from the sale of allowances may in general be decisive for the dominant firm to engage in predatory pricing. However, it is already difficult enough to prove predatory pricing *ex post*. It will be more difficult to predict whether a financial advantage from allocation plays a decisive role or not. However, as we have seen before, the profitability of the industry does matter: the higher the profits of the industry, the more difficult successful predation will be. This is the reason why we measure the likelihood of predatory pricing also with Indicator I. In other words, we interpret $I_b < 0.5$ for all firms as an indication for predatory pricing to remain unlikely, since the sector is too profitable, compared to the cost changes induced by emissions trading. If I_b exceeds 0.5, we will only evaluate the industry as "vulnerable" for predatory pricing. As discussed here, this is not a strict criterion, but serves rather as an indication.

6.3.5 Investment Advantage and Indicator II

The effect called investment advantage is difficult to evaluate. We have argued in Section 4 that investments/expenditures in activities such as R&D are considered, when a firm maximises its profits as well as all other expenditures/ investments for the direct manufacture of the product. However, the link between the expenditure for activities like R&D and the gain in market share is less rigorous than the links between marginal production costs, prices and optimal output. This difference requires a different approach. The simplest approach is to focus on the investments themselves rather than on their effects on the market share, which are difficult to estimate. In line with this reasoning, distortions of competition due to investment advantages would be proportional to the relation of the financial advantage from free allowances to the expenditures in one of these activities. We propose thus the following Indicator II:

$$II = \frac{c_q}{I} \tag{6}$$

In (6), I denotes total investments/expenditures of the firm under consideration in a certain activity such as R&D during a period with the same duration as the commitment period. Indicator II calculates the comparative cost advantage of a firm that can finance certain activities that are not directly related to output and price optimisation, such as publicity or R&D from free allowance revenues rather than from its income or loans. If the financial advantage of allocation equals the typical expenditures of the firm, Indicator II equals unity. However, there is no sharp critical threshold value, from which one could suppose the onset of distortions of competition. For instance, a firm with generous allocation may just pass the financial gift through to shareholders, without using it for any investments. We can only suggest interpreting Indicator II in the order of magnitude of 1 (rather than 0.1 or 0.01) as an indication of the significance of the cost advantage for the activity considered. In the sense of our definition, we would then call the firm with the stringent allocation vulnerable to distortions through the activity considered. If R&D expenditures are low in a sector, even very small differences in allowance allocation could result in values around unity for Indicator II. This does not, however, imply that the sector would suffer from distortions. Here, the fifth criterion of the first tier (see above) functions as a filter, requiring high expenditures/investments in the related activities, before Indicator II checks the relation between expenditures and allowance advantage.

6.4 Overview of the Two-tier Approach

Table 3 summarizes the conditions for the different distorting effects to happen and their various levels of likelihood.

6.5 Filling the Second Tier with Empirical Data

In the following chapters, we derive methodologies to fill Indicators I and II of the second tier with empirical data. The focus is on EU industry. However, data sources are not specialised and are available for most OECD countries.

6.5.1 Scenario Approach for 2010

For our analysis, we adopt the approach used by one of the predominant data sources, the studies that were commissioned by the European Commission in preparation of the proposal for the emissions trading directive. At the heart of these studies are scenario calculations for the year 2010, the central year of the first commitment period of the Kyoto Protocol. For this reason, we will also base our analysis on this year and assume that the EU emissions trading system will be up and running by that time. Furthermore, we choose the national industry level in Member States rather than the level of individual firms. This is a simplistic way to put Indicators I and II into operation and keep the data within manageable dimensions.

Taking into account the uncertainties surrounding the allowance price, a pragmatic approach is followed, assuming a broad range of prices that at least comprises one order of magnitude. This range was decided to be Eur 5–50/ton of CO_2, centred around Eur 20/ton. This range includes the case of the EU, significantly opening access to cheap allowances outside the EU and it covers the highest allowance prices predicted in models. The central value of Eur 20/ton of CO_2 corresponds to the estimates in EC studies and the above assumptions on allocation and emission sources.

Table 4 lists the different industries and their emissions in the EU. As can be seen, in terms of emissions, energy supply is by far the most important, followed

Table 3. The two tier approach. Likelihood of distortions of competition due to inhomogeneous allocation, conventions for the assessment

Prohibitive allocation and passive dumping	Extended lifetime	Predatory pricing	Investment advantage
(tier I.10–15 negative): Not relevant	(tier I.10–15 negative): Not relevant	(tier I.10–15 negative) or (tier (I.4 + I.6) or I.8 negative): Not relevant	(tier I.10–15 negative) or (tier I.5 low): Not relevant
(tier I.10–15 positive) and ($\pi \approx > c_a$) and ($c_q/(\pi - c_a) < 0.5$): Low	(tier I.10–15 positive) and ($\pi \approx < c_a$) and ($c_q\|/(\pi - c_a) < 0.5$): Low	(tier I.10–15 positive) and ($\pi < \approx c_a$) and ($\|c_q\|/(\pi - c_a) < 0.5$): Low	(tier I.10–15 positive) and (tier I.5 high) and (tier II, indicator II ≈ 1 or bigger): Vulnerable
(tier I.10–15 positive) and ($\pi \approx > c_a$) and ($0.5 < c_q/\pi < 1.5$): Medium	(tier I.10–15 positive) and ($\pi \approx < c_a$) and ($0.5 < \|c_q\|/\pi < 1.5$): Medium	(tier I.10–15 positive) and ($\pi < \approx c_a$) and ($0.5 < \|c_q\|/\pi$): Vulnerable	
(tier I.10–15 positive) and ($\pi \approx > c_a$) and ($c_q/\pi > 1.5$): High	(tier I.10–15 positive) and ($\pi \approx < c_a$) and ($\|c_q\|/\pi > 1.5$): High		

Tier I.4: market maturity, I.5: Investment in R&D, I.6: occurrence of predation in the past, I.8: concentration, I.10–15: geographical scope of competition (international vs. domestic).

Table 4. Industry sectors in the EU and CO_2 emissions

Industry	Total CO_2 emissions 1990/1995 [Mt]	Share [%]	Included in draft directive	Included in analysis of this paper
Electricity and heat generation generation)	1012	41.4%	Yes	Yes (electricity
Iron and steel (two routes)	196	8.0%	Yes	Yes (blast furnace route)
Glass, pottery, building materials	201	8.2%	Yes	Yes (cement)
Refineries	120	4.9%	Yes	Yes
Paper and pulp	29	1.2%	Yes	No
Non-ferrous metals	15	0.6%	No	No
Chemicals (incl. petrochemical)	108	4.4%	No	No
Food	46	1.9%	No	No
Other industry	718	29.4%	No	No
Total	2444	100.0%		

Source: modified data from (Capros *et al.*, 2001)

by building materials, steel and refineries. These four sectors were thus chosen for our analysis. However, since the two-tier approach requires a detailed level of analysis, we isolated within these industries those sub-sectors, for which products were most clearly separable. Regarding energy supply, we only chose electricity generation, and regarding building material we chose only the cement sub-sector. Both sub-sectors dominate their sectors. Within the steel sector, we only look into the classical blast furnace production route rather than in the electric arc route, since only the former directly emits CO_2, covered by the draft directive.

6.5.2 Empirical Data

For our analysis, we have chosen the six Member States (MS), in which the most important industries of the respective sectors are located (see Table 6). This limitation appeared to be adequate, since in most cases these 6 MS accounted for the bulk of the activity in the EU (60% for the oil refining industry, 82% for the steel industry and 96% for the cement industry).

The challenge in searching for adequate sources is that there is no single source covering all data needs. Furthermore, data are frequently inconsistent among different sources and are not available or spelled out at the required level of detail. Hence, information on several sources had to be mixed, leading to a certain degree of inaccuracy. We use the following principal sources as well as published articles:

- EC studies: these studies were carried out by consultant agencies on behalf of the European Commission in the context of the planned emissions trading scheme (bottom-up and top-down studies) and the IPPC directive. They are published by the European Commission.
- The PRIMES energy model of the EU, simulating numerically scenarios for all energy related industries in the EU, based on empirical data and policy inputs.
- EUROSTAT: the official statistical service of the EU.
- IPPC-BREFs: official Manuals on technologies and emission abatement in the framework of the IPPC-Directive of the EU.
- EC-merger decisions: EC decisions providing specific information on the market structure of the different industries.
- Associations: industry associations within the EU frequently have their own statistical services – data were used from EUROFER (Iron and Steel), CEMBUREAU (cement) and EUROPIA (the oil industry).
- Firms: corporate balance sheets and annual reports.

Table 5 summarizes the data types and sources used.

7 Empirical Results for Four EU Industrial Sectors

In this section, we summarise and discuss the results derived in the detailed empirical analysis of four industrial sectors: energy supply, building supplies, steel and oil refining. Our focus will be on the main results and their interpretation. The results for the two-tier approach for the four industrial sectors are

Table 5. Main data sources used for the empirical application of the two tier approach

Data type	Data source
Market structure	EC-merger decisions
Emissions, consumption, technologies	
SEC (specific energy consumption)	EC-bottom up studies, IPPC study
Trend improvement rate of SEC	PRIMES
SCO_2 (specific CO_2 emissions)	Own calculation, based on BREF, IPPC and EC-bottom up studies
Total CO_2 emissions of a sector	Own calculation, based on PRIMES for output and on SCO_2
Abatement measures (SEC)	EC-bottom up studies
Saving potential of a measure (SEC)	EC-bottom up studies
CO_2 saving potential of a measure	Own calculation, based on SEC and SCO_2 (specific CO_2 emissions)
Marginal abatement costs of a measure	EC-bottom up studies
Economic data	
Allowance prices	EC-bottom up and top down, own analysis
Gross operating surplus	EUROSTAT
Product prices	EUROSTAT, IEA, own investigation
Turnover	EUROSTAT, own calculations based on prices and output
Exits and entries of enterprises	EUROSTAT, BREF
Expenditures for R&D	EUROSTAT
Output base year and baseline 2010	PRIMES
Output 1990–1999	EUROSTAT, IEA, industry associations.

shown in Table 6. The resulting assessments with regard to distortions of competition are compiled in Table 7.

7.1 Tier I: Overview, Results and Discussion

The analysis carried out for the four sectors (energy supply, building supplies, steel and oil refining) revealed that there are some common characteristics, which are typical for all sectors examined. However, with regard to competition, each sector has its own characteristics and different means of how competition is exercised among the competing enterprises in different MS.

Table 6. Summary of the two tier approach for four EU industry sectors and selected Member States (MS)

	Steel manufacture	Cement manufacture	Oil refineries	Electricity generation
Number of Installations in the EU	~90 (BF/BOF route)	~250	~100	~1000
Tier I				
1. Privatisation	~95%	~95% (estimated)	~100%	Dominant State in BE, FR, IT, GR
2. Subsidies	less than 1€/ton steel, declining	Insignificant	Insignificant	Insignificant
3. Mark-up index (average 1970-1992)	(1.1 - 1.4)	n.a.	(1.07 - 1.19)	n.a.
4. Market maturity	Mature, saturated, overcapacity	Mature, saturated, overcapacity	Mature, saturated, overcapacity	Emerging market, overcapacity
5. R&D expenditures (high – low) [% of production value]	0.6 (low)	0.6 (low)	0.1 (low)	0.5 (low)
6. Occurrence of predatory conducts in the past	None	None	None	None
7. Exits and new entrants [per total installations, annually]	~10% exits, 5% entries	Stable, slight overall exits	Stable	Stable, barriers for new entries
8. Concentration on the European level (output capacity)	CR_1 28%, CR_3 50%, CR_5 61%	CR_1 21%, CR_3 56%, CR_5 70%	CR_1 19%, CR_3 41%, CR_4 48%	CR_1 18%, CR_3 36%, CR_5 48%
9. Internationalisation	Four biggest groups present in 2-4 big MS + several small MS, small companies only in one MS	Four biggest groups present in 2-4 big MS + several small MS, small companies in one MS	The four integrated majors present in all 4 big MS + NL+BE, small companies only in one MS	Three biggest groups present in 2-3 big MS + several small MS, small companies only in one MS.
10. EU internal trade intensity [imports of MS/consumption of MS]	~ 30-40%	~7% [exports/production]	~25% (16-42%, dep. on product)	~7% (physical, capacity constraint)
11. Transport costs, economic trade distances	Road: ~4,5-8€/100tkm, ~200-3000km (low-high cost product), more on train and ship	Road: 6-8€ /100tkm, ~ 500 km, several 1000km on ship	Road: ~ 100 - 150km, ~1000km barges, ~several 1000 km tankers	Domestic ~ 5-50€/MWh + 1-2€/Mwh cross border, whole EU
12. Deviation of prices among MS [%]	~10-20% (incl. VAT)	~ 20% (incl. VAT)	~ 10% (excl. VAT)	~ 20% (excl. VAT)
13. Price transparency	Weak	Very low	High	Low to fair
14. Product differentiation and customer service	Some	Insignificant	Insignificant	Insignificant
15. Geographical scope of competitive environment	EU	Regional	EU (indirect)	National (possibly EU in 2010)
Evaluation of Tier I	**Positive**	**Negative**	**Positive**	**Negative today, 2010 not clear**
Tier II				
MS selected for in depth analysis	BE, FR, GE, IT, SP, SV	BE, FR, IT, PO, SP, UK	BE, FR, IT, SP, UK	Different approach
1990 CO_2 emissions per ton output [kg CO_2/t]	2030 (EU), (1490NL-2230BE)	690 (EU), (660SV-726DK)	220 (EU)	~0.4, 0.7, 0.9 t/MWh (Gas, oil, coal)
2010 CO_2 per ton output after abatement under ET [kg CO_2/t]	1730 (EU)	590 (EU)	190 (EU)	n.a.
Prices per output, average (low - high price product) [€/t]	480 (300-660) (incl. VAT)	70 (incl. VAT)	245 (110-305) (excl. VAT)	50-90€/Mwh (incl. trans. & distr.)
Gross operating surplus/output (low-high price product) [€/t]	45 (28-64)	22	14 (13-15)	~2-10€/Mwh (generation only)
Relative gross profit margin, average 1990-1999	10%	30%	6%	3-16%
Maximum allowance costs[1] / ton product, allowance price 5, 20, 50€/t CO_2 [€/t]	9, 35, 87	3, 12, 30	1, 4, 10	2, 8, 20€/Mwh (Gas)
Indicator[2] I for middle product at a allowance price of 20 €/t CO_2, average 1990-1999.	~1.0	~0.6	~0.3	n.a.
Indicator[2] I for high cost product at a allowance price of 5, 20, 50 €/t CO_2, respectively, average 1990-1999.	~0.1, 0.6, 1.5	~0.1, 0.6, 1.4	~0.1, 0.3, 0.6	n.a.
Indicator[2] I for low cost product at a allowance price of 5, 20, 50 €/t CO_2, respectively, average 1990-1999.	~0.3, 1.3, 3.2	~0.1, 0.6, 1.4	~0.2, 0.9, 2.0	n.a.
Tolerable margins of inhomogeneous allocation[3] for allowance prices of 20 and 50 €/t CO_2, respectively.	~40 and 20 percentage points	n.a.	~50 and 20 percentage points	n.a.

Source: Own analysis, based on the discussed sources

1: Allowance costs in the case of the most inhomogeneous allocation: zero allocation to one firm and generous allocation to a competitor abroad. See table I for definition.

2: $I<0,5$ uncritical, $I>1,5$ critical. See equations (2) and (3) for the definition of indicator I.

3: Tolerable margins are expressed in percentage points of free allowance allocation per unit product output. See section 6.3.1 for the definition of tolerable margins.

Table 7. Likelihood for distortions of international competition due to maximum differences in domestic allowance allocation, evaluated for four industry sectors in the EU (differences between zero allocation and generous allocation[1])

Industry	Allowance prices [Eur/ton CO_2]	Prohibitive allocation and Passive dumping	Predatory pricing	Investment advantage (R&D)	Extended lifetime
Steel manufacture	5	Not relevant	Not relevant	Not relevant	Not relevant
	20	High (tinplate: medium)	Not relevant	Not relevant	Not relevant
	50	High	Not relevant	Not relevant	Not relevant
Cement manufacture	5	Not relevant	Not relevant	Not relevant	Not relevant
	20	Not relevant	Not relevant	Not relevant	Not relevant
	50	Not relevant	Not relevant	Not relevant	Not relevant
Refining of oil products	5	Not relevant	Not relevant	Not relevant	Not relevant
	20	Medium (fuel oil: high)	Not relevant	Not relevant	Not relevant
	50	High	Not relevant	Not relevant	Not relevant
Electricity generation in 2000[2]	5	Not relevant	Not relevant	Not relevant	Not relevant
	20	Not relevant	Not relevant	Not relevant	Not relevant
	50	Not relevant	Not relevant	Not relevant	Not relevant

1: See Table I for convention on terminology
2: Findings may change, if the future market environment becomes more competitive

Common points among the four sectors are the almost entire privatisation (apart from electricity generation in some Member States) and the absence of substantial subsidies in the form of State Aid. Furthermore, all sectors show overcapacities, due to different reasons, including past oil crises (refineries), new competing capacities outside the EU (steel), declining demand (cement) and a regulated past (electricity). Market dynamics in terms of new entries are generally low, being stable or with a slight tendency towards overall exits. Concentration is generally medium to high and tends to increase. Firms are trying to find optimum sizes to realise the advantages of economies of scale. Internationalisation is increasing as well, showing almost identical structures in the four sectors. Typically the market is split into the small firms and the major groups, the latter with concentration rates (CR_5) at around 60% on the European level. The major groups are typically present in several MS, including at least two and sometimes all four "big" MS. The small firms are typically only present in one MS.

As far as distortions of competition are concerned, predatory pricing and investment advantage could be evaluated as not relevant for all sectors already in the first tier. Expenditures in R&D are low (not exceeding 1% of the production value), so that distortions of competition due to investment advantage are not to fear. As for predatory pricing, concentration rates CR_1 do not reach dominant levels, and hence are insufficient for predatory pricing. The presence of mature and older industries and the fact that predation was not observed in the past in the EU, contributed to the negative evaluation of this effect in all four sectors.

Differences among the sectors relate to the geographic area of competition. In this regard, all four industrial sectors are distinct:

- Steel: for the steel sector, the scope of competition is EU wide and beyond. This is due to the fact that markets are liberalised, a significant wholesale market is in place, products can be differentiated to some extent and the relation of transport costs to product prices allows for transportation throughout the Union. The steel sector is the classic case of a competitive EU-internal market, where trade intensity is high and national boundaries have no implications for customers and steel manufacturers. The first tier is hence evaluated positively for the steel industry.

- Cement: for the cement sector, the geographical scope of competition is regional. It may at most comprise the border zones of neighbouring MS, if the cement plant is located in the border area. This is mostly due to the fact that transport costs are high, not allowing for road transport (accounting for 90% of the transport volume) to exceed distances of 500 km. The differences with the steel sector are obvious: one ton of steel costs around Eur 500, one ton cement costs not more than about Eur 70, but the transport costs are about the same for both products. Nevertheless, the high transport costs can still not explain the observed high price differences in the cement market, which are the highest of all four industrial sectors analysed. Prices for cement are as diverse as those for electricity. Whereas the latter industry has only recently been opened for competition and there are numerous obstacles for trans-boundary transportation of goods, which make big price differences no surprise, this is astonishing for cement, where in two neighbouring countries (France and Italy) prices differ by almost a factor of two. Even high transport costs should not frustrate exports from Italy to France. However, there are only insignificant exports from Italy (less than 3% of production). This had been also noted by the European Commission, when proceeding against the so-called cement cartel in 1994. Since then, however, the situation with regard to low trade intensity and big price differences seems to have remained unchanged in the EU. In conclusion, the cement market appears to be not competitive as far as EU internal competition of the cement manufacturers is concerned. Hence, the first tier is evaluated negatively for this industry.

- Oil refining: whereas the steel and cement industries are characterised by customers with the freedom of choice of their suppliers, this is not always the case for the oil refining market. The major products such as gasoline, diesel and fuel oil are sold mainly from peripheral depots or import terminals. Transport costs from the refinery or the cargo market to the depots and terminals are low, since transport means are typically pipelines or barges. However, the highest road transport costs of all industries analysed discriminate against any transport distance exceeding 150 km from these depots to consumers or retailers (filling stations). Hence, customers and retailers are bound to buy from local depots(which are usually owned by integrated oil companies), or from import terminals. The latter are controlled by both integrated oil companies and independents. Hence, for the market of refined oil products, the existence of a sufficient number of independently controlled import terminals, pipeline connections to the peripheral depots is a basic requirement for a competitive environment for oil refiners. Information on the extent of these independent capacities is incomplete. Available information suggests, nevertheless, that capacities are sufficient in the big MS and most small MS to allow big consumers and retailers a real choice between the cargo market and inland refineries. High import flows of refined products within the EU, high price transparency and the smallest price differences among all four sectors observed support this finding. Hence, in this sector, oil refineries in one MS compete with refineries in other MS indirectly via the cargo market. In conclusion, the geographical scope of competition among oil refineries comprises indirectly the whole EU and beyond. Hence, the first tier is evaluated positively for this industry.
- Electricity: due to its special nature (natural monopoly of transmission and distribution grids, public service obligations) and its regulated past, the electricity sector is different from the other sectors analysed. The other sectors are closely linked to the overall performance of the whole economy, whereas

economic cycles have traditionally little influence on the electricity sector. The first three sectors are characterised by market structures, where customers were either entirely free in their choice of the supplier on the EU level (cement and steel) or where customers could at least choose between a domestic supplier and an unspecified supplier from abroad (the cargo market in oil refining). In the electricity sector, though in the phase of liberalisation, this choice is still severely restricted for more than 50% of the demand in four MS, of which two are "big" MS. At what time these MS will open their markets remains to be seen. However, recent policy developments show that there are chances for an entirely liberalised EU electricity market by 2010. Besides market opening, there are other factors such as unbundling, high network tariffs, balancing regimes and dominant incumbents that still restrict competition even on the national level. This restriction of consumer choice and other obstacles also imply diminished competition among electricity generators. Since the sector is currently undergoing rapid changes subsequent to its liberalisation, it is difficult to evaluate the geographic area of competition. However, for the time being, the scope is clearly national at most. The prevailing high price differences among MS would normally be conducive for large trade flows. This is reinforced by the fact that the new system of transboundary tariffs allow in principal for economic electricity trading throughout the EU. However, the capacity of transboundary inter-connectors limits trades to the current low physical trade intensity, at least in the direction of the established trade flows. Hence, the first tier is evaluated negatively for electricity generation. In the future, this geographical area of competition may grow to the European level, depending on the further development of market opening and inter-connector capacities.

In conclusion, all four sectors show different characteristics of competition on the EU level. These are mainly grounded in different structures of the downstream market between the manufacture of the product and the final consumer, transport costs and political or physical restrictions. On the EU level, only the steel sector and the refining sector were found to be competitive. Cement was evaluated as not competitive and electricity generators may only become competitive in the future.

7.2 Tier II: Overview, Results and Discussion

Since we do not yet know the allocation plans of Member States, we can only apply Indicator I in the form of an *ex ante* sensitivity test. Hence, we test the effects of the maximum possible cost increase through allowance costs stemming from inhomogeneous allocation (zero allocation versus generous allocation).

Since the electricity market is currently in the process of ongoing liberalisation, only the other three sectors are quantitatively analysed.

Specific emissions differ widely among the products. Per ton of output, steel causes by far the most direct CO_2 emissions. Specific emissions for steel are about three times higher than those of cement and almost ten times higher than those per ton of mineral oil throughput. The emission reductions under emissions trading with a cost efficient industry target are similar for all three sectors and amount to a reduction of around 15% of the specific CO_2 emissions (an absolute reduction of 26% of CO_2 emissions from 1990 levels for the whole EU in 2010 has been evaluated as cost effective in the EC studies). The differences in gross profit margins are less pronounced. For a ton of average steel product we observe a profit margin of around Eur 45. This is about double the margin of one ton of cement and three times the margin of one ton of oil product.

The maximum cost increase for a firm under emissions trading is proportional to the assumed allowance prices times the specific emissions after abatement. At an estimated allowance price of Eur 20/ton of CO_2, a steel maker would have to spend Eur 35 in order to acquire all the allowances necessary to cover the emissions corresponding to the production of one ton of steel. For cement and average refined oil products this would only be Eur 12 and Eur 4, respectively.

From this data one can already see that steel making encounters much stronger pressure under emissions trading than cement making or oil refining. The quotients between the maximum expected cost differences (zero allocation – generous allocation) and profit margin for the three sectors relate to each other by a ratio of 4:2:1 for steel, cement and oil refining, and as do Indicator I (Table 6).

7.2.1 Evaluated Likelihood for Distortions of Competition

Table 7 compiles our results for the likelihood of the different distortions of competition, as evaluated based on the two-tier approach. As can be seen from the table, there are many cases for which distortions of competition due to differences in allocation can be ruled out. The effect of extended lifetime is evaluated as not relevant, since all the industries reviewed are too profitable for this effect to happen. However, we have to bear in mind that this assessment is made on the national rather than on the firm level. For individual firms, the picture may be different. This is, however, beyond the scope of this analysis. Predatory pricing and investment advantages were already negatively assessed in the first tier.

The effects of prohibitive allocation and passive dumping for the competitive steel and refinery sectors are relevant depending on the assumed allowance price. For the central value of Eur 20/ton of CO_2, the likelihood for these two effects are assessed to be medium and high for the refinery and steel industry, respectively.

Furthermore, we note a strong dependency of the likelihood through prohibitive allocation and passive dumping from the allowance price and product classes. Even for the most competitive sector (steel), at an assumed allowance price of Eur 5/ton of CO_2, Indicator I remains below 0.5 for all steel products and MS examined. Hence, the maximum cost increase is far lower than the profit margin and thus prohibitive allocation is not relevant. At an allowance price of Eur 20/ton of CO_2, only for the high value added tinplate, Indicator I remains below the critical 1.5 for all MS. For other steel products (hot rolled wide strip and narrow strip) distortions of competition are already likely at this allowance price. At an allowance price of Eur 50/ton of CO_2, markets for all steel products would be distorted with a high level of likelihood. For the oil industry, the structure of the results is similar, with the difference that the critical allowance prices are higher. Most vulnerable here is the low value fuel oil, while the vulnerability of gasoline and diesel is lower.

7.2.2 Tolerable Differences in National Allocation and Meaning for the EU Trading System

For the most vulnerable sector (steel), we calculate that at the maximum assumed allowance price of Eur 50/ton of CO_2 the absolute amounts of allowances allocated free of charge by national regulators should not differ by more than 20 percentage points in terms of the corresponding allowed specific emissions, in order to avoid these distortions. This tolerable margin could even be around 40 percentage points at an allowance price of Eur 20/ton of CO_2. Tolerable margins for the refinery industry would be somewhat larger.

These results have a direct bearing on the upcoming EU emissions trading directive. The distinction of the European Commission between allocation methodology (grandfathering versus auction) and allocation stringency is missing the point. If national regulators exceeded the tolerable differences of allocation stringency as outlined above, we predict that these allocations may have a prohibitive effect, driving some firms out of the market. This is possible – although regulators would allocate through grandfathering and not through an auction. This effect could be avoided by harmonizing the tolerable margins of national allocation at the EU level. This should not impose an undue burden on the subsidiary principle of the EU and the EU burden sharing agreement for the Kyoto Protocol, given the large tolerable margins that would provide national regulators with flexibility.

7.3 Conclusions

In conclusion we note that, in theory, output-optimising enterprises that receive free emission allowances would, in general, optimally strive for the same price and output decisions as those enterprises that have to pay for their allowances. The only exception is the case of combining an output or benchmark-based allocation system with a grandfathering system, in which case the allocation method drives the competing enterprises to different output decisions. This general independence of output and price decisions from the allocation method and stringency can be understood by the fact that free allowances entail opportunity costs, which have to be included in the optimisation calculus. However, in some cases, national allocations may be distributed so unevenly that enterprises will not behave like their competitors (prohibitive allocation), or that enterprises will be offered an option which does not exist otherwise (extended lifetime). Furthermore, in some cases, the distribution of allocations may allow enterprises to successfully pursue pricing strategies to force their rivals out of the market (passive dumping). These potential effects of distortion of competition exist despite the opportunity cost argument, which has been used by the European Commission in order to justify the non-harmonisation of allocation stringency in the draft emissions trading directive. Finally, there are the effects of financial positions, which can influence the competitive position of rivalling firms (investment advantage and more theoretically, predatory pricing).

Empirically, we find that European refiners and steel manufacturers do compete indirectly or directly on the EU level. For these two sectors, we evaluate at a central allowance price of Eur 20/ton of CO_2 a medium to high likelihood for distortions of competition through prohibitive allocation and passive dumping. The risk increases directly with the assumed allowance prices, for which Eur 20/ton of CO_2 are taken as a central estimate for the cost-efficient implementation of the EU Kyoto targets. However, we show also that this risk may be avoided, if allocations of national regulators are restrained to a tolerable allocation margin. This margin is quantified to a 40–50 percentage point difference in free allowances in terms of specific emissions of the product in both sectors. At higher allowance prices, the tolerable margin would be correspondingly smaller. The effects of extended lifetimes, investment advantage and predatory pricing are evaluated as not relevant, since neither market nor cost structures are found being conducive for the effects.

References

Bienaymé, C. (1998). 'Principes de Concurrence', Paris: Economica.

Blok, K., De Jager, D. and Hendriks, C. (2001). *Economic Evaluation of Sectoral Emission Reduction Objectives for Climate Change – Summary Report for Policy Makers*, Utrecht: ECOFYS, AEA Technology, NTUA.

BREF Cement (2000). *Reference Document on Best Available Techniques in the Cement and Lime Manufacturing Industries*, European Commission, Directorate-General Joint Research Centre, Seville, Spain: European IPPC Bureau, http://eippcb.jrc.es.

BREF Refineries (2001). *Reference Document on Best Available Techniques for Mineral Oil and Gas Refineries*, European Commission, Directorate-General Joint Research Centre, Seville, Spain: European IPPC Bureau, http://eippcb.jrc.es.

BREF Steel (2000). *Best Available Techniques Reference Document on the Production of Iron and Steel*, European Commission, Directorate-General Joint Research Centre, Seville, Spain: European IPPC Bureau, http://eippcb.jrc.es.

Brennan, Timothy J. (2000). 'The Economics of Competition Policy: Recent Developments and Cautionary Notes in Antitrust and Regulation', *Discussion Paper*, 0:7, Washington: Resources for The Future.

Capros, P. et al., (1999a). *European Union Energy Outlook to 2020*, Catalogue number CS-24-99-130-EN-C, ISBN 92-828-7533-4, Brussels: European Commission – Directorate General for Energy (DG-XVII).

Capros, P., Mantzos L., (1999b). *Energy System Implications of Reducing CO_2 Emissions, Analysis for EU Sectors and Member-States by using the PRIMES Version 2 Energy System Model*, Study on behalf of the European Commission, Final Report from ICCS/NTUA, Athens: National Technical University of Athens.

Capros, P., Mantzos L., D.W., Pearce, A., Howarth, C., Sedeem B.J. and Strengers, (2000). *Technical Report on Climate Change*, RIVM, EFTEC, NTUA and IIASA, Bilthoven: RIVM report 481505012.

Capros, P., Kouvaritakis, N., Mantzos, L., (2001). *Top-own Analysis of Greenhouse Gas Emission Reduction Possibilities in the EU*, Athens: National Technical University of Athens.

COM Performance (2001). 'Competitiveness, innovation and enterprise performance', The Competitiveness Report of the Innovation Scoreboard and the Enterprise Scoreboard, 2001 edition, Brussels: European Commission.

De Beer, J., Phylipsen, D. and Bates, J. (2001). *Economic Evaluation of Sectoral Emission Reduction Objectives for Climate Change. Economic Evaluation of Carbon Dioxide and Nitrous Oxide Emission Reductions in Industry in the EU – Bottom-up Analysis*, ECOFYS Energy and Environment, The Netherlands & AEA Technology Environment, Culham, UK.

EU Council (1989). 'Merger Control regulation', Council regulation 40/64/89, OJ L395, 1, 30.12.1989, last amended in OJ L180, 1, 9.07.1997.

Hall R. and Hitch C. (1939). 'Price Theory and Business Behaviour', *Oxford Economic Paper*, 2, Oxford.

IPPC (2000). *Study on Energy Management and Optimisation in Industry*, AEA Technology plc.

Kluwer (2002). *European Competition Law*, database regularly updated, The Netherlands: Kluwer Law International.

Koutstaal, P. (1997). *Economic Policy and Climate Change: Tradable Permits for Reducing Carbon Emissions*, Cheltenham, UK: Edward Elgar.

Martins, J.O., Scarpetta, S. and Pilat, D. (1996). 'Mark-up ratios in manufacturing industries, estimates for 14 OECD Countries', *Working paper 162*, Paris: OECD.

Milgrom, P. and Roberts, J. (1982). 'Predation, Reputation and Entry Deterrence', *Journal of Economic Theory*, 27.

Mulder, M. (2002). 'Economic effects of national emissions trading schemes; national dilemma's within a global issue', Paper prepared for the 25th Annual International Conference of the IAEE (International Association of Energy Economics), Aberdeen, Scotland, UK, June 26–29, 2002.

OECD (1989). *Predatory Pricing*, Paris: OECD.

Van der Laan and Nentjes (2001). 'Competitive distortions in EU environmental legislation: Inefficiency versus inequity', *European Journal for Law and Economics*, 11:2, 131–152.

Chapter 11

FRÉDÉRIC JACQUEMONT

The Kyoto Compliance Regime, the European Bubble: Some Legal Consequences

Under the climate change regime, the European Community (EC) and its Member States have decided to jointly reduce their greenhouse gas emissions under a common target during the first commitment period set by the Kyoto Protocol that lasts from 2008 to 2012. The European Community and its Member States are jointly committed under the Kyoto Protocol (KP) to cut their total aggregate anthropogenic carbon dioxide equivalent of greenhouse gas (GHG) emissions, a 'basket' of six gases, by 8% in 2012, relative to 1990 levels. This is the European Bubble[1]. While the Kyoto Protocol fixes a kind of global bubble for Annex I countries, namely the goal of an aggregate reduction of GHG by 5%, relative to 1990 levels, this global goal is broken down into differentiated national commitments. Article 4 of the Kyoto Protocol opens up the possibility to jointly fulfill these national targets, thanks to the insistence of the EC[2]. It allows a group of countries

1 Art. 3(1) Kyoto Protocol (KP). The six GHG listed in Annex A of the KP are carbon dioxide (CO_2), methane (CH_4), nitrous oxide (NO_2), hydrofluorocarbons (HFCs), Perfluorocarbons (PFCs) and sulphur hexafluoride (SF_6). Quantified emission limitations are listed in Annex B of the KP. The 'basket' approach means that all gases covered by the target would be considered together for the achievement of the target according to their carbon dioxide equivalent based on their global warming potential, rather than choosing a gas-by-gas approach where a target applies to each gas individually.

2 Art. 4(1) KP: "Any Parties included in Annex I that have reached an agreement to fulfil their commitments under Article 3 jointly, shall deemed to have met those commitments provided that their total combined aggregate anthropogenic carbon dioxide equivalent emissions of the greenhouse gases listed in Annex A do not exceed their assigned amounts calculated pursuant to their quantified emission limitation and reduction commitments inscribed in Annex B and in accordance with the provisions of Article 3. The respective emission level allocated to each of the Parties to the agreement shall be set in that agreement." For an historic of Article 4, see Oberthür and Ott (1999), 140–149.

MICHAEL BOTHE AND ECKARD REHBINDER (EDS.), Climate Change Policy, 351–406.

with emission reduction commitments, possibly together with a regional economic integration organization, to jointly meet those commitments by entering into an agreement that reallocates among the Parties their combined total assigned amounts calculated pursuant to Article 3 paragraph 1, and inscribed in Annex B of the Kyoto Protocol. Annex B of the Kyoto Protocol fixes the percentage for each Member State of the EC at 92%, and for the EC as a whole at 92% as well. As required by Article 4, paragraph 1, such agreement must set out the exact allocation of emission levels to each Party. This reallocation creates flexibility among the Parties that agree on limitation, reduction, or even increase targets for each of them according to their economic circumstances. This differentiated approach to sharing the burden of emission reduction aims to introduce fairness and cost-effectiveness within a relatively homogeneous group of countries such as the EC Member States, where environmental and economic dissimilarities still exist[3]. Different commitments are determined by taking into consideration the countries' different contribution to GHG emissions and different national economic development circumstances[4]. When they approved the KP, the Member States and the EC also communicated their national targets agreed among them at the European Council of 16 June 1998, as a result of the internal repartition of the European Bubble[5]. This joint commitment which binds the EC together with its Member States may have particular legal consequences both at the international and the European level which are not yet satisfactorily addressed.

The Kyoto Protocol prescribes legally binding quantified emission reduction targets and provides for certain consequences in the event of a Party not complying with its emission reduction or limitation commitment. Within the context of the Community, where Member States are jointly committed, the question arises who is internationally responsible, either the EC as a whole or only the failing Member States, or both. Who is to be sanctioned internationally, the Community as a whole, with the consequence of sanctioning all Member States, even those in compliance, or only the EC together with the individual failing Member State? The Kyoto Protocol gives no clear answer to this question and is subject to different interpretations. At the European level, although the EC has a long tradition of being jointly committed together with its Member States, very few cases have been decided by the European Court of Justice (ECJ) that could clarify the issue.

---

[3] For an overview of the advantages and disadvantages of differentiated environmental commitments, see Ringius (1997).

[4] Greece, Ireland, Portugal, and Spain, the so-called cohesion countries, were concerned that the Community's climate policy would negatively affect their development, and that it would be necessary to take into account their need for economic development. Ibid.

[5] See Council Decision 2002/358/CE of 25 April, 2002. For example: Belgium 92.5%, France 100%, Germany 79%, Greece 125%, see Annex I of this chapter.

Despite these legal uncertainties, the Community must secure its capacity to meet its target, and in order to do so it depends heavily on its Member States. The question thus arises whether there exist compliance systems, both under the KP and under Community law, which are efficient enough to achieve the treaty goal. The first part of this paper will describe the KP's compliance rules as set out in the Marrakech Accords. A second part will highlight the historic development of the joint commitment under the KP. Then the third part will analyze how the Kyoto regime addresses joint commitment, particularly the liability issue in relation to non-compliance. Finally, the last part will assess the Community response to being jointly committed under the Kyoto Protocol.

1 Compliance Rules under the Kyoto Protocol

In order to ensure that international legal rules are actually implemented, institutions and procedural rules have been developed, in particular the rules of state responsibility and the mechanisms of dispute settlement. However, in the field of international environmental law, recourse to state responsibility for breach of an obligation and traditional dispute settlement mechanisms are widely considered as inappropriate[6]. The KP corresponds to a trend, which can be seen in recent multilateral environmental agreements, that procedures to secure compliance should not be confrontational and punitive, but should rather consist of cooperative remedial and facilitative measures that address the causes of non-compliance.

Non-compliance is a very broad term that describes a situation in which a Party to an agreement does not implement it or does not fulfill obligations arising from the agreement[7]. For the purpose of this article, this section will concentrate on compliance mechanisms and rules that ensure respect for the substantive obligations of Annex I Parties to reduce or limit their emissions below certain levels. The following sections, thus, do not exhaustively address compliance rules and mechanisms adopted under the KP[8].

---

6 For an assessment of the traditional approaches to enforcement in international environmental law see Bothe (1996), 13–38. For comments on Draft Articles on State Responsibility of the International Law commission in relation to Multilateral Environmental Obligations see Peel (2001), 82–97.

7 See Sands (1995), 170.

8 For an overview of the Kyoto compliance regime and negotiations development, see Wang and Wiser (2002), 181–198.

1.1 The Link between Compliance and Inventory, Monitoring and Reporting Requirements

The incentive for Annex I countries to furnish reports rests on their conditional participation/or eligibility to the so-called flexible mechanisms, international emission trading (IET), joint implementation (JI) and the clean development mechanisms (CDM)[9]. These mechanisms offer Annex I Parties the possibility to reduce their emissions in a cost-effective manner, by working in cooperation with other partners. The CDM and JI are project-based mechanisms that allow Annex I Parties to gain emissions credits by investing in countries where the cost for emissions reduction is cheaper than at home. The CDM mechanism aims at the development of emission reductions projects in developing countries financed by Annex I Parties (developed countries). The JI mechanism facilitates such emission reductions projects between Annex I Parties. The IET should establish an international market that allows Annex I Parties to buy and sell emissions credits between them. Although the decision to participate is entirely voluntary some compliance requirements have been established which subject the participation of Annex I Parties in these mechanisms to certain conditions, the so called 'eligibility conditions'. The Parties concerned must fulfill seven conditions: (i) be a Party to the Protocol; (ii) have established its assigned amount; (iii) dispose of a national inventory system that estimate GHG emissions and removals; (iv) have in place a national registry; (v) submit annual inventories pursuant to Articles 5.2 and 7.2 of the KP; (vi) submit supplemental information on assigned amount pursuant to Articles 7.1 and 7.4 of the KP and; (vii) designate a national authority to ensure the supervision of JI and CDM projects. Depriving a Party of such tools for fulfilling its commitment would entail an additional cost in complying with reductions targets. Therefore, participation in the flexible mechanisms is used as a 'carrot' to induce compliance with some requirements. Furthermore, transparent and accurate information will secure that these mechanisms operate in an environmentally sound manner. In this case, participation by Annex I Parties in these flexible mechanisms depends mainly on their conformity with methodological and reporting requirements relating to the inventory, under Article 5, paragraphs 1 and 2, and to the registry, under Article 7, paragraphs 1 and 4 of the Kyoto Protocol[10]. Both measuring a country's emission levels based on a correct inventory, and tracking a Party's assigned amount units (AAUs) and the different credits resulting from its participation to these mechanisms through a registry are two essential elements of determining compliance. For this purpose, an

[9] Arts. 17, 6 and 12 KP respectively.

[10] Paragraph 5 Draft Decision-/CPM.1 (Mechanisms), FCCC/CP/2001/13/Add.2

Annex I Party, no later than one year prior to the commencement of the first commitment period, shall have in place a national inventory for estimating GHG emissions by source and their removal by sinks[11]. The methodologies used for estimating GHG emissions by source and removal by sinks shall be those recommended by the IPCC and will be agreed at COP 9 in order to secure the reporting of consistent, transparent, comparable, complete and accurate information[12].

Once the inventories are in place, Annex I Parties shall calculate their total assigned amount pursuant to Articles 3.7 and 3.8 and in accordance with the annex on modalities for the accounting of assigned amount (Decision 19/CP.7). Each Annex I Party has to submit, prior to the commencement of the commitment period, a report that contains all the information needed to calculate its assigned amount, and all relevant information that demonstrates the Party's ability to monitor, track and record transactions related to the flexible mechanisms. Furthermore, they will have to adopt and maintain a national registry pursuant to the modalities adopted under Article 7.4[13]. Each national registry would start with a country's initial AAUs and would be adjusted to reflect the issuance of credit units, as well as related transactions and cancellations. Transfers would be made directly between national registries, and transaction would be definitively recorded once made. In addition, each registry would cover removal by sinks through issuing credits or through the cancellation of these credits from the registry. The form of such registry would be an electronic record of the different credit units. Each credit unit would be labeled so as to identify the country of origin and carry a serial number and the date at which it was included in the registry. The different transactions would be recorded in a way to maintain this basic information, so that each credit retained could always be tracked back to the original seller. In order to supplement this national system, an independent transaction log to be held by the UNFCCC Secretariat would be used as the basis for the compliance assessment at the end of the commitment period by tracking and monitoring the operations of the Parties' registries[14]. It will allow automatic checks during transactions to verify whether a Party is eligible to use the mechanisms and to ensure that credit

[11] Art. 5.1 and Draft Decision-/CMP.1, and Annex FCCC/CP/2001/13/Add.3.

[12] Art. 5.2. An adjustment procedure is provided when inventory data submitted by the Annex I Parties are not consistent with IPCCC *Guidelines for National Greenhouse Gas Inventories*; see Draft Decision-/CMP.1 (Article 5.2), FCCC/CP/2001/13/Add.3. A description of national registries is required, see Decision 22/CP.7, Appendix (*Reporting of supplementary information under Art. 7.2*); FCCC/CP/2001/Add.3.

[13] See Draft Decision-/CMP.1 (*Modalities for the accounting of assigned amounts*) & Annex (*Modalities for the accounting of assigned amount under Art. 7.4 of the KP*); FCCC/CP/2001/13/Add.2.

[14] Para. 38, Point D of Section II, Annex to Draft Decision-/CMP.1; FCCC/CP/13/Add.2.

units are located in one registry, and are not being used more than once[15]. The public will be allowed to have access via internet to some information on a national registry, such as the list of units and their serial number held in each account[16]. Submission of supplementary information on assigned amounts pursuant to Article 7.1 by Annex I Parties and observance of guidelines for accounting of assigned amounts adopted pursuant to Article 7.4 are also eligibility requirements for all three mechanisms[17]. Finally, Annex I Parties shall every year report their inventory of GHG emissions and removal by sinks and their total amount of credit units detained within their national registry for the previous year[18].

All information forwarded will be reviewed and assessed by expert review teams, who will report to the COP/MOP and to the Secretariat[19]. They will also assess at the end of the Commitment period whether an Annex I Party has met its target. A special reinstatement review procedure is provided for Parties that were declared not eligible to the mechanisms[20].

1.2 Verification and the Review Expert Teams

Article 8 of the Kyoto Protocol provides that all information submitted by Annex I Parties shall be reviewed annually by expert review teams. The expert review teams will also assess at the end of the commitment period whether an Annex I Party has met its target. Expert review teams will act independently, be selected on the basis of competence and geographic considerations to reflect a balance between Annex I and non-Annex I Parties, and be coordinated by the Secretariat[21]. Where inventory data are incomplete an adjustment procedure is foreseen[22]. When the Party concerned disagrees with the expert review team on an adjustment, the case is forwarded to the Compliance Committee. After completion of the review process the expert review teams will report to the Secretariat, the Party concerned, the COP/MOP and the Compliance Committee.

---

15 Para. 42, Point D of Section II, Annex to Draft Decision-/CMP.1; FCCC/CP/13/Add.2.

16 Point E of Section II, Annex to Draft Decision-/CMP.1; FCCC/CP/13/Add.2

17 See Decisions 19/CP.7, 22/CP.7 and 22/CP.8.

18 Art. 7.1 & Decision 22/CP.7 (*Guidelines for the preparation of the information required under Art. 7 of the KP) Appendix (Reporting of supplementary information under Art.7.1*), FCCC/CP/2001/13/Add.3.

19 Art. 8 KP, and Decision 23/CP.7 and Draft Decision-/CMP.1 Art. 8); FCCC/CP/2001/13/Add. 3.

20 Appendix II to Decision 23/CP.7; FCCC/CP/2001/13/Add.3.

21 Section E, Annex to Draft Decision-/CMP.1 (Article 8), FCCC/CP/2001/13/Add.3.

22 Sub-section 4 to Section D, Annex to Draft Decision-/CMP.1 (Article 8), FCCC/CP/2001/13/Add.3.

1.3 The Compliance Committee

The Compliance Committee is a standing body, composed of twenty members (ten for each branch of the Committee) elected by the COP/MOP based on a fair geographical distribution between the five UN regional groups (involving a fair share between Annex I and non-Annex I countries, plus a member for each branch from small island developing states) and their competences relating to climate change. This geographical representation rule was strongly defended by the G77/China and opposed by the Umbrella Group in Bonn. The Umbrella Group feared that this procedure could be used for political reasons other than reasons related to compliance issues, which made the group more reluctant to adopt any legally mandatory regime[23]. The Compliance Committee can address cases of non-compliance by two different means: by offering the possibility to restore compliance, and by sanctioning this state. The structure of the Compliance Committee reflects this double differentiation in the treatment of non-compliance. The Compliance Committee consists of two branches, the Facilitative and the Enforcement Branch.

The Facilitative Branch has the general aim to assist all Parties, developed and developing, in their implementation of the Kyoto Protocol. Regarding Annex I countries, it will give individual advice and formulate recommendations on submitted national inventory, and in meeting eligibility requirements[24]. It will particularly serve for Annex I Parties as an early warning when they have difficulties in meeting their emissions targets.

The Enforcement Branch serves as a quasi-judicial forum for determining whether an Annex I Party (i) has met its target, (ii) has complied with its monitoring and reporting requirements and (iii) has complied with the eligibility requirements for participating in the flexible mechanisms[25]. It decides on the adjustments of inventories as well as on the correction of the accounting of assigned amounts in the event of a disagreement between the expert review team and the Party involved[26]. Finally, when the enforcement branch finds that a Party has failed to comply with one of these obligations, it will apply the appropriate and predetermined consequences[27]. The enforcement procedure can be triggered by a Party regarding another Party, or by the Party itself, or on the basis of the assessment of the expert review

---

23 Annex, Section II para. 3 & 6, Decision 24/CP.7, FCCC/CP/2001/13/Add.3.

24 Sections IV & XIV, Annex to Decision 24/CP//2001/13/Add.3.

25 Para. 4, Section V, Annex to Decision 24/CP/2001/13/Add.3.

26 Para. 5, Section V, Annex to Decision 24/CP/2001/13/Add.3.

27 Para. 6, Section V, Annex to Decision 24/CP/7, FCCC/CP/2001/13/Add.3.

teams forwarded by the Secretariat to the Climate Change Convention and to the KP[28].

1.4 Consequences of Non-Compliance

When the Party is not in compliance with its reporting obligations under Articles 5 paragraph 1 or paragraph 5, or under Article 7, paragraph 1 or paragraph 4, the Enforcement Branch declares the state of non-compliance and requires the Party involved to submit a precise plan indicating how and by when the Party intends to restore compliance with such reporting requirements[29].

Regarding the eligibility requirements for the flexible mechanisms, the Party shall be suspended from using these mechanisms when the Enforcement Branch has determined that an Annex I Party does not meet one or more of the eligibility requirements[30]. No appeal to the COP/MOP is provided once the decision of suspension has been taken. Whether a particular transaction is restricted depends on the eligibility requirement for the particular flexible mechanisms[31]. A Party may be thus excluded from one of these flexible mechanisms, but still enjoy the use of the others. However, the consequence of being suspended from the EIT may be that no transactions relating to any credit units resulting from JI or CDM can be carried out until eligibility is restored[32]. A Party may submit a request to reinstate its eligibility[33].

Finally, when an Annex I Party does not comply with its emissions target specific consequences will be applied by the Enforcement Branch: (i) each excess ton of emissions x 1.3 will be deduced from that Party's AAUs in the second commitment period. That rate may be increased for subsequent commitments periods. (ii) The Party concerned will develop a detailed compliance action plan describing the measures that it will undertake for complying with the subsequent target; the enforcement branch will review the plan and will assess whether the action plan will work or not. (iii) Finally the Enforcement Branch shall suspend the Party's right to transfer assigned amounts under Article 17.

---

[28] Furthermore, reports of the COP/MOP, or the other branch of the Compliance Committee can initiate such procedure. See section VIII, Annex to Decision 24/CP/2001/13/Add.3.

[29] Paras. 1, 2 & 3; Section XV; Annex; Decision 24/CP/7; FCCC/CP/2001/13/Add.3.

[30] Para. 4, Section XV, Annex to Decision 24/CP/7, FCCC/CP/ 2001/13/Add.3.

[31] Para. 4, Section XV, Annex to Decision 24/CP/7, FCCC/CP/2001/13/Add.3.

[32] Para. 2, Annex, Decision 18/CP.7, FCCC/CP/2001/13/Add.2.

[33] Para. 2, Section X, Annex, Decision 24/CP/7, FCCC/CP/2001/13/Add.3

1.5 Reinstatement Procedures

Before such consequences apply, reinstatement procedures concerning the reinstatement of the Party's eligibility to participate in the three flexible mechanisms, and a delay period to secure compliance with its target are at the disposal of the Parties concerned. At the request of a Party, the reinstatement to the flexible mechanisms will be examined in collaboration with the Expert Review Team, or by the Enforcement Branch[34]. Furthermore, in order to comply with their targets, Annex I Parties concerned may dispose of an additional period of hundred days, after the date for the completion of the expert review process for the last year, to acquire credits[35].

1.6 General Procedures

Rules are aimed at securing due process, transparency, and legal certainty depending upon which basis a branch takes a decision. They allow Parties to be represented, they provide the opportunity for them to make comments, and they permit competent intergovernmental organizations and NGOs to submit factual and relevant technical information. Transparency is secured by making information, comments and the final decision available to the public, bearing in mind relevant confidentiality rules[36]. Special procedural rules have been developed for the Enforcement Branch in order to take into account the time constraint[37]. This should ensure that compliance problems in relation to the flexible mechanisms and emissions reduction targets are dealt with efficiently in a relatively short time and that the market maintains confidence in the whole system.

1.7 The Right of Appeal

After the Enforcement Branch has found that a Party has exceeded its target, that Party may appeal to the COP/MOP "if it believes it has been denied due process"[38]. This last wording gives a broad possibility to the Party concerned to seize the COP/MOP. The COP/MOP can override the Enforcement Branch's decision by a majority of three quarters of the members present and

---

34 Section X, Annex to Decision 24/CP.7 and Appendix II to Decision 23/CP.7; FCCC/CP/2001/13/Add.3.

35 Section XIII, Annex to Decision 24/CP.7; FCCC/CP/2001/13/Add.3.

36 Para. 7, section VIII, Annex, Decision 24/CP.7; FCCC/CP/2001/13/Add.3.

37 See Sections IX & X, Annex, Decision 24/CP.7; FCCC/CP/2001/13/Add.3.

38 Paragraph 1, Section XI, Annex to Decision 24/CP.7; FCCC/CP/2001/13/Add.3.

voting. In this event, the COP/MOP shall refer the matter of the appeal back to the Enforcement Branch[39]. The decision of the Enforcement Branch shall stand pending the decision on an appeal. It shall become definitive if no appeal is made against it within a period of 45 days after the Party has been informed of the decision of the Enforcement Branch[40].

1.8 The Role of COP/MOP

The role of the COP/MOP, which is the political body of the KP, is rather limited in the compliance procedure. It considers the reports of the Compliance Committee on its work, provides general guidance to the Committee and adopts decisions on administrative and budgetary matters.[41]

1.9 The Legally Binding Nature of the Compliance Rules and Their Adoption

"Any procedures and mechanisms under this Article entailing binding consequences shall be adopted by means of amendment to this Protocol" (Article 18 KP). On the one hand, this requirement reflects the inability of the Parties represented in Kyoto to agree on the consequences of non-compliance[42]. On the other hand, this formal procedural requirement demonstrates that states want to secure a high degree of legal certainty for rules which may encroach on their sovereign rights. At the Marrakech COP, compliance rules and mechanisms were enclosed in an Annex to Decision 24/CP.7, but not in the form of a draft COP/MOP decision like other agreed rules, in order to preserve the recourse to the amendment procedure. According to Decision 24/CP.7, it will be the prerogative of the first COP/MOP to agree on the legal form of the procedures and mechanisms relating to compliance[43]. Thus, there is no guarantee that compliance mechanisms will be in fact adopted as provided in Article 18 KP, although this procedure is recommended by Decision 24/CP.7[44].

Whether the first COP/MOP will adopt the compliance regime in the form of an amendment to the Protocol, or just as a COP/MOP decision is of little practical relevance[45]. Under recent multilateral environmental

[39] Paragraph 3, Section XI, Annex to Decision 24/CP.7; FCCC/CP/2001/13/Add.3.
[40] Paragraphs 2 and 4, Section XI, Annex to Decision 24/CP.7; FCCC/CP/2001/13/Add.3.
[41] Section XII, Annex to Decision 24/CP.7; FCCC/CP/2001/13/Add.3.
[42] On the negotiating history of this article see Oberthür and Ott (1999), 217–221.
[43] Point 7, Decision 24/CP.7, FCCC/CP/2001/13/Add.3.
[44] Para. 2, Decision 24/CP.7, FCCC/CP/2001/13/Add.3.
[45] For Lefeber (2001), 25–54, the second sentence of Art. 18 is a "superfluous addition".

agreements, most *ad hoc* compliance procedures have been established by a decision of the respective COP[46]. Depending on the relevant enabling provisions, this does not necessarily put into question the legally binding character of these procedures[47]. Adopting the compliance rules and mechanisms in the form of an amendment to the KP entails a cumbersome procedure. Though consensus is not required, it is the preferred means of adoption, which entails a risk of reopening discussions on the core rules of the regime[48]. The latter critique may also apply to the draft decisions to be adopted at the COP/MOP, but this procedure avoids at least the lengthy ratification process, which, in addition, would result in a situation where some Parties may be bound by the amendment and others not. In the context of the KP, where all Annex I Parties have legally established targets, enforcement rules have to apply equally to all Parties to secure confidence in the system.

1.10 Conclusion

The Kyoto Protocol is unique among the modern multilateral environmental agreements as it contains a very elaborate and complex regime of compliance rules and mechanisms. Conditioning the use of the flexible mechanisms on the establishment of prior accurate GHG inventories and of a registry to track the credit transactions, including an annual report on these matters, is an efficient incentive for Parties to comply. In addition, independent expert review should ensure that the factual evaluation of the reports and data is done in an objective manner. This approach is supplemented by an independent transaction log, thus giving confidence that the Kyoto system is able to detect and prevent non-compliance. Further procedural rules are designed to facilitate compliance reinstatement by the Facilitative Branch in a co-operative manner. A quasi-judicial body, the Enforcement Branch performs the legal evaluation of non-compliance cases. However, the general approach adopted by this compliance regime is mostly to reinstate and facilitate compliance rather than to impose negative consequences for non-compliance. These non-compliance consequences are rather weak. Strong penalties are lacking. The ‘penalty’ of withholding total

---

46 See Marauhn (1996), 730.

47 See Bothe (1996), 31. See also, Lavranos (2002), 44–50. However, COP decisions are still not regarded as being legally binding by some scholars or their legal form is considered as undetermined, see Sands (1995), Lang, W. (1996), 683–695.

48 Article 20 KP requires that “the Parties shall make effort to reach agreement on any proposed amendment to this Protocol by consensus; If all efforts at consensus have been exhausted, and no agreement reached, the amendment shall as a last resort be adopted by three-fourths majority vote of the Parties present and voting at the meeting […]”.

assigned amounts of the subsequent commitment period the equivalent of the preceding commitment period's excess tons of emissions multiplied by 1.3 does not really deter a strategy which amounts to borrowing. Where a government finds that a policy of emission reductions is too costly for its economy, it may opt for not complying and for borrowing by postponing reductions to future periods. The interest rate for this 'loan' is rather low. That State will, thus, over time accumulate debts, thus putting into jeopardy the viability of the system and also its own ability to restore compliance. The suspension of transfers under Article 17 is only a means to limit overselling[49]. Finally, the requirement of developing a compliance action plan does not lead very far. It may stay at the elaboration stage, as the Enforcement Branch, despite its name, does not dispose of any real means to enforce its implementation by the Party concerned. As there are few meaningful possibilities to compel a Party that did not reach its target to do so, some multilateral environmental agreements provide for trade restrictions in order to ensure compliance from[50]. However, such trade sanctions are limited and controversial as to their compatibility with the World Trade Organization rules, and the Kyoto Protocol does not envisage such a tool yet. However, one may not disregard the political and psychological effects of a Party being declared to be in a state of non-compliance with its binding reduction target.

As to the particular case of the EC, in the event of non-compliance, the EC as a Party to the KP is liable to face non-compliance consequences as well. As the EC is committed jointly with its Member States to achieve a common reduction target in the form of the so called 'Bubble', it must be analysed how these non-compliance consequences would apply to both levels or to either one of them. Before addressing this problem, however, the history of the European Bubble has to be analysed in the next section.

2 The Political Background of the EU Bubble Concept

2.1 The Bubble Concept at the European Level

The joint commitment approach can be traced back to the beginning of the nineties when the EC had to address, during the Rio negotiations, its future commitment under the United Nations Framework Convention on Climate Change (UNFCCC) to stabilize CO_2 emissions at 1990 levels by the year

[49] For a full assessment of different proposals for compliance and their efficiency regarding International Emissions Trading, see Haites and Missfeldt (2002).

[50] For example the Montreal Protocol on Substances that deplete the Ozone Layer, the Basel Convention, and CITES.

2000[51]. The concept of burden sharing, with different targets for different members, derived partly from previous European experience in the field of acid rain policy, where the Commission, in determining sulphur dioxide emission ceilings had to take more into account Member States' specific situations[52]. In the field of climate change, the Commission failed to propose a formal coordinated top-down approach that would have included the different Member Sates' targets in a binding legal instrument of Community law[53]. Member States were reluctant to transfer competences, especially in the energy sector, to the EC. To the contrary, a bottom-up approach was adopted, leaving to the Member States the possibility to set their own targets individually, which would somehow result in an emission stabilization at the Community level in line with the stabilization target under the Climate Change Convention adopted at Rio[54]. However, the EC decision to ratify the UNFCCC implicitly acknowledged a common emission stabilization target:

> The European Economic Community and its Member States declare that the Commitment to limit anthropogenic CO_2 emissions set out in Article 4 paragraph 2 of the Convention will be fulfilled in the Community as a whole through action by the Community and its Member States, within the respective competence of each. In this perspective, the Community and its Member States reaffirm the objectives set out in the Council Conclusions of 29 October 1990, and in particular the objective of stabilization of CO_2 emissions by 2000 at the 1990 level in the Community as a whole. The European Economic Community and its Member States are elaborating a coherent strategy in order to attain this objective[55].

Consequently, the issue of burden sharing arose again during the negotiations for the adoption of the Kyoto Protocol where binding emission reduction targets had to be adopted. In order to show a leading role in climate change at the international level, the Community pushed for a common ambitious reduction target position to be adopted before COP 3,

[51] See *Community Policy Target on Greenhouse Issue*, communication from the Commission of 16 March 1990 (COM90/496).

[52] See Council directive 88/609/EEC on the limitation of emissions of certain pollutants into the air from large combustion plants. In its preamble (point 7) it is stated: "Whereas in establishing the overall annual emission ceilings for existing large combustion plants due account has been taken of the need for comparable effort, whilst making allowance for the specific situations of Member States [...]". See also Haigh (1996), 163.

[53] See Haigh (1996), 163, 173.

[54] Art. 2 (b) UNFCCC.

[55] Declaration by the European Economic Community on the Implementation of the United Nations Framework Convention on Climate Change; (94/69/EC) Council Decision of 15 December 1993 concerning the conclusion of the United Nations Framework Convention on Climate Change, Annex C.

with a view to influence other Parties at the ongoing climate change protocol negotiations to do the same[56].

Thus, the Commission had to identify the Member States' individual reduction targets, and start negotiating a formal agreement among them as well. Three strategies were debated: (i) for economic reasons, the strategy of adopting equal targets for all Member States was not considered as being a rational approach. The infrastructure of power production varied too drastically from one Member State to the other[57]. (ii) The adoption of a different set of targets for each Member State would have undermined the position of the EC as a key actor and was not in line with the EC declaration to fulfill commitments as a whole[58]. (iii) Finally, adopting a common emission reductions target at Community level, yet allowing different emission commitments at Member State level, would ensure that all Member States, especially the so-called 'cohesion countries', would participate in the Kyoto Protocol[59].

Due to the efforts of the Dutch presidency at the Environment Council of 2 March 1997, Member States agreed on their respective national emission targets under the so-called Burden Sharing Agreement (BSA). The Commission and the Member States established a common negotiation position of a 15% emission reduction target covering three GHG for the EC as a whole to be attained by 2010[60]. This success in adopting an internal differentiation of emission reductions covered by a common target was a temporary proposal only, which might change during the negotiations[61]. The total emissions reduction of the Member States as agreed under the BSA represented in fact barely 10% reduction, while 15% was proposed for the EC as a whole[62]. If the 15% reduction target were agreed in Kyoto, the remaining 5% gap would have been redistributed amongst Member States afterwards. The internal agreement reached was considered as a negotiating position, which did not lead to a unilateral commitment on the part of the

56 See Gerhardsen (1998).

57 Sweden and France depend heavily on carbon-free nuclear and hydropower, when Danish and German electricity production was mainly coal-based.

58 See n.5 above.

59 In order for these countries (Greece, Ireland, Portugal and Spain) to catch up economically they are allowed to increase their emissions while improving the energy efficiency of their economic activities. See Oberthür and Ott (1999), 141 and Gerhardsen (1998), 9.

60 Carbon dioxide (CO_2), methane (CH_4), and nitrous oxide (N_2O).

61 The basis for the differentiation used a triptique approach elaborated by the University of Utrecht, which divides national economies into three emissions contributor sectors: domestic, energy intensive industry, electricity. See Ringius (1997), 20–23. However, the Dutch proposal based on the triptique did not correspond with what Member States see as an acceptable national reduction targets. Gerhardsen (1998), 18.

62 In fact, it represented a total reduction of 9.2%.

EC. The 15% target was seen as an important political signal in order to influence the adoption of strong commitments on the part of the United States and Japan during the Kyoto negotiations, but the final figure would depend on these industrialized countries' agreed targets.

2.2 The European Bubble at Kyoto

At the international negotiations on the Kyoto Protocol the Community faced criticisms and difficulties to make its 'Bubble' concept acceptable[63]. This was to a large extent due to the fact that the EC had to overcome internal struggles in agreeing on individual targets for the Member States, a fact which prevented the community from clearly communicating the Bubble concept to its negotiating partners at an early stage. The EC was accused, by the US and its coalition, of adopting double standards by advocating for itself a certain flexibility, while proposing a flat-rate target of 15% for all Annex I Parties[64]. In addition, the Community's proposal failed to address clearly the future enlargement of the EU to the Central and Eastern European countries. The initial European proposal allowed the burden-sharing agreement to be renegotiated until five years before the expiration of the commitment period. This language gave rise to suspicions among Conference participants that the EC had the intention to take advantage of the 'hot air' of certain countries which were to join the EU, like Germany had already benefited from a decrease in emissions as a result of the economic breakdown of East Germany. The EC would implement its target with relatively low costs and thus enjoy an unfair competitive advantage compared to other industrialized Parties where similar reductions could only be reached at a higher cost.

Another major critical aspect of the Bubble concept was raised by Australia, and not addressed by the EC when it made its proposal. This was the question of responsibility for jointly fulfilling the obligations and the ensuing problem whom to turn to in case of non-compliance. Australia made its acceptance of the European Bubble and the EC participation conditional on a declaration stating clearly the division of competence between the organization and its Member States, subject to the approval by the COP/MOP[65].

---

[63] On the EC as a negotiating Party at Kyoto see Sjöstedt (1997/98), 225–256.

[64] The US was part of an informal coalition, the so-called 'JUSSCANNZ' at Kyoto, an acronym for United States, Switzerland, Canada, Australia, Norway and New Zealand, joined sometimes by Iceland and South Korea, which had for common purpose an opposition to stringent targets, thus mainly opposed to the EC. The so-called 'Umbrella group' joined by Russia and the Ukraine but left by Norway and Switzerland has replaced this former group.

[65] FCCC/AGBM/1997/MISC.1/Add.2, p6.

The Kyoto process ended with a success for the EC regarding the issue of joint fulfillment. The common target adopted is 8% instead of 15% as proposed by the EC, covering a basket of six gases instead of three as envisaged by the BSA, by 2012 rather than 2010[66]. The European Bubble was accepted against some major concessions in order to respond to some of the critiques previously mentioned. The possibility to create a 'Bubble' was extended to all Annex I Parties to the KP and not left to the exclusive use by the Community. Any agreement reallocating the quantitative reduction commitments among the Parties to this agreement shall be notified to the UNFCCC Secretariat at the time of ratification. It will then be forwarded for information to the other Parties to the Kyoto Protocol[67]. Once notified, the agreement cannot be renegotiated, a fact which entails certain consequences regarding the accession countries.

2.3 The Formalization of the Burden Sharing Agreement

The BSA was renegotiated at the European Council in June 1998 to take into account the additional fluorocarbonated GHG, and the level of the reduction target adopted at Kyoto. The Council conclusions acknowledged the political agreement achieved in these terms: "the Council has now agreed the determination of the contributions of Member States to 8% reduction and the commitment of each Member State", whose target figures are attached in an annex[68]. Furthermore, it provided for the legal formalization of this political agreement: "the terms of this agreement will be included in the Council Decision on the conclusion of the Protocol by the Community"[69]. Thus, the final agreement, containing each Member State's emission level as allocated and the overall Community target was notified on 25 April 2002, in the annex to the EC Decision approving the Kyoto Protocol, to the UNFCCC Secretariat as required by Article 4, paragraph 2 of the Kyoto Protocol[70]. Once forwarded to the Secretariat, the arrangement shall operate during the first commitment period, a period of five years, without any alteration.

[66] Less than in the initial European proposal, but of a basket of six gases instead of three, see n. 1 and 11 above.
[67] Art. 4.2 of the KP.
[68] Point 2 of Council conclusions, 16–17 June 1998, DN: PRES/98/25. See annex I to this chapter.
[69] Ibid.
[70] See Annex II in (2002/358/CE) Council Decision 25/04/2002, concerning the approval, on behalf of the European Community, of the Kyoto Protocol to the United Framework Convention on Climate Change and the joint fulfilment of commitments thereunder. The Council decision was forwarded to the Secretariat on 12 June 2003, Agreement Between the European Community and its Member States under Article 4 of the Kyoto Protocol, Doc. FCCC/CP/2002/2.

During that period, it thus excludes the accession countries from the 'Bubble'[71]. Therefore, the accession of the ten countries to the EU in May 2004 does not alter the different targets agreed between the 15 old members in June 1998. Those new entrants having their own target under the KP will be committed to that individual target as set out in the Annex B of the KP[72]. Article 4, paragraph 4 KP, in its last sentence together with paragraph 3, leaves the door open for future 'Bubble' arrangements with the new EU members for further commitments beyond 2012[73]. Nevertheless, the particular feature of a an enlarged Community of 25 members, where the EC is jointly committed together with the 15 members Parties to the BSA to reduce by 8% its GHG emissions, and of new entrants with targets excluded from that joint commitment arrangement, raises some questions in terms of responsibility. This will be further discussed in the next section.

3 Joint Commitment

Being jointly committed means that both the Member States and the Community participate as Parties to the treaty with a view to achieving their targets together. It is the special feature of the EC that it is an integration organization with an autonomous legal system, where the EC competence to regulate certain areas complements or even totally substitutes the respective competence of the Member States. In the field of the environment, both the EC and the Member States have concurrent competences[74]. Hence, the fulfillment of obligations under international environmental conventions requires the participation of both the EC and the Member States as Parties. These conventions, where both the Community and some or all of its Member States have become Party, are called mixed agreements[75]. This is

---

[71] Art. 4(3): "Any such agreement shall remain in operation for the duration of the commitment period specified in Article 3, paragraph 7", & Art. 4(4): "If Parties acting jointly do so in the framework of, and together with, a regional economic integration organization, any alteration in the composition of the organization after adoption of this Protocol shall not affect existing commitments under this Protocol. Any alteration in the composition of the organization shall only apply for the purpose of those commitments under Article 3 that are adopted subsequent to that alteration."

[72] Malta and Cyprus do not have any reduction targets under the KP, and are not Annex I Parties under the UNFCCC, see Fernández Armenteros and Massai in this volume.

[73] Article 4.3 KP: "Any such arrangement shall remain in operation for the duration of the commitment period specified in Article 3, paragraph 7." Art. 3.7 lays down the duration of the first commitment period from 2008 to 2012.

[74] See Art. 174.6 TEC.

[75] For detailed discussions see Ehlermann (1983), 3–21; Schermers (1983), 23–33; Dolmans (1985); Neuwahl (1991), 717–740; Bourgeois (1998), 83–98; Garzon Clariana (1998), 15–26; Leal Arcas (2001), 483–513.

the case especially with the UNFCCC and the Kyoto Protocol whose subject matter encompasses, and has consequences in, diverse areas (agriculture, trade, internal market, transport, energy, environment), where the Community has either exclusive or concurrent competences. It is said that many environmental conventions have been concluded as mixed agreements more for political and diplomatic reasons than for legal ones[76]. The internal distribution of competence between the EC and its Member States as regards the implementation of the obligations arising out mixed agreements, however, cannot be underestimated. As both the EC and its Member States tend to insist on their respective competences in relation to the subject matter of a particular treaty, the attempt to identify in detail the responsibility that each level has for fulfilling particular obligations is difficult and contentious[77]. However, in order to secure a full and uniform implementation of all obligations resulting from mixed agreements, the ECJ requires a close cooperation between the Member States and the institutions of the Community, which applies to the whole life of the agreement[78]. This flows from the principle of unity of the EC and its Member States vis-à-vis third States at international level. Under the KP, the term 'jointly committed' refers more generally to a group of Parties that decides to achieve together a common reduction target rather than to the specific feature of the joint participation of the EC and its Member States under the KP. How the KP addresses and organizes such joint participation of the EC and its Member States under its regime will be the subject of the following paragraph.

3.1 Joint Commitment under the Kyoto Protocol

Article 3 paragraph 1 KP allows Annex I Parties to 'individually' or 'jointly' fulfill their quantified emission limitation and reduction commitments. Article 4 KP provides for the Parties willing to fulfill their emission reductions jointly to do so by forming a 'Bubble' equal to their total emission limitation and reduction commitments and then re-allocating their

[76] See Lang, J.T. (1986), 157–176; Neuwahl (1996), 667–687; Granvick (1999).

[77] See Heliskoski (2001).

[78] Moreover, the ECJ repetitively states in the same words that: "where it is apparent that the subject matter of an agreement or convention falls in part within the competence of the Community and in part within that of its Member States, it is essential to ensure close cooperation between the Member States and the Community institutions, both in the process of negotiation and conclusion and in the fulfillment of the commitments entered into. That obligation to cooperate flows from the requirement of unity in the international representation of the Community. The Community institutions and the Member States must take all necessary steps to ensure the best possible cooperation in that regard"; See Ruling 1/78 [1978] ECR 2151, paras. 34 to 36; Opinion 2/91[1993] ECR I-1061, para. 36; Opinion 1/94 [1994] ECR I-5267, para. 108; and Opinion 2/00 [2001] ECR I-09713, para. 18.

commitments among themselves through a formal agreement. The terms of this agreement must be notified to the Secretariat on the date of the ratification of the Protocol. In turn, the Secretariat shall inform other Parties of the terms of the agreement, which shall remain in operation without any alteration for the duration of the commitment period. In order to respond to the need for legal clarity in determining the responsibility with regard to the binding targets established by the Parties at Kyoto, Article 4 addresses the responsibility of each Party to such arrangement in case of a failure to achieve the common reduction commitment.

3.2 Determining Responsibility

As to the responsibility for failure to jointly fulfill a common target, Article 4 distinguishes between two situations. Article 4, paragraph 5 addresses the case where Parties jointly committed without a regional organization fail to achieve their overall target. In this case, each of them is responsible for its own target as stipulated by the agreement between those Parties[79]. The non-fulfillment of the collective target triggers the individual responsibility of each Party for achieving its agreed individual target. Thus, this Article lays down an individual state responsibility as a general rule. However, Article 4 paragraph 6, which deals with Parties acting together with a regional economic integration organization, itself a Party to the KP, namely the EC, adopts a joint responsibility rule. It reads as follows: "If Parties acting jointly do so in the framework of, and together with, a regional economic integration organization which is itself Party to this Protocol, each Member State of that regional economic integration organization individually, *and together with* the regional economic integration organization acting in accordance with Article 24, shall, in the event of failure to achieve their combined level of emission reductions, be responsible for its level of emissions as notified in accordance with this Article"[80]. Should the common reduction target of 8% not be met, both the Member States individually in relation to their BSA target, together with the Community, would be held liable. Therefore, the international responsibility of the Community will be at stake. The wording 'in the event of failure to achieve their combined level of emission' could be interpreted *a contrario*. As long as the EC and its Member States deliver their combined reduction target of 8%, and do so despite the failure of some Member States to fulfill their individual targets while other Member States offset their failure, both the EC and the failing

[79] Art. 4.5: "In the event of failure by the Parties to such an arrangement to achieve their combined level of emission reductions, each Party to that arrangement shall be responsible for its own level of emissions set out in the agreement."
[80] Emphasis added.

Member State appear not to be liable under the KP. This is at least a possible interpretation, but the text of Article 4 paragraph 6 is not clear and its consequences are, therefore, uncertain. This leads to three problems: (i) the *a contrario* interpretation of Article 4, paragraph 6 KP might induce some Member States Parties to the BSA to use the 'Bubble' as a 'profit sharing', i.e. for gaining a profit from efforts made by the other Member States only. (ii) Is the EC liable in respect of the new Member States, which are not Parties to the BSA, but nevertheless Member States of the EC? (iii) Finally, what are the consequences of non-compliance within the joint liability context?

3.2.1 The EC and the Accession Countries: A Joint Liability?

We have seen that Article 4, paragraph 4 KP precludes the new Member States from entering into the European 'Bubble' arrangement once the EC and its 15 Member States have ratified the Protocol. Thus, the EC is committed to its 8% target, while at the same time the new Members are committed to their individual reduction targets as set in Annex B of the Protocol – e.g. Poland with 6%. From an international point of view, the new entrants with reduction commitments are seen as third Party in relation to the joint fulfillment arrangement although they are Members of the EC. As seen, Article 4, paragraph 6, establishes a joint liability for the EC and its Member States if they act jointly towards a common target. On the other hand, the Protocol is silent on the liability of the EC within the context of an enlarged Community of 25 with Member States having their own targets and being excluded from joint fulfillment. This can be explained by the fact that this paragraph was introduced at the request of the EC in order to appease the opposition to the 'Bubble' concept that came from other Parties at Kyoto. It was formulated only to address this particular problem. The KP is the first mixed environmental agreement that creates a joint liability of the EC and its Member States[81]. Thus, the enlarged EC of 25, together with the 15 Member States Parties to the BSA, are jointly responsible and the accession countries, through their accession, are drawn into the responsibility of the EC. But if a new Member State fails to fulfill its own individual reduction commitment, should the new Member State be jointly liable together with the EC? Though this will probably not occur in practice, as these new members are well

[81] With the exception of the United Nations Convention on the Law of the Sea (UNCLOS), which states: "Any State Party may request an international organization or its Member States which are State Parties for information as who has responsibility in respect of any specific matter. The organization and the Member States concerned shall provide this information. Failure to provide this information within a reasonable time or the provision of contradictory information shall result in joint and several liability." See Annex IX Article 6, para. 2 UNCLOS III (1982).

below their reduction targets, the case could at least be raised theoretically. This falls within the broader problem of the international responsibility of the EC for mixed agreements in general, which is not the object of this paper[82]. Nevertheless, through their accession to the EC, the new entrants are under Community law jointly committed with the EC to fulfill their reduction commitment as set in Annex B of the KP. This is the consequence of the fact that the Community and the Member States have to act in close cooperation to fulfill these environmental obligations. Neither level can alone deliver a complete performance over the whole breadth of the KP obligations, as the legal powers to fulfill the obligations under the treaty are shared between the Community and the Member States. Furthermore, the internal division of competences under Community law cannot be opposed to third Parties[83]. Thus, the new Member States cannot rely on the fact that they have lost certain powers in favor of the Community in order to justify non-compliance. The most convenient construction concerning this problem is that the Community and its Member States are co-responsible for the performance of their obligations, and thus jointly liable under the Protocol when either level fails to deliver that performance[84]. This conclusion is in line with the principle of unity, the Community and the Member States constituting 'one and the same Party' vis-à-vis third States[85].

3.2.2 Joint Fulfillment: A Profit Sharing?

The joint responsibility is only triggered if the Parties acting jointly with an organization fail "to achieve their total combined level of emission reductions". Thus, one can conceive the BSA working as a 'profit sharing', where over-compliance by one Member State offsets another Member's non-compliance, with the result that the overall 8% target is attained. Article 4, paragraph 6 KP requires Member States to attain their original individual target only if the common target is not attained. Nor does it establish any

[82] On this aspect see Björklund (2001), 373–402.

[83] See Okowa (1995), 169–192. And Art. 27 of the 1986 Vienna Convention on the Law of Treaties between States and international Organizations or between International Organizations. Download at: http://www.un.org/law/ilc/texts/trbtstat.htm.

[84] In the doctrinal debate regarding the EC responsibility under international law for mixed agreement, authors agree on a joint liability of the EC together with its Member States. Tomuschat (1983), 125–132. For Dolmans (1985), non-performance for obligations under mixed agreements where both the Community and Member States have concurrent competence results in a joint and several responsibilities. For advocate General Jacobs: "Under a mixed agreement the Community and Member States are jointly liable unless the provisions of the agreement point to the opposite", Opinion of Advocate General Jacobs of the ECJ in Case C-316/91, European Parliament v. Council, [1994] ECR I-625, para. 69. And Björklund (2001) for a critical comment.

[85] See Macleod, Hendry and Hyett (1996), 329.

rule as to the use which Member States may make of any surplus in order to facilitate compliance by another Member State or the EC. The only element required is the formal agreement stipulating the different agreed targets among the Member States, to be forwarded for information only to the other Parties to the KP. Such a possibility of using the joint fulfillment possibility as a profit sharing may create the incentive for some Member States to adopt a free rider option to the extent that they suppose other Member States will offset the non-compliance by the former. Under the Protocol, the only limit to such option is the possibility for Annex I Parties to bank their excess AAUs for the following Commitment period, or to transfer their excess AAUs under Article 17 to Parties other than the Member States. Article 3, paragraph 13 of the KP allows to carry over to subsequent commitment periods any surplus AAU[86]. In this context, if a Member State decided to bank or transfer its remaining AAUs while another Member State is in non-compliance, this could lead to a situation of EC non-compliance, which would entail joint liability.

The new Member States are not involved in the system of joint liability under Article 4 as they are not Parties to the Bubble. Nevertheless, as Member States of the EC, they are part of the European system. Whether and how their emission credits could be transferred in order to contribute to other Member States' compliance is unclear[87]. This uncertainty results from the difficulties encountered by the EC in discussing the Bubble concept as a separate compliance issue during the Kyoto and subsequent negotiations, thus leaving the topic for internal discussion at the European level[88].

3.2.3 Joint Liability and Consequences of Non-Compliance Under the Protocol

As pointed out previously, the joint commitment entails joint liability. What does joint liability mean in practice when applied to the EC and its Member States? There is no international practice relating to the specific issue of joint responsibility of the EC and its Member States for mixed agreements from which one could deduce any relevant rules of international law[89]. Nor does a clear definition of joint liability exist in international law[90]. Recourse to the Community declaration of competences would also not be very helpful in

[86] This right is known as banking. Some restrictions have been imposed at Marrakech regarding the possibility to bank other Kyoto credit units, see Decision 19:CP.7 para. 16 & Annex, section F.

[87] See Fernández Armenteros and Massai in this volume.

[88] See Oberthür and Ott (1999). One of the criticisms raised by Australia was the possibility for the EC to benefit from hot air through accession countries joining the Bubble. Dealing with excess AAUs was a touchy issue.

[89] Björklund (2001), 386.

[90] Ibid.

order to determine the responsibility of each entity, the EC and the Member States, in the performance of their commitments. Usually, the wording of the Community declarations of competence under mixed agreements is rather vague. Furthermore, pursuant to the wording of Article 4, paragraph 6, the relevant question is whether the common reduction commitment has been fulfilled, and not which level is competent to take the relevant measures. The expression in Article 4, paragraph 6, "each Member State [...], and together with the organization" excludes a joint and several liability of all other Member States (which are in compliance of their notified target) where, due to the failure of one Member State, the Community is unable to reach the aggregate common target. It rather indicates that only the failing Member State individually and jointly with the Community would be responsible for non-compliance. Thus, it is suggested that the Member States which are in compliance with their BSA targets should not be affected. However, if the Community were sanctioned, the consequences may affect all Member States. As regards the role of the Compliance Committee, especially its Enforcement Branch, it will simply conclude that the relevant Member State and the EC as a whole are in non-compliance and will determine and apply the consequences of non-compliance. Thus, the responsibility for these consequences and remedial action will have to be decided on the basis of Community law at the European level. In the light of such a consequence, the Community would be certainly entitled to some kind of internal compensation from the defaulting Member States. At least, it would not be prevented from taking some steps against Member States whose omissions risk to engage the Community's responsibility[91].

3.3 Other Protocol Provisions with Regard to the Joint Participation of the EC and its Member States

Regarding other provisions particularly relevant for the joint participation of the EC and its Member States, the KP (Article 24, paragraph 2), similar to the UNFCCC, contains a standard clause allowing a 'regional economic integration organization' to become a Party jointly with its Member States, subject to the condition that they are already Parties to the UNFCCC. For this matter, "the organization and its Member States shall decide on their respective responsibilities for the performance of their obligations under this Protocol" and they are "not entitled to exercise rights under this Protocol concurrently"[92]. This leaves it to the organization to decide with its Member

[91] See Ehlermann (1983), 20.

[92] Art. 24.2 KP. Under the UNFCCC such clause is enshrined in Art. 22.2 of the Convention.

States on the allocation of responsibilities for the obligations that would be assumed. In addition, the KP contains a clause requiring that in its instruments of acceptance the organization shall declare the extent of its competence with respect to matters governed by the Protocol[93]. The reason for this requirement is to secure that either Member States or the Community have competence to fulfill the obligations arising from the Protocol[94]. The organization, in matters within its competence, shall vote with a number of votes equal to the number of their Member States that are Parties to the Protocol, and shall not exercise its right to vote if any of its Member States exercises its right to vote and vice versa[95]. Most conventions contain such a clause in order to prevent the double voting on a subject matter. It is the principle of 'alternative voting'. These requirements are similar to those found in most of the multilateral environmental agreements allowing the EC and its Member States to jointly participate in the regime established by them[96].

One ambiguity in the formulation of Article 24, paragraph 2, should be mentioned with regard to the EC and its Member States relating to the exercise of rights under the KP. Article 24, paragraph 2, of the Kyoto Protocol states that when both the organization and its Member States are Parties to the Protocol "(t)he organization and the Member States shall not be entitled to exercise rights under this Protocol concurrently". The meaning of 'rights' is unclear. Does it refer to the "respective responsibilities for the performance of their obligations" to be decided between the Member States and the EC under the KP, as stated in the preceding sentence of Article 24.2? Or does it refer to other rights, and, if so, which? For Temple, commenting on a similar clause of the Vienna Convention on the Ozone Layer, the word 'right' does not refer at all to the 'responsibilities' accepted by the organization and its Member States[97]. It is rather the application of the rule of alternative voting rights to other procedural rights under the Convention[98].

---

93 Art. 24.3 KP.

94 See Temple (1986), 157–176.

95 Art. 22.2 KP.

96 See Lang, J.T. (1986).

97 Ibid.

98 For Lacasta, Barata, Cavalheiro and Dessai (2001) it refers to procedural rights: "*However, it does not seems that the intention in drafting this particular sentence was to restrict the role of the organization, but rather to clarify that procedural rights under the Protocol, such as the right to vote, cannot be exercised both by Member States and the European Community. See also Article 22.2*". However, they do not explain why it is mentioned twice, and within the context of Article 24, to which procedural aspect it refers. One must remember that during the Kyoto negotiations, Australia wanted a clear statement on the division of competence between the EC and its Member States as to who is entitled to exercise or execute rights or obligations, against the acceptance of the European Bubble

Nonetheless, he does not put forward substantial arguments to support this interpretation, and especially why such alternative rule of voting right is placed there. If the word 'right' refers to the respective responsibilities for the performance of the obligations under this Protocol, this has to be decided at Community level as stated by Article 24, paragraph 2. This ambiguity has been used by some Member States to deny to the Community the possibility offered by the Protocol of using the three flexible mechanisms alongside with the Member States. This would reserve the possibility to fulfill part of their greenhouse gas emissions reduction obligations through the implementation of EIT, JI and CDM to the exclusive use by the Member States. Such interpretation would deprive the Community as a Party to the KP of important instruments to ensure the fulfillment of the common reduction target. It would result in an awkward situation where the Community as a Party is bound to achieve its reduction obligation, but deprived of the right to do it by directly using the flexible mechanisms itself.

The wording of an agreement shall be interpreted in good faith in accordance with the ordinary meaning to be given to the terms of the treaty in their context, such as its preamble, and in light of its object and purpose[99]. The object of the KP is that Annex I Parties shall, individually or jointly, ensure that their aggregate GHG emissions do not exceed their assigned amounts calculated pursuant to their quantified emission limitation and reduction commitments inscribed in Annex B, with a view to reducing their overall emissions of such gases by at least 5% below 1990 levels in the commitment period 2008 to 2012[100]. For the purpose of meeting its commitment, the KP allows any Party included in Annex I to benefit from the use of the flexible mechanisms. As Annex I Party with a reduction commitment inscribed in Annex B, the EC should therefore be entitled under the KP to benefit from such instruments. The word 'concurrently' usually means simultaneously. In the context where obligations have to be fulfilled, such as a reduction commitment, rights designed to facilitate the fulfillment of this obligation could not be possessed concurrently if they conflict with such aim. This would leave it to the Member States and the EC to determine whether such 'conflict' would arise. The possibility of undertaking CDM and JI reduction projects in parallel would not necessarily result in a conflict with the aim of reducing GHG emissions. A conflict of rights, if any, could only occur during the additional period where both the EC and a Member State are declared in a state of non-compliance and benefit from an additional period of hundred days to purchase credits in order to reinstate

concept. Thus, this sentence may be a reminder of this attempt. However, instead of clarifying the situation of the EC, it introduces an ambiguity.

[99] See Art. 31 of the 1986 Vienna Convention on the Law of Treaties between States and International Organizations or between International Organizations.

[100] Art. 3.1 of the KP.

compliance. If several Member States were found in a state of non-compliance and competed together with the EC for purchasing credits, such competition may result in a price increase and, thus, conflict with their interest of complying at a lower price. It is less a legal than a political problem how to deal with such a competition at the Community level. As regards international law, the EC seems fully entitled to participate in the so-called flexible mechanisms as long as it fulfills the eligibility requirements for such participation.

3.4 Conclusion

The EC succeeded in obtaining the possibility of a joint commitment concerning a common reduction target and, in order to attain it, of forming a 'Bubble' with its Member States. This success has a price; the KP establishes a joint responsibility of the EC and its Member States in relation to their individual targets as agreed under the BSA, if the overall reduction target is not fulfilled. Furthermore, the manner in which this joint responsibility is formulated may induce some Member States to consider the 'Bubble' as a profit sharing device. They may be tempted to rely on other Member States' possible over-compliance to offset their own failure or their 'business as usual' strategy. The result of such a strategy may be that the Community would fall into non-compliance and would have to face the consequences of this non-compliance as provided by the Kyoto compliance regime. This would ruin the original idea of the 'Bubble', which allows all Member States with different targets to participate jointly with the EC towards a common reduction target, and thus would compromise the international credibility of the Community. In the light of such a bleak scenario, the EC is entitled to take early internal actions and measures to prevent Member States from adopting a free riding attitude. As regards to the new Member States, the KP considers them as third states in relation to this joint commitment arrangement, since the EC and its fifteen Member States have ratified the KP before the accession of these new Members. Finally, the EC as a Party to the KP is fully entitled to participate in the three flexible mechanisms as long as it fulfills the eligibility requirements as provided by the Kyoto regime.

The most important legal consequence that faces the EC is to be held liable. Because of the uncertainties in the interpretation of Article 4, paragraph 6, the EC has adopted some mechanisms to ensure that Member States comply with their targets. These measures will be analyzed in the next section.

4 Compliance with the Kyoto Protocol under Community Law

In the Conclusions of the Council meeting of March 2002, the Member States and the Community agreed on a coordinated ratification of the Protocol by both the Member States and Community, with a view to simultaneously deposit their instruments of ratification before June 2002[101]. With the aim of giving a strong signal to the international community concerning the European commitment to climate change policy, and with a view to bring about a prompt entry into force of the Protocol before the Johannesburg World Summit held in September 2002, the coordinated European ratification process was successfully achieved, and all Member States as well as the Community deposited their instruments of ratification or of approval at the United Nations Headquarter in New York by 31 May 2002. As to the driving force the European ratification was expected to constitute for other Annex I Parties, their ratification has proved to be more difficult than expected. It took more than two additional years for the Kyoto Protocol to enter into force. The problem of the entry into force has had a strong impact on the debate regarding the opportunity to pursue an emission reduction policy at the European level to achieve the targets set out in the BSA.

The Community approved the KP by a Council decision of 25 April 2002, which, along with other purposes, transforms the BSA into a legally binding instrument[102]. The proposal for the decision was initially based on Article 174, paragraph 4, of the EC Treaty[103]. Following the Opinion of the ECJ on the legal basis for the conclusion of mixed environmental agreements, the Commission changed the original legal basis. Although this change does not affect the qualified majority voting procedure, the decision was adopted unanimously on the basis of Article 175, paragraph 1[104]. Nevertheless, some Member States, in the light of the substantial impact the emission reduction obligation has on their energy policies, demanded the adoption of the decision on the basis of Article 175, paragraph. 2, which would have required unanimity.

---

[101] See Council conclusions of the 2413rd Environment Council meeting, Brussels, 4 March 2002, C/02/47.

[102] Council Decision of 25 April 2002 (2002/358/CE).

[103] See proposition 11 of the Proposal for a Council decision concerning the approval, on behalf of the European Community, of the Kyoto Protocol […] and the joint fulfilment of commitments thereunder, COM(2001) 579, 2001/0248 (CNS).

[104] Opinion 2/00 on the Cartagena Protocol on Biosafety, 6 December 2001, [2001] ECR I-09713.

4.1 Council Decision of 25 April 2002

The Decision both approves the KP and organizes the joint commitment at Community level by imposing obligations upon Member States, and by empowering the Commission to take further decisions through the comitology procedure. In this aspect, the Decision of 25 April 2004 differs from other decisions concerning the approval of international agreements[105]. Article 1 of the Decision approves the Protocol. The BSA, which establishes the respective emission levels allocated to the Member States, is included in Annex II and forwarded to the UNFCCC Secretariat as required by Article 4, paragraph 2 KP[106]. In accordance with Article 24, paragraph 3 KP, a declaration of competence is incorporated in Annex III of the Decision, which vaguely states that in accordance with Article 175, paragraph 1, of the EC Treaty, the EC has the competence to enter into international agreements and to implement the resulting obligations, and it cites the different objectives stated by Article 174, paragraph 1. It emphasizes, without giving further precision, that the Community and its Member States through action within their respective field of competence will fulfill under Community law the emission reduction commitment under the Protocol. Finally, it states that in accordance with the requirement established by Article 7, paragraph 2 concerning supplementary information, the Community will provide such information on a regular basis. The vagueness of the declaration of competence is upheld by the ECJ:

> It is further important to state, as was correctly pointed out by the Commission, that it is not necessary to set out or determine, as regards other Parties to the Convention, the division of powers in this respect between the Community and the Member States, particularly as it may change in the course of time. It is sufficient to state to the other contracting Parties that the matter gives rise to a division of powers within the Community, it being understood that the exact nature of that division is a domestic question in which third Parties States have no need to intervene. In the present instance the important thing is that the implementation of the Convention should not be incomplete[107].

Furthermore, for obvious political reasons, Member States and the Community prefer to avoid disputes about the division of competences

[105] See Decision (94/69/EC) concerning the conclusion of the UNFCCC, contains two Articles plus three Annexes, containing the text of the Convention, a declaration of competence, and a declaration of joint fulfilment, while Decision (2002/358/CE) contains 7 Articles plus three Annexes, the text of the Protocol, the BSA, and a declaration of competence.

[106] See n. 78 above.

[107] Case 1/78, Draft Convention of the International Atomic Energy Agency, [1978] ECR 2151, para. 35.

concerning obligations arising from a mixed agreement, which might also result in freezing, in relation to the agreement, the competence of each level[108]. The decision organizes the coordinated ratification process with a view to achieve the simultaneous deposit of the instruments of ratification of the KP[109]. Regarding the relation between the Member States and the Community, the preamble of the Decision states: "The Community and its Member States have an obligation to take measures in order to enable the Community to fulfill its obligations under the Protocol without prejudice to the responsibility of each Member State towards the Community and other Member States to fulfilling its own commitment"[110]. In other words, both the Community and the Member States, being jointly committed, must take measures. Member States are under a Community obligation of loyalty towards the Community and towards each other to attain their individual BSA targets in order to enable the Community as a whole to comply. The Decision excludes the possibility of using the BSA as a profit sharing by recalling that each Member States is committed to its individual target under Community law. This proposition is legally enshrined in Article 2 of the Decision, which recalls that Member States and the EC are jointly committed with regard to the provision of Articles 3, paragraphs 1 and 4, of the Protocol and with regard to the provision of Article 10 of the EC Treaty. For that purpose, their respective emission levels are to be found in Annex II. Thus, both the Community and its Member States shall take the necessary measures to comply with their emission levels as set out in Annex II, which shall be determined in accordance with Article 3 of the Decision[111]. To do so, the respective Member States emission levels have to be determined in relation to their respective base year[112]. The reduction and limitation targets under the BSA are expressed in a percentage of base year emission levels, and will have to be calculated and recorded in terms of tons of carbon dioxide equivalent pursuant to Article 3, paragraphs 7 and 8 KP and pursuant

[108] See Heliskoski (2001).
[109] Art. 6 Council Decision 2002/358/CE.
[110] Preamble, point 12 Council Decision 2002/358/CE.
[111] "The European Community and its Member States shall fulfil their commitments under Article 3(1) of the Protocol jointly, in accordance with the provisions of Article 4 thereof, and with full regard to the provisions of Article 10 of the Treaty establishing the European Community. The quantified emission limitation and reduction commitments agreed by the European Community and its Member States for the purpose of determining the respective emission levels allocated to each of them for the first quantified emission limitation and reduction commitment period, from 2008 to 2012, are set out in Annex II. The European Community and its Member States shall take the necessary measures to comply with the emission levels set out in Annex II, as determined in accordance with Article 3 of this Decision", Art. 2 of the Council Decision (2002/358/CE).
[112] For the three fluorocarbon gases the Parties may either chose 1990 or 1995 as a base year, see Art. 3.8 KP.

to the Annex to Decision 19/CP.7 on modalities for the accounting of assigned amounts, taking into account the methodologies for estimating anthropogenic emissions by sources and removals by sinks referred to in Article 5, paragraph 2 KP, adopted at COP 9[113]. Parties jointly committed under Article 4 shall use the respective emission level allocated to each of them in the agreement instead of the percentage established for it in Annex B to the KP[114]. The calculation of the assigned amount in terms of carbon dioxide equivalent is an eligibility requirement for the participation in the flexible mechanisms, which shall be submitted in a pre-commitment period report to demonstrate that the Annex I Party is able to monitor, track and record the mechanisms' transactions. To do so, the Commission, on the basis of Member States' inventories of emissions, assisted by the Climate Change Committee composed of Member States national experts[115], will determine in a decision, adopted through the comitology procedure[116], the assigned amount of each Member State for the first commitment period by 31 December 2006[117]. This date is in conformity with the requirement to submit the pre-commitment period reports to the UNFCCC Secretariat before 1 January 2007[118]. However, the calculation of the Member States assigned amount may re-open some political discord. The BSA targets agreed in June 1998 frustrated some Member States, and they may be tempted during the conversion process to readjust their share of assigned amounts in their favor. Such national interests had been left aside during the adoption procedure of the Council decision in order to rescue the international Kyoto process in the face of the American opposition to the KP, and to strengthen the credibility of the Community. Furthermore, the BSA was incomplete when adopted, based on provisional data with respect to GHG base year emissions, and did not take into account the impact of removal by sinks as no modalities to do so were set at that time. Both the Commission and the Council were aware that some adjustments were needed by stating that the calculation of the

---

[113] See decision 11/CP.7 on methodologies for estimating anthropogenic emissions by sources and removal by sinks.

[114] Para. 5(c), section B of Annex to Draft Decision-/CMP.1 on modalities for accounting of assigned amounts, FCCC/CP/2001/13/Add.2.

[115] Established under Council Decision 1999/296/EC amending Council Decision 93/389/EEC, for a monitoring mechanism of Community CO_2 and other greenhouse gas emissions as amended by Council Decision 1999/296/EC of 26 April 1999. Now repealed by Decision 280/2004/EC concerning a mechanism for monitoring Community greenhouse gas emissions and for implementing the Kyoto Protocol.

[116] Council Decision 1999/468/CE of 28 June 1999 laying down the procedure for the exercise of implementing powers conferred on the Commission.

[117] Articles 3 & 4.2 Council Decision 2002/358/CE.

[118] See Draft Decision-/CMP.1 (*Modalities for the accounting of assigned amounts*) & Annex (*Modalities for the accounting of assigned amount under Art. 7.4 of the KP*) para. 2, FCCC/CP/2001/13/Add.2.

Member States' emission levels: "shall include any necessary adjustments in order to secure that the sum of the assigned amounts calculated for the individual Member State equals the aggregate assigned amounts calculated for the Community and that the overall balance between the contributions of the Member States to the achievement of the Community's commitment, as agreed in the Council Conclusions of 16–17 June 1998, is preserved [...]"[119]. Nevertheless, with regard to the uncertainties that impede the entry into force of the KP, some Member States started questioning the European strategy to implement the BSA alone[120]. The economic cost of such emission limitation and reduction policy would be too high in comparison of the global environmental benefit in terms of climate change[121]. Therefore, they put into question to be legally bound by the BSA if the KP did not enter into force. As stated by the Council when adopting the Decision: "The allocation of emission levels between Member States (burden sharing), which forms part of the decision, reflects individual commitment to be undertaken by the Member States as Parties to the Protocol"[122]. Such statement makes it clear that Member States conceive their BSA targets above all as an obligation arising from the KP. The legal effect of the BSA being contained in a Council decision will be discussed in the following paragraph.

4.2 The Legal Effect of the Burden Sharing Agreement.

The BSA was first a political agreement between the Member States, which was then embodied in a legal instrument, a Council decision. There has been a vivid debate on the question whether any legal consequences derive from the decision if the Kyoto Protocol did not enter into force. Usually, the Council would approve international agreements in the form of a regulation. Recently however, decisions have been used for concluding agreements. They have to be distinguished from a normal decision under Article 249 EC Treaty. They are decisions *sui generis* whose only purpose is expressing the Council's intention to approve the agreement[123]. They serve as a bridge to introduce the international agreement into the Community law in order to give effect to its provisions within the Community legal order when it enters into force. It is when the international agreement enters into force that the

[119] Joint statement by the Council and the Commission concerning Article 3, and Annex II, Brussels of 12 April 2002, Doc. 7510/02 ADD1, 2.

[120] See Le Boucher (2004).

[121] The EC account for 20% of the global emission, a reduction of 8% alone would not have environmental benefit globally. Ibid.

[122] Statement by the Council of 12 April 2002, Doc. 7510/02 ADD1, 6.

[123] See Macleod, Hendry and Hyett (1996), 21.

decision becomes binding on Member States as a Community obligation[124]. It is, thus, less the form of the act than its content that determines its legal status and effect. In the Council Decision of 25 April 2002, the approval of the KP is contained in Article 1. By itself, the Article does not yet give legal effect to the Protocol within Community law, as long as the Protocol is not in force. Although, as stated by the ECJ in Opinion 1/91, the Council act approving an agreement can stipulate that the agreement's provisions apply within the Community as from an earlier date than that of the entry into force[125]. Such an express stipulation cannot be found in the Council Decision of 25 April 2002. Other Articles in the Council Decision have legal effect, such as Article 6, which organizes the coordinated ratification procedure of the Member States, or Articles 3 and 4 that govern the conversion procedure for the calculation of Member States' assigned amount. As to the BSA, the Decision transforms the political agreement of June 1998 into a legal Community instrument, which the Council considers to be legally binding, without giving further precision[126]. Being the act of approving the KP, the Decision was forwarded to the UNFCCC Secretariat[127].

As already stated, such act is not legally binding *per se*. One has to look at its provisions. The formalization of the BSA in legal terms is found in Article 2 of the Decision, which states that the different "emission limitation and reduction commitments agreed by the European Community and its Member States for the purpose of determining the respective emission levels allocated to each of them [...] are set out in Annex II", and that "(t)he European Community and its Member States shall take measures to comply with the emission levels set out in Annex II [...]". For some authors, Article 2 lays down an obligation binding the Member States under Community law which is independent of the entry into force of the KP[128]. This may be formally correct. However, one must not neglect the object, the purpose, and the context of the decision. The object of the decision is to approve the KP and to organize the joint commitment under Article 4 of the KP at the Community level for the purpose of enabling the Community to fulfill its quantified emission reduction commitment under Article 3, paragraph 1[129]. As such, secondary Community provisions have been established under the Decision with the aim to technically prepare, and legally organize the relationship between the Community and its Member States for the entry into force of the KP. These provisions are intimately linked to, and

124 Case 181/73 Haegeman [1974] ECR 449, para. 5.
125 Opinion 1/91 on European Free Trade Association, [1991] ECR 6079; para. 37.
126 Conclusion of the Environment Council 16–17 June 1998; 9402/98 (Presse 205).
127 See n. 78 above.
128 See M. Pallemaerts, 'La Communauté européenne comme Partie contractante au Protocole de Kyoto', in Aménagement-Environnement (2003), 16–28.
129 See Preamble, points 7, 8, 9 & 10 of Council Decision 2002/358/CE.

anticipate, the entry into force of the Protocol. If the KP did not enter into force, the main object of the Decision was frustrated. In the context of a Decision deprived of its main object, the KP, the purpose of the BSA would also be put into question. Member States have agreed on reduction and limitation targets under the BSA because they are Parties to the Protocol, and they have been more than reluctant to accept them as an independent Community obligation[130]. Furthermore, the Decision is addressed to the fifteen Member States only; the new Member States are not part of the BSA. If the BSA had been fulfilled as a Community obligation without the KP being in force, there would have been the awkward situation where fifteen Member States were committed to a reduction and limitation target, while others would have no binding target under international law, nor under Community law. By the same token, such a situation would also affect the distribution of allowances to entities under the European emission trading scheme with respect to its second phase (2008–2012). This allocation of allowances must be based on objective and transparent criteria, including the KP targets for the new entrants[131].

In the face of these problems, the legal basis for the Community to enforce the implementation of the BSA by the fifteen old Member States would be rather doubtful. Whether Member States and the Community want to give effect to this agreement if the Kyoto Protocol does not enter into force, is another question, which is mainly political. But this agreement should then be negotiated with the participation of all Member States. In order to give legal effect to such agreement in the whole Community, a different type of legal instrument would be needed.

The Council Decision of 25 April 2002 responds to the requirement of Article 4 of the KP. It establishes a joint commitment under Community law. This is also governed by Article 10 of the EC Treaty, which states that:

> Member States shall take all appropriate measures, whether general or particular, to ensure fulfillment of the obligations arising out of this Treaty or resulting from action taken by the institutions of the Community. They facilitate the achievement of the Community's tasks. They shall abstain from any measure which could jeopardize the attainment of the objectives of this Treaty.

Thus, when the KP enters into force, all Member Sates and the Community would be bound under both international and Community law to

[130] See Pallemaerts (2003), 28.

[131] See Art. 9 and Annex III of the 2003/87/EC Directive, and Communication from the Commission on guidance to assist Member States in the implementation of criteria listed in Annex III, COM(2003) 830 final at section 2.1.1.

jointly fulfill their reduction and limitation targets[132]. Member States are under an obligation of loyalty towards the Community and towards each other in fulfilling their reduction and limitation targets. Conversion into and calculation of Member States and Community assigned amount will be operated under the responsibility of the Commission assisted by the Climate Change Committee. It is not stated clearly under the Decision who will hold the AAUs, the Community or the Member States, once the Commission will have determined the respective emission levels allocated to the EC and to each Member State in terms of tons of carbon equivalent. When adopting the Decision, both the Council and the Commission recognized that some additional legislation would be needed to ensure joint implementation with regard to the BSA, including arrangements to facilitate preventive action in accordance with Article 10 EC Treaty "to ensure that the Community complies with its commitment under Article 4 of the Protocol, in a manner which ensures that Member States can benefit from taking early action to reduce their emissions in excess of their allocated emission level [...] and have an incentive to do so"[133].

4.3 The Decision on the Monitoring Mechanism

With the development of the reporting requirements by the Marrakech Accords, in relation to GHG inventories, especially with regard to the modalities for accounting anthropogenic emissions by sources and removal by sinks, registries, and other information, with the view to participating in the flexible mechanisms, a new monitoring mechanism instrument was adopted by the Council on 11 February 2004 (hereafter the Monitoring Decision)[134]. The current Monitoring Decision replaces the monitoring mechanism put in place by Decision 93/389/EEC[135]. This mechanism aims at providing a harmonized emission forecast at both Member States and Community level, by monitoring all anthropogenic GHG, and thus evaluating progress towards fulfilling the reduction and limitation emission commitments. It implements UNFCCC and Kyoto requirements as regards national programmes, GHG inventories, national systems and registries of the Community and its Member States. It is meant to ensure timely reporting

---

[132] New Member States with a target under the KP would also be committed under Community law, as the KP regime, when entered into force, would become Community law, see Chapter 12.

[133] Joint statement by the Council and the Commission, Doc 7510/02 ADD 1, p.4 & 5.

[134] Decision No 280/2004/EC of the European Parliament and of the Council of 11 February 2004 concerning a mechanism for monitoring Community greenhouse gas emissions and for implementing the Kyoto Protocol, OJ L 49 of 19 February 2004.

[135] OJ L 167, 9 July 1993.

by the Community and its Member States to the UNFCCC Secretariat in a complete and accurate form[136]. Both the EC and the Member States are as Parties to the Kyoto Protocol subject to the same reporting requirements, and their inventories will be reviewed under Article 8 to assess whether they are in compliance or not. The Preamble emphasizes that "it is appropriate to provide for effective cooperation and coordination in relation to obligations under this Decision, including in the compilation of the Community greenhouse gas inventory, the evaluation of progress, the preparation of reports, as well as review and compliance procedures enabling the Community to comply with its reporting obligation under the Kyoto Protocol [...]"[137].

4.3.1 Evaluation of Community and Member States' Progress

Therefore, reliable GHG emissions inventories with accurate data are crucial for the Commission to make a quantitative review of the progress made by the Member States and the EC as a whole in attaining their emission targets. Furthermore, compliance with the guidelines for calculating and reporting Greenhouse Gases, as well with the inventories guidelines, are eligibility criteria for participation in the KP flexible mechanisms[138]. Thus, the purpose of the monitoring mechanism is both to monitor progress made by the Member States in reaching their individual BSA targets and to enable the Commission to assess the progress of the EC as a whole in meeting its commitments and to comply with its reporting obligations under the UNFCCC and the KP. The architecture of the Community monitoring mechanism reflects the interdependence between the EC and its Member States.

The Commission, assisted by the European Environment Agency, in consultation with the Member States, based on their national programmes, evaluates annual progress made by the Community and whether Member States' progress is sufficient to fulfill their commitments[139]. The result of that assessment shall be annually reported to the European Parliament and the Council[140]. Consequently, the Decision imposes upon each Member State

---

[136] Art. 1 Decision 280/2004/EC.

[137] Recital 13 Decision 280/2004/EC.

[138] See para. 5 of Draft Decision-/CMP.1 (Mechanisms), FCCC/CP/2001/13/Add.2, p. 4.

[139] Art. 5 Decision 280/2004/EC. This annual assessment shall be based on information given by the national programmes of the Member States, by national registries, and by reports of the Member States submitted pursuant to Article 21 of the Directive 2003/87/EC on European Emission Trading System (ETS). The annual report of the Member States pays attention to arrangements for the allocation of allowances, operation of registries, application of monitoring and reporting guidelines with regard to operating entities. Such report shall be sent to the Commission by 30 June 2005.

[140] Art. 5.2 Decision 280/2004/EC.

and upon the Commission the obligation to devise and implement national and Community programmes for limiting or reducing the emissions of six GHG[141]. By 15 March 2005, Member States shall report to the Commission on their national programmes, and then on a biannual basis. National reports shall include (i) information on actual progress and (ii) information on projected national GHG emissions by sources and removal by sinks for the years 2005, 2010, 2015 and 2020, including policies and measures to be implemented and their quantitative effects on GHG emissions[142], (iii) information on the use of the flexible mechanisms and on their 'supplementarity' at the national level, as the KP requires that the use of the flexible mechanisms shall be supplemental to domestic actions[143]. The Commission shall prepare a special report to demonstrate the progress achieved by 2005, taking into account sinks according to Article 3, paragraph 3 KP, to the UNFCCC Secretariat as required by Article 3 paragraph 2 KP, at the latest by 1 January 2006[144]. The Commission may adopt additional reporting obligations with a view to fulfilling requirements to demonstrate progress under Article 3, paragraph 2 KP[145]. This report is a critical element for assessing Member States' and, thus, Community progress in relation to their targets. The same obligation with regard to the special report on the demonstration of progress, to be forwarded to the UNFCCC Secretariat, applies to each Member State[146].

In addition, the Commission shall, in cooperation with the Member States and assisted by the European Environment Agency, annually compile a community GHG inventory and a Community GHG report to be circulated to the Member States as a draft, on 28 February of each year, in order to be completed, and then to be published and submitted to the UNFCCC Secretariat by 15 April each year in conformity with Article 7, paragraph 1 KP and subsequent Kyoto decisions[147]. The Member States, with a view to enabling the Commission to prepare its own annual inventory reports, must report their national GHG inventory by 25 January of each year. National reports shall contain information on complete GHG emissions offsetting emissions and removal of sinks during the year before the previous year (year of the report –2). During the period between 28 February and 15 March, Member States can provide to the Commission additional data to fill gaps or to remove major inconsistencies. Complete national GHG

---

[141] Art. 2.1 Decision 280/2004/EC.
[142] Art. 3.2 Decision 280/2004/EC.
[143] Art. 2.2 & 3.2 (a)(vi) & 3.2(d) of the 280/2004/EC Decision. And Draft Decision-/CMP.1 on Mechanisms, para. 1, FCCC/CP/2001/13/Add.2.
[144] Art. 5.3 Decision 280/2004/EC.
[145] Art. 5.6 Decision 280/2004/EC.
[146] Art. 5.4 Decision 280/2004/EC.
[147] Art. 4.1 Decision 280/2004/EC.

inventories shall be communicated by 15 March 2005. The EC has less than four months to prepare its own annual inventory, which may be rather short in the light of the previous experiences[148]. The EC greenhouse gas inventory is the result of compiled annual Member States inventories, supplied when necessary by, and compared with, information from the European Environmental Agency[149]. The Community GHG inventory system depends on the implementation and on the quality of national inventory systems, as the main responsibility for reporting annual emissions lies upon the Member States. Then the Commission, through the comitology procedure, assisted by the Climate Change Committee, shall adopt at the latest by 30 June 2006 a Community inventory system with two aims: (i) it shall secure accurate, compatible, complete and national inventories consistent with the Community GHG inventory system, and (ii) it shall establish a procedure for estimating data missing in the national inventory[150]. While Member States shall have in place at the latest by 31 December 2005 their national inventory system for the estimation of anthropogenic emissions of CO_2 by source and removal by sinks, the first complete Community GHG inventory should be established by 2006, which will allow the Community to fulfill its reporting requirements under Article 5, paragraph 1 KP, taking into account that Annex I Parties shall have in place a national inventory system no later than one year prior to the start of the first commitment period.

The tight time frame of less than four months which the Commission has at its disposal for preparing its own report raises two issues: (i) the need for a timely report from the Member States, and (ii) the fact that the time period is too short to undertake a deep and comprehensive review of each Member State inventory in order to solve inconsistencies. Compliance by Member States with reporting schedules is still poor, which may, when the Kyoto Protocol enters into force, compromise the Community's ability to meet its reporting obligations under the Protocol[151]. To address these issues, two options are available: to sanction late submission or to supply missing data. The current monitoring mechanism opts for the second option. The Commission, together with the Climate Change Committee and through the comitology procedure, may take appropriate measures to implement the provisions for reporting, such as filling gaps before handling the Community

148 See Hyvarinen (1999).

149 Art. 4, paras. 1 & 3 Decision 280/2004/EC.

150 Art. 4.3 Decision 280/2004/EC.

151 Seven Member States submitted their year 2000 GHG inventories to the Commission on time by 31 December 2001, other Member States reported their 2000 GHG inventories by April 2002 and data on national policies and measures even later. See Third Progress Report under Council Decision 93/389/ EEC, COM5 (2002) 702 final.

report[152]. Furthermore, special Articles of the Decision deal with the reports to be submitted to the UNFCCC Secretariat within the additional period as set in the Marrakech Accords. The Community and each Member State shall submit such report, and the Commission, through the comitology procedure, may establish additional requirements for reporting[153]. Following the submission of the report on the demonstration of progress by 2005, the Commission shall review the extent to which the Community and its Member States are making progress towards achieving their emission levels under the BSA, and the extent to which they are meeting their commitments under the KP. In the light of this assessment, the Commission may make proposals to the European Parliament and the Council to ensure that the Community and its Member States comply with their emission levels and that all their commitments are met[154]. Thus, the Monitoring Decision gives the Commission the possibility to adapt the monitoring mechanisms to the circumstances, and when needed to adopt additional measures that secure compliance by the Community with its reporting requirements and reduction commitments.

4.3.2 Ensuring Full Information and Cooperation between Member States and the Community

Article 8 of the Monitoring Decision requires full and effective cooperation between the Community and the Member States regarding the compilation of, and the report on, the annual Community GHG inventory, the report on the demonstration of progress by 2005, and adjustments under the review and compliance procedures established according to the Kyoto Protocol. To secure consistency, Member States shall submit, in their national report forwarded to the UNFCCC Secretariat by 15 April each year, identical information to the Commission as provided by the Monitoring Decision procedure. Thus, Member States are subject to a double reporting requirement: under Community law and under the international climate regime. The Decision recalls that the Community and its Member States are all Parties to the UNFCCC and its Protocol, and are each responsible for reporting, establishing and accounting for their assigned amounts and for

[152] Art. 4.1 Decision 280/2004/EC. The Commission, with the help of the European Environmental Agency, may work with the Member States to agree on estimates of missing data, in order to ensure the completeness of the inventory for the Community as a whole. If there is no agreement the Commission identifies the missing data in its annual EC inventory report. According to Art. 3.3 of Decision 280/2004/EC, the Commission implements the provisions for the reporting of the information referred to in paragraphs 1 & 2.

[153] Art. 5 paras. 5 & 6 Decision 280/2004/EC.

[154] Art. 10 Decision 280/2004/EC.

*Table 1: Annual process of submission and review of MS inventories and compilation of the EC inventory**

Element	Who	When	What
1. Submission of annual inventory by MS	Member States	15 January annually	– Anthropogenic CO_2 emissions and CO_2 removals by sinks, for the year n-2 – Emissions by source and removals by sinks of the other greenhouse gases; [1)]
2. Initial check of MS submissions	European Commission (incl. Eurostat), assisted by EEA	up to 28 February	– Initial checks (by EEA) – Comparison of energy data in MS IPCC Reference Approach with Eurostat energy data (by Eurostat and MS)
3. Compilation and circulation of draft EC inventory	European Commission (incl. Eurostat), assisted by EEA	28 February	– Draft EC inventory (by EEA), based on MS inventories and additional information where needed – Circulation of the draft EC inventory on 1 March
4. Submission of updated or additional data by MS	Member States	up to 15 March	Updated or additional data submitted by MS [2)]
5. Final annual EC inventory	European Commission (incl. Eurostat), assisted by EEA	15 April	– Submission to UNFCCC of the final annual EC inventory. This inventory will also be used to evaluate progress as part of the Monitoring Mechanism
6. Additional review of MS submissions and EC inventory	European Commission (incl. Eurostat), assisted by EEA	June to December	– Additional review aimed at improving the next annual MS and EC inventories – In November Eurostat makes available to MS energy balance data (1990 to inventory year)

1) In accordance with Art. 3(1) and 3(2) of Council Decision 280/2004/EC

2) updating is limited to the following situations: to remove major inconsistencies, to fill major gaps or to provide essential additional information.

* Source: European Community

maintaining their eligibility to participate in the Kyoto flexible mechanisms[155]. Any adjustment or change regarding the Member States' national inventory reports occurring before their submission to the UNFCCC Secretariat, or made thereafter by the Kyoto expert review teams, should be timely reported to the Commission. If a Member States is nevertheless subjected to an adjustment procedure for its report being incomplete or inconsistent with the requirement established pursuant to Article 5, paragraph 2 KP, the Community, as a consequence, will sooner or later be affected and will face an adjustment procedure as well. Thus, an information and cooperation procedure must be in place between the Commission and the Member States to solve the adjustment issue together. Furthermore, an appropriate coordination between the Member State concerned and the Commission is necessary before the Member State undertakes any action in relation to any adjustment applied by the expert review teams because any changes in the national inventory affect the Community inventory, and thus the overall compliance by the Community. Thus, the Commission may also lay down procedures and time scales for such cooperation and coordination[156].

4.3.3 National and Community Registries, Initial Assigned Amounts and Eligibility for the Flexible Mechanisms

With a view to their participation in the Kyoto flexible mechanisms, Annex I Parties need to establish national registries for accurately accounting for the issuance, holding, transfer, acquisition, cancellation and withdrawal of the different Kyoto credits and for carrying-over allowed excess credits to the next period[157]. Furthermore, the compliance assessment of Annex I Parties is based on the comparison between the quantity of valid credits for the commitment period held in the registry and the aggregate anthropogenic carbon dioxide equivalent emissions of the GHG during the commitment period. The balance determines whether the Annex I Party is in compliance or not[158]. Thus, not only emissions but also emission rights and credits need to be monitored through a registry in order to assess compliance at the end of the commitment period. Article 6 of the Monitoring Decision requires that the Member States and the Community shall establish and maintain registries in the form of a standardized electronic database[159]. Each national registry must incorporate the registries created for the European Emission

---

[155] Recital 14 Decision 280/2004/EC.

[156] Art. 8.3 Decision 280/2004/EC.

[157] See Art. 7.4 of the KP and Decision 19/CP7.

[158] Art. 3 paras. 7 & 8 KP and Sections D & E, Draft Decision-/CMP.1 on modalities for accounting of assigned amounts, FCCC/CP/2001/13/Add.2.

[159] Art. 19 Directive 2003/87/EC on emission trading.

Trading Scheme (ETS). Information regarding all transactions under the national registry should be provided in real time to the Community Central Administrator designated under the ETS[160]. Thus, the Central Administrator would be aware of transactions outside the scope of the ETS, which are relevant for meeting the requirements of the KP governing the use of its flexible mechanisms. The Kyoto regime requires that each Party designate an organization as its registry administrator to maintain the national registry of that Party[161]. Once the national registry is in place, assigned amount units and other Kyoto credits shall be documented in such registry.

As prescribed in Decision 2002/358/CE, the Commission shall determine the respective emissions levels and corresponding assigned amount allocated to the EC and to each Member State in terms of tons of carbon dioxide equivalent at the latest by 31 December 2006. While each Member State has to report on its emission level and assigned amount to the UNFCCC Secretariat, Member States shall endeavor to report simultaneously[162]. As regards the choice of the base year for the calculation of fluorocarbonated gases, contrary to the initial proposal for the Monitoring Decision, nothing is said in the current Monitoring Decision.[163] Initially, for the calculation of the initial assigned amount, the Commission had proposed 1995 as the base year for the group of fluorinated gases. The Commission proposition was amended in order to maintain the flexibility allowed by the KP which gives Annex I Parties the possibility to choose a base year between 1990 and 1995 and, thus, to allow Member States the same flexibility with the aim of giving fair credit to those sectors that have undertaken early action to reduce these emissions already before 1995[164]. As a result and contrary to the Commission wish, there is no common base year for the calculation of fluorinated gases levels within the Community. When different assigned amounts are calculated, initial assigned amounts units, consistent with their BSA emission limit and reduction target, will be issued in the Member States national registries[165]. It means that initially, the EC registry will be empty.

---

[160] Art. 6.2 Decision 280/2004/EC.

[161] Para.18, section II, Decision-/CMP.1 on modalities for accounting of assigned amounts, FCCC/CP/2001/13/Add.2.

[162] Art. 7.1 Decision 280/2004/EC.

[163] The proposal chose 1995 as base year for the three groups of fluorinated gases for the calculation of the initial assigned amount in order to ensure accuracy and consistency of reporting. Moreover, a majority of Member States prefer 1995, see Proposal for a Decision of the European Parliament and of the Council for a monitoring mechanism of the Community greenhouse gas emissions and the implementation of the Kyoto Protocol, COM(2003) 51 Final.

[164] See report of the European Parliament on the proposal for a European Parliament and Council Regulation on a monitoring mechanism of Community greenhouse gas emissions and the implementation of the Kyoto Protocol, doc A5-0290/2003.

[165] Art. 7.3 Decision 280/2004/EC.

Although this choice was not clear in the Community Decision to approve the KP, it is now stated that "(p)ursuant to Decision 2002/358/EC, the Community is not to issue assigned amounts"[166]. This shows the unwillingness of the Member States to surrender units to the Community, which might have reduced their possibilities to transfer credits within the EITS pursuant to Article 17 KP. As a result, it also stresses that the Member States and not the Commission are individually responsible for meeting their emissions limitations or reductions targets under Community law. However, the Community will be able to fill its registry with credits issued from CDM and JI projects[167].

Finally, the Commission and the Member States have agreed on the possibility for the Community to participate in parallel with the Member States in emission reduction projects under Articles 6 and 12 KP. Such emission reduction projects financed by the Community would enhance the Community policy to reduce GHG emissions outside its own borders. Thus, Article 24, paragraph 2 KP has been interpreted by the Commission as well as by the Member States as not being an impediment to such 'concurrent' participation. Nevertheless, the elusive wording used in the Preamble of the Monitoring Decision on this matter shows how cautiously the Community envisages such possibility: "The Community registry may be used to hold emission reduction units and certified emission reduction generated by projects funded by the Community [...], and be maintained in a consolidated system together with Member States"[168]. Why such credits resulting from Community projects should "be maintained in a consolidated system together with Member States' registries" is not clear. It may suggest that the Community should only hold its registry in a consolidated system together with Member States in order to accommodate Member States susceptibility by showing that it is not acting concurrently but in a manner complementary to the Member States. Furthermore, the Preamble of the Decision states: "The purchase and use of emissions reduction units and certified reduction units by the Community should be subject to further provisions to be adopted by the European Parliament and by the Council on a proposal of the Commission"[169]. Thus for the time being, the Community does not decide anything with regard to the use of such credits. Nevertheless, the following recital recalls that the Member States have a Community obligation to take action to fulfill their BSA target, and states: "Provisions laid down on the use of emission reduction units and certified emission reduction held in the Community registry should take into account Member States' responsibilities to fulfill their own commitments in accordance with

166 Preamble, recital 16, Decision 280/2004/EC.

167 Recital 10 & 11 of the preamble, Decision 280/2004/EC.

168 Recital 10 of the preamble.

169 Recital 11 of the preamble, Decision 280/2004/EC.

Decision 2002/358/EC"[170]. This suggests that such credits may be used for the purpose of Member States compliance, and rules to regulate such purpose are postponed to a later date. It should be noticed that nothing is said on the possibility for the EC to participate in an IETS under Article 17. This silence may be interpreted to the effect that the EC does not envisage such a tool at all, or that Member States deny the Community the right to use this possibility concurrently with them, or that there are some technical impediments. Regarding the technical aspect, two observations can be made: (i) in relation to the so-called 'supplementarity' principle, and (ii) with regard to the commitment period reserve that is needed under Article 17.

The Marrakech Accords provide that the "use of the Mechanisms shall be supplemental to domestic actions and domestic action shall thus constitute a significant element of the efforts made" by each Annex I Party in meeting its reduction or limitation commitments[171]. It was the result of a compromise between the EC, which during the Kyoto regime negotiations favored a ceiling on the use of the flexible mechanisms, and other industrialized Parties that fought against any quantitative cap. For the EC, industrialized countries are to take significant action at home to meet their reduction commitments and use the KP mechanisms to meet only part of these commitments. A qualitative assessment of whether Annex I Parties meet the supplementary condition was agreed, which requires those Parties to submit information about their utilization of the mechanisms and domestic actions in accordance with Article 7 KP for review under Article 8. The same information should be submitted in the report on demonstrable progress under Article 3, paragraph 2 KP. No criteria are so far established in the Kyoto regime on the basis of which 'supplementarity' could be assessed. Within the Community, some Member States, such as the Netherlands, have stated that they would achieve half of their reduction commitment through domestic actions, and the other half way would be done through the use of the flexible mechanisms. Other Member States did not decide on such a concrete division. One may question whether a fifty-fifty division could still be considered as supplementary. The Monitoring Decision requires that Member States report on the supplementary character of their use of the flexible mechanism, but no concrete threshold is set. Thus, supplementarity at Community level would be the result of the aggregate information on supplementarity provided by Member States. If the Community makes use of the flexible mechanisms, it would have to take into account that its own use of the flexible mechanisms does not conflict with the supplementarity principle, taking into account the use which Member States have already made of the flexible mechanisms. In its proposal for a directive on linking JI

---

[170] Recital 12 of the preamble, Decision 280/2004/EC.

[171] Para. 1, Draft Decision-/CMP.1 Mechanisms, FCCC/CP/2001/13/Add.2.

and CDM credits within the European emission-trading system, the Commission states that "this proposal finds a balance between the goal of promoting JI and CDM on the one hand and the concern for their supplementarity to domestic emission reduction measures on the other, taking into account that this measure by itself cannot guarantee supplementarity as it does not affect the use that Member States may make of the Kyoto flexible mechanisms"[172]. Thus, the threshold for the introduction of CDM and JI credits within the ETS triggers a Commission review on supplementarity, but it does not affect the use of these mechanisms made by the Member States. It rather ensures the supplementarity of the flexible mechanisms within the EC as a whole. Therefore, the Commission, when developing procedural regulations with a view to purchase project related reduction credits, must take into account their compatibility with the supplementarity principle at Community level in the light of Member States' use of the flexible mechanisms.

As established under Decision 18/CP.7 on emission trading under Article 17 of the KP, each Annex I Parties shall establish and maintain a commitment period reserve, which should not drop below 90% of the Party's assigned amount[173]. The reserve can hold any Kyoto units valid for that commitment period, and shall be maintained in the Party's registry[174]. The rationale behind this reserve is to prevent over-selling. Therefore, a Party shall not make a transfer that would result in these holdings falling below the required level of the commitment period reserve[175]. In the pre-commitment report that determines Annex I Parties' capacity to participate in flexible mechanisms, a Party shall submit the calculation of its commitment period reserve[176]. As under the decision mentioned earlier assigned amounts shall only be issued in the Member States' national registries, the Community, at the beginning of the first commitment period, cannot hold and calculate its period reserve for the pre-commitment period report. Practically, one may question whether it is at all possible to constitute a commitment period reserve by obtaining other Kyoto units adding up to 80% of the total assigned amounts of all fifteen Member States which are Parties to the BSA. This would constitute a serious obstacle to the Community participating in the IET under Article 17 KP. Pursuant to the negotiating text as adopted in

[172] Explanatory memorandum, proposal for a Directive of the European Parliament and of the Council, amending the Directive establishing a scheme for greenhouse gas emission allowance trading within the Community, in respect of the Kyoto Protocol's project mechanisms, COM(2003) 403 final.

[173] Para. 6, Annex, Draft Decision-/CMP.1 Article 17, FCCC/CP/2001/13/Add.2.

[174] Para. 7, Annex, Draft Decision-/CMP.1 Article 17, FCCC/CP/2001/13/Add.2

[175] Para. 8, Annex, Draft Decision-/CMP.1 Article 17, FCCC/CP/2001/13/Add.2.

[176] Para. 8(a), Annex, Draft Decision-/CMP.1 Modalities for accounting assigned amounts, FCCC/CP/20001/13/Add.2.

Bonn (2001), the constitution of a reserve period was originally considered as an eligibility requirement to transfer and /or acquire credits under Article 17 of the KP. At Marrakech, however, the text on the reserve period was deleted from the eligibility requirements section in order to be treated in separate paragraphs under the Annex of the Draft Decision-/CMP.1 (Article 17)[177]. The text of the Marrakech Accords stipulates that "a Party shall not make a transfer which would result in these holdings being below the required level of the commitment period reserve"[178]. One may argue that the reserve requirement is limited to transfers as stipulated in the text, i.e. to selling and, thus, does not hinder the Community to buy under Article 17. It would leave room for an Annex I Party, which does not hold a period reserve but fulfills Article 17 eligibility requirements, to participate in the IET in buying credits only. Nevertheless, this interpretation is contentious. In the face of this uncertainty, the Community seems to prefer to abstain from using Article 17 for itself, which may explain the silence of the Monitoring Decision on this aspect.

4.3.4 Compliance Arrangements

A preventive compliance instrument is established by the Monitoring Decision. Article 7, paragraph 2 of the Decision requires that, after completion of the review of their national inventories under the KP, Member States would have to withdraw assigned amount units and other Kyoto credits equivalent to their net annual emissions for each year of the first commitment period. With respect to the last year of the commitment period, such withdrawals of units shall take place prior to the end of the additional period. This tool would restrict the danger of over-selling, and complement the commitment period reserve. One criticism is that this option alone does not prevent a situation in which a Member State simply has no more KP units in the last commitment period year because the Member State did not limit its emissions[179]. This tool does not prevent a business as usual attitude.

4.4 Conclusion

The Monitoring Decision imposes on Member States an obligation under Community law to report to the Commission. This obligation should ensure that the Community as a Party to the KP could fulfill its own reporting duties. For that purpose, the Community depends essentially on the Member

[177] Paragraphs 6, 7, 8, 9 and 10, Annex, Draft Decision-/CMP.1 (Article 17), FCCC/CP/2001/13/Add.2.
[178] Paragraph 8, Annex, Draft Decision-/CMP.1 (Article 17), FCCC/CP/2001/13/Add.2.
[179] See Haites and Missfeldt (2002).

States' diligence in reporting to it. In the light of previous reporting experience, the four months period available to fill gaps or iron out inconsistencies may be too short. The Commission may supply missing data, and can develop, together with the Climate Change Committee, new rules to respond to such a situation. The Monitoring Decision allows the Commission enough flexibility to secure cooperation from the Member States in relation to a reinstatement procedure, and with regard to the additional period established under the KP. By 2006, the Community should have an inventory system with a registry in place, and Member States should have issued their assigned amount units in their national registries, allowing the Community to report demonstrable progress and to fulfill the eligibility requirements of the KP. The Community will be able to undertake JI and CDM reduction projects in order to obtain credits. The Commission, when developing procedural regulations with a view to purchase project related reduction credits, must take into account their compatibility with the supplementarity principle at Community level in the light of Member States' use of the flexible mechanisms. Although the use of such credit units is not clearly established, they may be used to facilitate compliance, either at Member States level or at that of the Community. The concrete shape of this possibility shall be determined in the future. As regards the possibility for the Community to participate in emission trading under Article 17, there is legal uncertainty as to whether the Community could entirely make use of such possibility, as it cannot yet create a commitment period reserve. In the light of the 2006 demonstrable progress report, the Commission will decide whether further reduction actions and/or further compliance tools are needed. Currently, the only preventive compliance instrument is the annual surrender of credit units. While limiting over-selling, it may not prevent the case that a Party has no credit units to withdraw at the end of the commitment period.

If a Member State falls into non-compliance, it can be asked whether the infringement procedure laid down by Articles 226 and 228 of the EC Treaty may be used as compliance tool within the context of commitment reductions.

5 The Community Infringement Procedure

According to Council Decision 2002/358/CE, Member States Parties to the BSA have a Community obligation to either limit or reduce their GHG emissions. Furthermore, international agreements concluded by the Community become an integral part of Community law and are, as from the date of their entry into force, binding on the Member States and on the

Community institutions.[180] Both the accession States with a KP reduction target and the other Members face a double obligation: (i) under international law, as Parties to the Protocol; and (ii) in the form of a Community obligation[181]. According to Article 10 EC Treaty, Member States shall take steps for ensuring that the obligations arising from such agreements are fulfilled in order to enable the Community to honor its obligations. Article 211 EC Treaty assigns to the Commission the function of a watchdog of the application of Community law. Therefore, if a Member State does not comply with its reduction target as agreed under the BSA, the Commission could bring the case before the European Court of Justice (ECJ) under Articles 226 and 228 EC Treaty. This infringement procedure has been successfully used, for instance, to enforce compliance by Member States in the field of the environment.[182]

5.1 The Procedure

An infringement of a Community obligation by a Member State is brought to the attention of the Commission by different means: by another Member State, by an individual from a Member State, by a question or petition from the European Parliament, or by any other source of information. If the Commission is of the opinion that the case constitutes an infringement of the Community law, the procedure that follows has three stages: (i) the pre-litigation procedure (Article 226 paragraph 1); (ii) litigation before the ECJ (Article 226 paragraph 2); (iii) enforcement of the Court's decision (Article 228).

5.2 The Effectiveness of the Infringement Procedure Regarding Non-Compliance with the BSA

Failure to comply with the BSA target can only be verified at the end of the first commitment period by 2012 and later. The individual inventory review process under the Kyoto Protocol will usually be completed one year after reception of the report. Then, a non-complying Member State has at its

[180] See Case 181/73 Haegeman [1974] ECR 449; Case 12/86 Demirel [1987] ECR 3719 and Art.300.7 EC.

[181] "Member States must fulfill an obligation not only in relation to the non Member State concerned but also above all in relation to the Community which has assumed the responsibility for the due performance of the agreement." Case Hauptzollamt Mainz v. Kupferberg [1982] ECR 3641.

[182] See the Greek cases, Case 45/91 Commission v. Hellenic Republic [1992] ECR I-02509, and case 387/97 Commission v. Hellenic Republic [2000] ECR I-05047

disposal a hundred days as an additional commitment period, which starts at the date set by the COP/MOP for the completion of the expert review process for the last year of the commitment period. Thus, a Party to the KP can only be found in non-compliance after this additional commitment period. What does this mean for the question when a Member State must be considered not to be in compliance with its BSA reduction or limitation obligation under Community law? Under the Monitoring and the KP Approval Decisions no answer is given. Article 7, paragraph 2 of the Monitoring Decision, which deals with the particular case of annual withdrawal of credit units to prevent over-selling, suggests that it is the decision of the Enforcement Branch after completion of the review process under the KP that also entails non-compliance under Community law. Indeed, Article 7 paragraph 2 states: "Member States shall following the completion of the review of their national inventories under the Kyoto Protocol [...], withdraw assigned amount units [...]". The wording "following the completion of the review of their national inventories under the Kyoto Protocol" suggests that the decision on non-compliance according to the Monitoring Decision depends above all on the relevant decision taken within the KP process. The latter decision is taken by the Compliance Committee. Thus, the breach of the international obligation results in a breach of the Community obligation. This is in line with the ECJ ruling[183]. The breach of Community law becomes final at the end of the additional period and after completion of the review process for such additional period, i.e. when the Kyoto Compliance Committee declares a Member State in non-compliance. At this stage, what would be the reason for the Commission to start an infringement procedure? Advocate General Ruiz-Jarabo Colomer in his opinion on the Greek case, stated that the infringement procedure under Article 228 "must likewise be regarded as a means of obtaining ultimate compliance and not as a penalty which serves to punish a Member State for its unlawful conduct or still less a form of compensation for the damage caused as a result of the delay in compliance"[184].

The infringement procedure is too lengthy to prevent a Member State's non-compliance at early stage. In the Greek case, from the date 1987, when individuals brought to the attention of the Commission the breach of the Community obligation by the Greek government, until the second judgment was delivered by the ECJ which imposed a daily penalty on the Greek government in 2000, thirteen years had passed[185]. When the ECJ might hand down a judgment on non-compliance with a BSA obligation, it would be too

[183] See cases supra notes 188 & 189.

[184] See Opinion of the Advocate General Ruiz-Jarabo Colomer delivered on 28 September 1999; Case C-387/97.

[185] See n. 192 above.

late. The Member State in question would be in a state of non-compliance, and the infringement procedure would not be able any more to obtain ultimate compliance with regard to the Member State's reduction or limitation commitments. This shows the inadequacy of the procedure within the Kyoto context[186]. Such a compliance regime for climate change has to be preventive and provide an opportunity to reinstate Member States into compliance at an early date.

As seen above, the review by the Commission of the first progress reports of the Member States by 2005, where progress under the Kyoto Protocol is to be demonstrated, should trigger a regulatory compliance procedure when found necessary. Based on these reports, the Commission should assess the progress made by Member States just before the commencement of the first commitment period, and make proposals to adopt measures as appropriate. Such measures could include recommendations to take additional action or to elaborate national compliance plans as well as proposals on the use of Kyoto units held by the Community or on preferential trade arrangements. These possibilities will be reviewed in the following section.

6 Additional Compliance Measures

6.1 National Compliance Action Plans

On the basis of the Member States' progress towards their target, the Commission may require that those Member States lagging behind adopt within a definite timeframe certain policies and measures to remedy the situation. Such a compliance plan should be followed by detailed progress reports on its implementation. However, the exact effect of policies and measures and the moment when they result in real emission reductions are difficult to determine. This strategy would only be useful if adopted before the commencement of the first commitment period on the basis of the first progress reports to be submitted by 2005. It is not helpful for reestablishing compliance within a short period of time.

6.2 The Use of the Flexible Mechanisms and of a Credit Pool

As envisaged under the Monitoring Decision, the EC is willing to purchase credits through JI and CDM projects. The EC can thus obtain credits and create a compliance reserve or a pool of credits to ensure that either a defaulting Member State attains its target or the Community as a whole

[186] For a similar conclusion, see Lacasta, Barata, Cavalheiro Dessai (2001).

achieves its overall target. The Community will have its own registry; a special account for holding those credits can be thus established within this registry. Credits from the pool could either be used during the commitment period or at the end, either at the request of a Member State, or imposed upon it as a compulsory means at the recommendation of the Commission. The use of this pool could start at the end of the commitment period when the Commission can already forecast that the Member State would not comply, or during the additional period given by the Kyoto Protocol. The latter option is more consistent with the aim of this instrument. Modalities for triggering this option and conditions regarding the price for a non-complying Member State using this compliance tool would have to be fixed.

The funding of those projects could be done either through the EC budget or through an *ad hoc* climate fund set up by a Council decision, financed by the Member States and managed by the Climate Change Committee where the Commission and the Member States are represented. Such financing should be additional to the official development assistance to respond to developing countries preoccupations[187]. The source of the financing would determine the modalities for using the credits available for compliance. Nevertheless, the use of such credits by Member States could be offset by a withdrawal from the Member State's assigned amount units for a subsequent commitment period. This formula would have the advantage of not favoring non-complying Member States and secure the amount of credits held in the reserve pool for next commitment periods. Within the context of an *ad hoc* fund financed by the Member States, the withdrawal of credits by a non-complying Member State could be done under the condition that it returns an equivalent amount of credits back to the pool. Unused credit units could be banked for next commitment periods, distributed back to the Member States, or cancelled.

The use by the Community of JI and CDM projects for constituting a reserve pool can promote the integration of environmental policy into EU external policies and contribute to the Community strategy on sustainable development, by furthering the host countries' sustainable development and promoting the transfer of environmentally sound technologies.

6.3 Preferential Trading within the EU

The idea behind a preferential trading scheme is the following: when a Member State finds that at the end of the commitment period it does not hold enough credits to offset its emissions, while other Member States hold more credits than they need for this purpose, the latter countries would transfer

[187] See Decision 17/CP.7, FCCC/CP/2001/13/Add.3.

their excess credits. This would prevent the Member State holding excess Kyoto units from selling them in the emission trading system established under Article 17 or to carry them over under Article 3, paragraph 3 KP. This could be justified by the obligation of loyalty pursuant to Article 10 of the EC Treaty, as specified by the Kyoto Protocol approval decision. However, this preferential trading in favor of non-complying Member States would be done under the condition that the latter hold the credited amount transferred by the seller, in order to preserve its right of banking as well, and against a price high enough to discourage a free riding behavior. This possibility should be available for the non-complying Member State during the last stage, i.e. during the additional period, if no more credits can be bought on the international market. This preferential trading could be extended beyond the borders of the Community to countries such as Russia or Ukraine, which are expected to remain well below their target and, thus, to dispose of a large amount of credits, the so-called 'hot air'. Regarding the latter, agreements between the Member States and those countries could secure an amount to be put at the disposal of the Member States against a higher price than can be obtained on the market. Advantageous as this might be for these countries, it would entail the political risk of undermining the position of the European countries and of the Community that have defended the environmental integrity of the Protocol by advocating a limited use of the flexible mechanisms.

6.4 Surrender of Excess Kyoto Units to the Commission

It has also been suggested that Member States holding excess Kyoto units should be obligated to hand them over to the Community level as a consequence of their duty of Community loyalty under Article 10 EC Treaty. This obligation would mainly concern the accession countries that dispose of 'hot air', with the ensuing political consequences. In order to preserve this right established by the Kyoto Protocol, banking could also be done at the Community level. The amount of credits taken from an over-complying Member State at the first commitment period could be given back to this Member State at the subsequent commitment period and an appropriate interest would have to be paid. This amount would be deducted from the total amount allocated to the non-complying Member State for the subsequent period. Among all these possibilities, the compliance solution that is politically the most acceptable, and that preserves most Member States' rights, is the constitution of a credit pool by the Commission, whose credits would be obtained through reduction projects under JI and CDM financed by the Community. This seems to be the preferred option of the Community, which can be deduced from the Monitoring Decision.

7 Conclusion

The main legal consequence deriving from the fact that the Community and the Member States are jointly committed under the Kyoto Protocol is their joint liability. As to the accession countries, their relationship to the Community in matters of the Kyoto Protocol is uncertain, as the Community is jointly committed with its fifteen old Member States only. One may argue that from a Community law perspective, once the Protocol enters into force, the EC and the accession countries, which are Parties to the KP, will be co-responsible in achieving their commitments. How this type of joint responsibility could operate with regard to the consequences of non-compliance to be applied by the Kyoto compliance regime is also uncertain, as no real precedent exists. This question is left to the Community and its Member States. In addition, uncertainties exist as to the possibility for the Community to meet all requirements in relation to its participation as a full actor under the international emission trading system as laid down by Article 17.

When the Protocol enters into force, all Member States will also be jointly committed under Community law, and thus, bound by a loyalty obligation towards the Community and towards each other to individually fulfill their emission reduction and limitation commitment under the Burden Sharing Agreement. Emission reduction and limitation commitments resulting from the Burden Sharing Agreement could be considered to apply independently of the entry into force of the Protocol. This would, however, result in the awkward situation that only fifteen Member States would be so committed. This effect should be avoided by some type of new internal arrangement in which all Member States Parties to the Protocol should participate.

As regards compliance, the consequences provided for by the Protocol are not dissuasive enough to prevent free riding. The Community's ability to comply with its own commitment under the KP depends mainly on the Member States' compliance. Until today, the Community has not adopted preventive compliance measures that would anticipate Member States non-compliance. The Community has rather established a procedure of cooperation with the Member States. It allows the Community to be prepared for the operation of the Protocol. It envisages that the Community participates in JI and CDM projects. The operation of such participation, however, are deliberately left open for the time being. Furthermore, additional preventive compliance measures have been postponed to the future in the light of Community progress. It is suggested that among possible preventive measures, a credit pool should be the preferred option.

Annex I. Member States' commitments in accordance with Article 4 of the Kyoto Protocol and the Burden Sharing Agreement [188]

Member State	*Commitments under the Burden Sharing Agreement**
Austria	–13
Belgium	–7.5
Denmark	–21
Finland	0
France	0
Germany	–21
Greece	+25
Ireland	+13
Italy	–6.5
Luxembourg	–28
Netherlands	–6
Portugal	+27
Spain	+15
Sweden	+4
United Kingdom	–12.5

European Community's Commitment in accordance with Article 3.1 of the Kyoto Protocol

European Community	–8

* Percentage change in emissions of the six Greenhouse Gases for the first Commitment period 2008–2012 relative to 1990 base year levels.

[188] In Council Decision 2002/358/EC on the approval by the EC of the Kyoto Protocol the different commitments of the Member States are expressed as percentage change from the base year. In 2006 the respective emission levels shall be expressed in terms of tonnes of carbon equivalent (Art. 3 Council Decision).

References

Anderson, M. (2001). *Report to CAN on Articles 5, 7 and 8 section of the Marrakech Accords to the Kyoto Protocol.* London: Vertic. http://www.climatenetwork.org/pages/InternationalNegotiations.html.

Baron, R. (2001). *OECD and IEA information paper on the Commitment Period reserve.* Paris: OECD. http://www.oecd.org/env/cc.

Bothe, M. (1996). 'The Evaluation of Enforcement Mechanisms in International Environmental Law – An Overview', in Rüdiger Wolfrum (Ed.), *Enforcing Environmental Standards: Economic Mechanisms as Viable Means?*. Berlin: Springer Verlag, 13–38.

Bourgeois, J.H.J. (1997). 'Mixed Agreements: A New Approach ?', in Bourgeois, Dewost and Gaiffe (eds.), *La Communauté européenne et les accords mixtes*. Bruxelles: Presses Interuniversitaires Européennes, 83–98.

Björklund, M. (2001). 'Responsibility in the EC for Mixed Agreements – Should Non-Member Parties care?'. *Nordic Journal of International Law*, 373–402. The Hague: Kluwer Law International.

Dolmans, M. (1985). *Problems of Mixed Agreements: Division of powers within the EEC and the Rights of the Third States*. The Hague: T.M.C Asser Institute.

Ehlermann, C.D. (1983). 'Mixed Agreements – A List of Problems', in O'Keeffe, D. and Schermers, H.G. (eds.), *Mixed Agreements*. Rijkuniversiteit Leiden: Europa Instituut, 3–21.

Garzon Clariana, G. (1998). 'La Mixité: le droit et les problèmes pratiques', in Bourgeois, Dewost, Gaiffe (eds.), *La Communauté européenne et les accords mixtes*. Bruxelles: Presses Interuniversitaires Européennes, 15–26.

Gerhardsen, M. (1998). 'Who governs the environmental policy in the EU? A study of the process towards a common climate target'. *CICERO Policy Note*, 1998:4. http://www.cicero.uio.no/media/98.pdf

Granvick, L. (1999). *The Treaty-Making Competence of the European Community in the Field of International Environmental Conventions*. Helsinki: Finnish Society of Environmental Law.

Haigh, N. (1996). 'EC Climate Change and Politics', in O'Riordan, T. and Jäger, J. (eds.), *Politics of Climate Change – A European perspective*. London: Routledge, 155–185.

Haites, E. and Missfeldt, F. (2002). 'Limiting Overselling in International Emissions Trading I: Costs and Environmental Impacts of Alternative Proposals'. *UNEP Working Paper*, 10. Copenhagen: UNEP. http://www.unprisoe.org/workpapers.htm

Heliskoski, J. (2001). *Mixed Agreements as a Technique for Organizing the International Relations of the European Community and its Member States*. The Hague: Kluwer Law International.

Hyvarinen, J. (1999). 'The European Community's Monitoring Mechanism for CO_2 and other Greenhouse Gases: the Kyoto Protocol and other Recent Developments'. *Review of European Community and International Environmental Law (RECIEL)*, 8: 191–197.

Lacasta, N.S., Barata, P.M., Cavalheiro, G. and Dessai, S. (2001). 'Options for a System Within the European Community to Ensure Compliance with the EC Target for the Reduction of Greenhouse Gas Emission under the Kyoto Protocol and Individual Member States'. Study to the European Commission under Contract B4-3040/40/2001/324138. Lisbon: Euronatura.

Lang, J.T. (1986). 'The Ozone Layer Convention: A New Solution to the Question of Community Participation in 'mixed' International Agreements'. *Common Law Review*, 23: 157–176. The Hague: Kluwer Law International.

Lang, W. (1996). 'Compliance Control in International Environmental Law: Institutional Necessities'. *Zeitschrift für ausländisches Öffentliches Recht und Völkerrecht*, 56/II:683–695. Berlin: Kohlhammer.

Lavranos, N. (2002). 'Multilateral Environmental Agreements: Who Makes the Binding Decisions?'. *European Environmental Law Review*, 11: 44–50. Oxford: Blackwell.

Leal-Arcas, R. (2001). 'The European Community and Mixed Agreements'. *European Foreign Affairs Review*, 6: 483–513. The Hague: Kluwer Law International.

Lefeber, R. (2002). 'From The Hague to Bonn and Beyond: A Negotiating History of the Compliance Regime under the Kyoto Protocol'. *Hague Yearbook of International Law*, 14: 25–54. The Hague: Nijhoff.

Macleod, I., Hendry, I.D. and Hyett, S. (1996). *The External Relations of the European Communities.* Oxford: Clarendon Press – European Community Law Series.

Manin, P. (1997). *Les Communautés Européennes, l'Union Européenne* (3rd edn). Paris: Pédone.

Marauhn, T. (1996). 'Towards a Procedural Law of Compliance Control in International Environmental Relations'. *Zeitschrift für ausländisches Öffentliches Recht und Völkerrecht*, 56/II: 696–731. Berlin: Kohlhammer.

Neuwahl, N.A. (1991). 'Joint Partnership in International Treaties and the Exercise of Power by the EEC and its Member States: Mixed Agreements'. *Common Law Review*, 28: 717–740. The Hague: Kluwer Law International.

Neuwahl, N.A. (1996). 'Shared Powers or Combined Incompetence? More on Mixity'. *Common Market Law Review*, 33: 667–687. The Hague: Kluwer Law International.

Oberthür, S. and Ott, H.E. (1999). *The Kyoto Protocol International Climate Policy for the 21st Century*. Berlin: Springer.

Okwa, P. (1995). 'The Community and International Environment Agreements'. *Yearbook of European Law*, 15: 169–192. Oxford: Clarendon Press.

Ott, H.E. (1996). 'Elements of a Supervisory Procedure for the Climate Regime'. *Zeitschrift für ausländisches Öffentliches Recht und Völkerrecht*, 56/II: 732–749. Berlin: Kohlhammer.

Pallemaerts, M. (2003). 'La Communauté européenne comme Partie contractante au Protocole de Kyoto'. *Revue d'Etudes Juridiques*, Numéro spécial: 16-28.

Peel, J. (2001). 'New State responsibility Rules and Compliance with Multilateral Environmental Obligations: Some Case Studies of How the New Rules Might Apply in the International Environment Context'. Review of European Community and International Environmental Law, 10: 82–97. Oxford: Blackwell.

Ringius, L. (1997). 'Differentiation, Leaders and Fairness – Negotiating Climate Commitments in the European Community'. *CICERO Report*, 8. Oslo: CICERO http://www.cicero.uio.no/media/99.pdf, p8.

Sand, P.H. (1996). 'Institution-building to Assist Compliance with International Environmental Law: Perspectives'. *Zeitschrift für ausländisches Öffentliches Recht und Völkerrecht*, 56/II: 774–795. Berlin: Kohlhammer.

Schermers, H.G. (1983). 'A Typology of Mixed Agreements', in O'Keeffe, D. and Schermers, H.G. (eds.), *Mixed Agreements*. Deventer: Kluwer Law and Taxation, 23–33.

Sjöstedt, G. (1998). 'The EU Negotiates Climate Change – External Performance and Internal Structural Change'. *Cooperation and Conflict*, 32/33: 225–256. Stockholm: Nordic Committee for the Study of International Politics.

Tomuschat, C. (1983). 'Liability for Mixed Agreements', in O'Keeffe, D. and Schermers, H. G. (eds.), *Mixed Agreements*. Deventer: Kluwer Law and Taxation, 125–132.

Wang, X. and Wiser, G. (2002). 'The Implementation and Compliance Regimes Under the Climate Change Convention and its Kyoto Protocol'. *Review of European Community and International Law*, 11: 181–198. Oxford: Blackwell.

Werksman, J. (1996), 'Compliance and Transition: Russia's Non-Compliance Tests the Ozone Regime'. *Zeitschrift für ausländisches Öffentliches Recht und Völkerrecht*, 56/II: 750–773. Berlin: Kohlhammer.

Wiser, G. (2001). 'Report on the Compliance Section of the Marrakech Accords to the Kyoto Protocol'. Washington: CIEL. (http://www.climatenetwork.org/pages/International Negotiations.html).

Chapter 12

MERCEDES FERNÁNDEZ ARMENTEROS
AND LEONARDO MASSAI*

Emissions Trading and Joint Implementation: Interactions in the Enlarged EU

1 Introduction

Parties to the UNFCCC and the Kyoto Protocol have the possibility to fulfill part of their greenhouse gas emissions reduction obligations through the implementation of the so-called flexible mechanisms. This prospect was confirmed by the Marrakesh Accords, which added definitive rules – the so-called eligibility criteria – in order to regulate the modalities for participation in the flexible mechanisms[1]. As the Clean Development Mechanism's (CDM) project rules have been already established[2], the official start allowing the possibility to issue and register the emissions reduction units generated by the implementation of the mechanisms (ERUs, AAUs or RMUs)[3] concerning International Emissions Trading (IET) and Joint Implementation (JI) will be in 2008[4]. This slight difference in timing concerning the setting of the regulatory framework of the different mechanisms is cause of uncertainty for Parties to the UNFCCC and to the

* At the time of writing Leonardo Massai was research fellow at the Environmental Law Research Centre of the University of Frankfurt, Germany

1 The Marrakesh Accords are the result of the COP 7 negotiations carried on in Marrakesh from 29 October to 9 November 2001. See UNFCCC decision 15/CP.7 UNFCCC document FCCC/CP/2001/13/ADD.2.

2 See UNFCCC decision 17/CP.7, UNFCCC document FCCC/CP/2001/13/ADD.2.

3 These units may be in the form of assigned amount units (AAUs), removal units (RMUs), emission reduction units (ERUs) and certified emission reductions (CERs), see Article 3.10/11/12 of the Kyoto Protocol.

4 See UNFCCC decision 16/CP.7 and 18/CP.7, UNFCCC document FCCC/CP/2001/13/ADD.2.

MICHAEL BOTHE AND ECKARD REHBINDER (EDS.), Climate Change Policy, 407–450.

Kyoto Protocol with regard to the definition of climate change national policies. As a sort of precursor for the forthcoming entry into force of JI projects and IET, some domestic Emissions Trading[5] schemes as well as Activities Implemented Jointly (AIJ) are already on track or under preparation in several countries[6].

Central and Eastern European Countries (CEECs)[7] – the group of states which have joined the EU on 1st May 2004 as agreed in the Athens European Council on 15/16 April 2003[8] – are single Parties to both the UNFCCC and the Kyoto Protocol, while the EC and its old 15 member states are jointly committed to those treaties as allowed respectively by Article 4.2 (a) and 12.8 of the UNFCCC[9] and by the procedure indicated in Article 4.1 of the Kyoto Protocol.

Starting from the fact that EU enlargement will not affect the EC "joint" position in the international climate regime in the first commitment period[10], this chapter will investigate the future implementation, constraints and possibilities of Joint Implementation projects as well as the application of the

---

[5] In the context of this article, IET refers to the International Emissions Trading system designed under the Kyoto Protocol, whilst EATD refers to the Emission Allowances Trading Directive. ET refers to Emissions Trading intended as new instrument of environmental policy.

[6] It is important to remember that the Kyoto Protocol has not been ratified yet by the number of Parties required for its entry into force as indicated in Article 25 of the Protocol. The implementation of those mechanisms will receive strong support upon the Protocol's entry into force.

[7] The Central and Eastern European Countries (CEECs) are among those countries classified under the UNFCCC as Economies in Transition (EITs) to a market economy. Ten of these countries joined the EU by May 2004 and are Annex I Parties to the UNFCCC: the Czech Republic, Cyprus, Estonia, Hungary, Latvia, Lithuania, Malta, Poland, Slovakia and Slovenia. At the international level, the term CEECs refers not only to the above countries but also to Romania, Bulgaria, Croatia, Belarus, Russia and Ukraine. In the context of this article, the terms CEECs and new member states refer exclusively to the eight Annex I countries that have already joined the EU, unless otherwise specified. Malta and Cyprus are non-Annex I Parties to the Convention and they are not considered in this paper though they have also joined the EU.

[8] On 16 April 2003 at the Athens European Council, CEEC representatives signed the Accession Treaties to the EU fixing 1 May 2004 as the official date for EU enlargement. The Accession Treaties must be ratified by the all EU-25 member states.

[9] The EC is Party to the UNFCCC as a "regional economic integration organization". See Articles 1.6 (definitions), 14.2 (settlement of disputes), 18.2 (right to vote), 20 (signature), 22 (ratification) and 23 (entry into force) of the UNFCCC.

[10] Under Article 4.4 of the Kyoto Protocol, "[I]f Parties acting jointly do so in the framework of, and together with, a regional economic integration organization, any alteration in the composition of the organization after adoption of this Protocol shall not affect existing commitments under this Protocol. Any alteration in the composition of the organization shall only apply for the purposes of those commitments under Article 3 that are adopted subsequent to that alteration".

EU Emission Allowance Trading Directive (EATD) in the new member states[11].

Thus, taking into consideration the EU climate policy's latest developments, a set of different problems are raised by the accession process when looking at the implementation of the Kyoto Protocol flexible mechanisms: the correlation between EU and international rules; the different position and obligations of CEECs and the EC in the international climate arena; and, the influence of the *acquis communautaire*[12] in the design of CEECs climate policies.

2 Acceding Countries within the UNFCCC and the Kyoto Protocol

All CEECs have ratified the UNFCCC and are Annex I Parties[13]. CEECs are also members of the Annex I Expert Group, an ad hoc group composed of Annex I government officials, which reflects the EU new member states concerns[14].

CEECs are furthermore considered within the UNFCCC and the Protocol as "Economies in Transition" (EITs): states that are in the process of making the transition to a market economy. Article 4.6 of the UNFCCC and Articles 3.5 and 3.6 of the Kyoto Protocol allow "the Parties included in Annex I undergoing the process of transition to a market economy" a "certain degree of flexibility" in the implementation of their commitments. This acknowledgement was of vital importance especially when CEECs decided to negotiate and choose a different base year than the one accorded to most

---

[11] The relevance of IET will be also considered as it is linked with the implementation of EATD in Europe.

[12] The *acquis communautaire* covers the whole EU institutional and regulatory setting, which acceding countries are required to fulfil as a condition for membership to the Union. It requires the harmonization of CEEC legislation with EU rules through both the transposition of EC law into the national systems in CEECs and its full implementation .

[13] Under Article 4.2 of the UNFCCC, Annex I countries are committed to reduce their greenhouse gas emissions to 1990 levels by the year 2000. Practically, they are the OECD countries excluding Mexico, the designated Economies in Transition countries, and Turkey. The Czech Republic, Slovakia and Slovenia have been added to Annex I list only recently through an amendment based on a decision taken at COP 3 as the result of the dissolution of Yugoslavia and of Czechoslovakia. The amendment entered into force on 13 August 1998 through UNFCCC decision 4/CP.3 adopted at COP 3.

[14] The group aims to support these countries' efforts in order to address and develop climate change policies through, for instance, the enhancement of the exchange of information at the government level.

other Annex I Parties[15]. Likewise, since 2000, CEECs have jointly participated in the international negotiations. During COP 6, the Central Group 11 (CG11) – composed of CEECs together with Bulgaria, Romania and Croatia, with Malta and Cyprus as observers – was set up with the aim to establish a common strategy towards the implementation of the Kyoto Protocol. Advantages and disadvantages of the participation in the flexible mechanisms were analysed in the group to foster a close collaboration with the EU position during the negotiations. The group was recently dissolved because of their EU accession[16].

All CEECs have ratified the Kyoto Protocol, under which they are committed to emissions limitations as Annex B Parties[17]. Three main reasons have induced these countries to ratify the Protocol. Firstly, the ratification of the Protocol became part of the *acquis communautaire* when the EC submitted its instrument of ratification in June 2002[18]. Thus, together with the World Summit on Sustainable Development, held in Johannesburg in September 2002, the ratification of the Protocol by the EC gave strong political impetus for ratification by each CEEC. Furthermore, according to the latest greenhouse gas emission information, CEECs will be able to meet their targets easily as most of them are already in compliance with their reduction commitments and will remain in compliance as their economies grow[19]. This is due to the fact that CEEC reduction commitments were adopted using base years before which their economies collapsed. Finally, CEECs are very keen to use the flexible mechanisms, which represent an opportunity to gain economic and environmental benefits, and the ratification of the Kyoto Protocol is one of the preconditions for participating in these instruments[20].

---

[15] Bulgaria and Poland have chosen 1988, Hungary the average between 1985–1987, Romania 1989 and Slovenia 1986. The choice of a different base year for the calculation of their emissions commitments is due to the political and economic changes that occurred in the region during those years and which allow the maximization of the difference with the aggregate emissions as to the year 2000.

[16] During the eighteenth sessions of the subsidiary bodies to the UNFCCC (SB18), held in Bonn from 4 to 13 June 2003, CEECs delegates decided to join the EU negotiating group.

[17] They are those Parties to the Kyoto Protocol with quantified emission limitation or reduction commitments – the so-called individual emissions targets – and included in Annex I to the UNFCCC, having assumed legally binding commitments for the period 2008–2012 as indicated in Article 3.1 of the Kyoto Protocol.

[18] The EU officially submitted its instrument of ratification to the Secretariat on 31 May 2002, see Council Decision 2002/358/EC concerning the approval, on behalf of the EC, of the Kyoto Protocol to the UNFCCC (OJ L 130, 15.5.2002, p.1).

[19] Only Slovenia and likely Lithuania will represent an exception compared to the emissions average of the other CEECs, as they could have extra absolute emissions at the end of the first commitment period.

[20] Massai (2002).

The EC and its old member states are jointly committed to the Kyoto Protocol through the EU's burden sharing agreement (BSA)[21], which has already been adopted and submitted to the UNFCCC together with the EC's instrument of ratification. The obligations under the BSA will be binding from both the international and European perspectives for the first commitment period once the Kyoto Protocol enters into force. The BSA (or "EU bubble") distributes the global EC reduction commitment among the member states, while CEECs each have independent obligations and are not bound by this joint instrument[22]. Considering Article 4.3 of the Kyoto Protocol, which states that "any such an agreement shall remain in operation for the duration of commitment period", the position of CEECs in the international climate regime is diverging from the one to which the EU-15 member states agreed, as new member states cannot be included in this agreement at least for the first commitment period. Along with this important clarification, it must be said that no rules at the European or international levels prevent existing member states from engaging in JI projects in Central and Eastern Europe, or to trade emission allowances with CEECs as foreseen by the IET. The EU bubble indicates the "contributions of each member state to the overall Community reduction commitment" and the submission of the EC instrument of ratification creates obligations for the EC and its existing member states only in terms of state responsibility[23]. No specific problem is caused by the accession of new states and CEECs will be simply subjected to the same responsibilities with regard to their individual targets. Consequently, ERUs generated through JI projects or allowances traded by entities through the EATD can be used by EU-15 member states within the bubble in order to fulfil their international obligations. If we look at Article 10 of the EC Treaty[24], one may even go further and see that CEECs will have an obligation to ensure that the EC as a whole meets its targets.

---

[21] Document 9702/98 of 19 June 1998 of the Council of the European Union reflecting the outcome of proceedings of the Environment Council of 16–17 June 1998. See also Annex I to the Council Decision 2002/358/CE, supra note 14.

[22] The EC is in any case responsible as a whole and CEECs – not part of the bubble agreement – will be also responsible for their individual targets once they enter the EU.

[23] See point 12, preamble of the Council Decision 2002/358/CE, supra note 18.

[24] "Member States shall take all appropriate measures, whether general or particular, to ensure fulfillment of the obligations arising out of this Treaty or resulting from action taken by the institutions of the Community. They shall facilitate the achievement of the Community's tasks. They shall abstain from any measure which could jeopardize the attainment of the objectives of this Treaty." Article 10 of the EC Treaty.

Within the international climate regime, Malta and Cyprus[25] have a different legal status compared to the rest of the new member states. Both countries have ratified the UNFCCC and the Kyoto Protocol, but are still considered non-Annex I Parties. Looking at the international climate regime, several questions may arise once these two states officially become members of the EU. Firstly, from a political point of view, the EC, which is a Party to the UNFCCC and the Kyoto Protocol as a "regional economic integration organization" (Article 4.4 of the Kyoto Protocol), will embrace two states formally considered as developing countries. Legally speaking, although Article 4.4 of the Kyoto Protocol makes clear that "any alteration in the composition of the organization after adoption of this Protocol shall not affect existing commitments under this Protocol", some inconveniences will appear when considering for instance the obligations to meet the eligibility criteria for the participation of these two countries in JI projects or in the IET[26]. It is still unclear and particularly complicated to envisage how a change of the legal status of these two countries could rapidly occur. These countries present several shortcomings related to the greenhouse gas emissions information obligations required by the UNFCCC. In order to prepare its national communications, Malta set up a National Board on Climate Change Affairs in 1999. While experts are currently preparing Cyprus' first national communication, information on Cyprus is still vague[27]. What is certain is that the positions of Malta and Cyprus will have to be clarified at the start of the discussions for the second commitment period, since all EU member states should have the same legal status so as to act jointly within the international climate regime via Article 4 of the Kyoto Protocol.

The fact that CEECs, as well as Malta and Cyprus, joined the EU does not affect in any sense the status these countries have within the international climate regime, at least for the first commitment period. This implies that

---

[25] Cyprus has ratified both the UNFCCC and the Kyoto Protocol, and has prepared a National Strategic Plan in order to fulfil its obligations. Cyprus is not an Annex B Party to the Protocol and is currently considered in the list of developing countries. Given their upcoming accession to the EU, the status of Cyprus in both the UNFCCC and the Kyoto Protocol will have to be changed also for political reasons and included in the Annex I list. Cyprus' emission reduction obligations may be included in the second Kyoto commitment period starting after 2012, but no decisions have been taken so far on this. An Ad-Hoc Committee on Climate Change has been established by the national government with the aim to design a strategy to reduce greenhouse gas emissions.

[26] Respectively Article 6 and 17 of the Kyoto Protocol.

[27] It must be remembered that non-Annex I Parties have a different timetable for the submission of their national communications to the UNFCCC Secretariat. Each non-Annex I Party shall submit its initial communication within three years of the entry into force of the Convention for that Party, or of the availability of financial resources (except for the least developed countries, who may do so at their discretion). Article 12.5 of the UNFCCC.

CEECs remain Economies in Transition (EITs), and Malta and Cyprus remain non-Annex I Parties for the moment. Through the accession to the EU, CEECs have certainly showed and certified some structural changes – one of the conditions for accession to the EU was the fulfilment of political, economic and legal criteria as indicated by the accession conditions of Article 49 EC Treaty – and when negotiations over the second commitment period start in 2005 a change of their status is envisaged. The same criteria will be applied to Malta and Cyprus, which currently cannot participate in JI or in IET. This is so even if the shift from being a non-Annex I to an Annex I Party is much more complicated due to difficulties regarding compliance with several monitoring and reporting obligations as required by the Protocol and its rules.

Although it is legally possible, the chances that existing member states will invest in CDM projects in Malta and Cyprus are very remote as there is practically no interest or space to implement such activities in these two small countries[28].

3 The Accession Process and Climate Policy

In contrast to the previous enlargements where the adoption of the *acquis communautaire* was just a condition for accession, this time, the implementation of the *acquis* was verified before accession took place[29]. The European Commission has annually undertaken an assessment of the degree of compliance with the *acquis communautaire* by the different CEECs focusing on each of the 30 negotiating chapters of the Europe Agreements, which are the first legal instruments of approximation[30].

The EU accession requirements and the *acquis communautaire* obligations represent additional driving forces for strengthening and recovering national structures dealing with environmental protection and therefore also for climate policy and the EU's international obligations.

However, *de facto* integration has not occurred so far. Indeed, the absence of discussion about climate policies between the EU Commission and new member states before accession can be observed in the different yearly reports that the Commission has published concerning the Approximation of the Environmental *acquis*. Several reasons have provoked

---

[28] For the time being, the same applies to JI.

[29] For a deep analysis on the particularities, obstacles and eventual solutions of the accession of Eastern and Central States to the EU, see Mayhew (2000).

[30] The Europe Agreements provided the means whereby the European Union offers the associated countries the trade concessions and other benefits normally associated with full membership to the EU. They prepare the way for the EU and the partner countries to converge economically, politically, socially and culturally.

this non-appearance of climate policies within the different phases of the enlargement.

The disregard for climate change is evident considering for instance the Guide to the Approximation of EU Environmental Legislation[31], where climate change does not appear as an individual chapter in contrast to other environmental areas. Likewise, in the EU Commission's annual assessments of the CEEC progress towards meeting the EU *acquis* requirements, climate policies are rarely mentioned. In the case of climate change, the real concern is that the *acquis communautaire* is and continues to be a moving target.

Obviously, one of the reasons for this lack of concern is the fact that the EU climate policy is still very much a work in progress and as such, when accession negotiations were undertaken, there was basically only one piece of legislation representing the "Climate Change *acquis*"[32]. Only recently, and thanks to the inputs coming from the Kyoto Protocol ratification, EU institutions have begun to promulgate regulatory acts on this issue[33].

In addition, the absence of discussions at the EU level on the potentialities of Joint Implementation and Emissions Trading both for EU member states and acceding countries is linked to the particular nature of these new instruments. Indeed, JI and ET can be included within the category of the market-based instruments. Yet, during the accession negotiations, the EU ignored environmental economic instruments as tools to implement the environmental *acquis* in the acceding countries. Rather, during the pre-accession period, emphasis has been put on the traditional command and control regulations that have until recently been the approach followed by the EU's environmental policy[34]. Only lately has the EU embraced environmental/economic approaches – among which Joint Implementation and Emissions Trading are included – in the context of climate policies, entailing a change of philosophy in the formation of European environmental law and policy. Certainly, the introduction of these new policy instruments will have an effect on the implementation of the

---

31 See European Commission (1997).

32 Council Decision 99/296/EC amending Council Decision 93/389/EEC for a Monitoring Mechanism of Community CO_2 and other Greenhouse Gas Emissions, which assesses improvements about emissions reduction scheduled by the Kyoto Protocol, OJ LL 117/35 of 5th May 1999.

33 These regulatory acts include, among others, the proposal for a new Community greenhouse gas monitoring mechanism, the compromise between the European Parliament and the Council on the EATD and the proposal for a directive linking JI/CDM projects to ET.

34 See Sofia Initiative on Economic Instruments (1999), 7 "Current EU environmental legislation is dominated by climate change measures. The accession negotiations center on those and seem to leave little room for flexible mechanisms ..." "While official documents often refer to environmental charges as possible instruments to address EU needs to approximate financial legislation, few mention the use of economic instruments in a flexible manner that would reduce implementation costs through incentive impacts".

environmental *acquis* by CEECs, providing new member states with a mandate for adopting instruments, such as JI and ET. Nonetheless, the EU should embark on a more active approach encouraging CEECs to adopt environmental economic instruments in such a way that a basic step both for the EU and CEECs would be the elaboration of a clear EU strategy for the development of JI and ET in new member states, which at least since the adoption of the EATD did not exist.

Having said that, it should be recognised that in the last phases of the elaboration of the EATD, representatives from acceding countries have been involved in the discussions. The same is valid for the current negotiations for the directive linking ET and JI.

When dealing with the implementation of JI and ET in CEECs, the reader could be easily misled as to the nature of the legal regime applicable to these mechanisms. Firstly, it is once again important to stress that climate policy, more than any other environmental issue, is currently still very much in evolution at international, European and domestic levels. Rules have still to be agreed on both regarding JI and ET internationally, while at the European level the implementation of JI projects is affected by many existing and forthcoming regulations and the EATD has only recently become a reality.

Although the intention of this chapter is to investigate some parts of the *acquis communautaire* that might be relevant for the implementation of JI and Emissions Trading among CEECs, it is not intended to offer an exhaustive view of the whole approximation process. Indeed, the *acquis communautaire* that could affect the design of ET and JI in the acceding countries is extremely vast, consisting not only of environmental, but also of energy and transport legislation. Furthermore, competition, state aid and internal market policies had to be integrated by CEECs during the pre-accession period and consequently are going to shape the development of JI projects and Emissions Trading. Apart from the existing EU legislation on climate[35], very significant pending proposals recently adopted by the EU will directly affect the scope of both JI and ET. The extent to which such new legislation will shape climate policies in new member states will be exposed in the forthcoming paragraphs.

The task of clarifying which and to what extent EU legislation will mould the implementation of JI and ET in CEECs, is of primary importance for these states – although the other member states are also enduring the same kind of uncertainties. Hence, to draw a general picture of the juridical framework relevant for the development of JI and ET appears complicated. Primarily the environmental *acquis* – which will be by far the most significant in terms of shaping JI and ET – emerges as one of the hardest

---

35 Ratification of the Kyoto Protocol – see supra note 18 – and Council Decision 99/296/EC – see supra note 32.

chapters since environmental policy regulations have remarkably increased in recent years becoming one of the EU's most extensive areas of law, and making its transposition and implementation an arduous task[36]. Furthermore, European environmental law is under continuous evolution and ongoing reviews and new regulatory frameworks are added frequently. Also, during the negotiations with the EU, most CEECs have obtained transitional and derogation periods for certain areas and regulations, which differ significantly from one country to another[37]. Likewise, and this is also applicable to the existing EU member states, both the interpretation and enforcement of community legislation by the acceding countries will diverge, which leads to great uncertainty. Probably, the most illustrative case – as will be explored in the next point – is represented by the Integrated Pollution Prevention and Control (IPPC) Directive[38] and its requirements concerning Best Available Technology (BAT) on the way in which it will affect the establishment of emission reduction baselines[39].

Finally, one should consider that at the time of the signature of the Accession Treaties in April 2003, the very core of the EU climate change regulation – EATD, Linking Directive, Registries Directive and review of the Monitoring Directive – was still not yet adopted, leaving these instruments outside the "environmental *acquis*" designed by those treaties.

4 Eligibility Criteria and Flexible Mechanisms

As for any other Annex I Parties, CEECs must comply with the Protocol's eligibility criteria in order to participate in the flexible mechanisms. Before engaging in the flexible mechanisms, Annex I Parties must submit national communications and national inventories of greenhouse gas emissions to the

---

[36] See European Parliament (2000).

[37] It is important to stress that transitional measures do not relate to new installations in line with the EU position that all new investments should comply with the environmental *acquis*.

[38] Council Directive 96/61/EC concerning integrated pollution prevention and control, OJ L 257, 10/10/1996, p.26–40.

[39] Another example of how transition periods can affect the development of JI in acceding countries was provided by an ERUPT-2 project in Slovakia. The project, the first of this kind in Slovakia, concerned landfill gas recovery operations at eight large landfills in Slovakia. A consortium of three companies are implementing the project and between 2003 and 2006. The landfills are planned to be equipped with methane extraction equipment and the methane emissions from the site will be flared. Slovakia acceded the EU in 2004, but does not have to fully comply with the EU landfill directives until 2012. The EU directive on landfills regulates the capture of methane emissions as well as electricity generation. From 2007 power generators are to be implemented at the sites, at which point the gas will be utilized for electricity production. See JIQ (2002), 10.

Convention Secretariat[40]. Furthermore, they must comply with the Kyoto Protocol's reduction obligations and present data according to the decided guidelines. The Protocol and the Marrakesh Accords[41] have set up international rules for emissions monitoring, government reporting and review of information[42], and an accounting system for transactions under the flexible mechanisms and for sinks activities[43].

Concerning the submission of national communications to the UNFCCC Secretariat, EITs were awarded a longer timeframe compared to the other Annex I Parties, requiring that the second national communications must only be presented by the majority of these countries by 1998 instead of 1997. The third set of communications was requested by 30 November 2001, but not all Parties have so far complied with this requirement. Slovenia is currently drawing up its third national communication and has skipped the preparation of its second. Romania and Lithuania are also preparing their third version. On greenhouse gas inventories[44], no data were submitted in 2002 by Lithuania or Slovenia[45].

As can easily be observed, the situation on information is far from positive as not all CEECs have submitted their national communications and greenhouse gas inventories and data are sometimes missing and often of poor quality. CEECs also face problems when updating national inventories due to demanding timeframe and difficulties in the estimation of data uncertainty. Deficiencies concern the lack of internal structure in terms of country focal points, staff, agencies, and specialised institutions, as well as the presence of barriers and obstacles when acceding to documents.

The problem of capacity-building within CEECs was recognized by the international community, firstly at COP 5[46], and then at COP 7, where guidelines requiring Annex II Parties to provide financial and technical

---

[40] See Article 4.1 and 12 of the UNFCCC.

[41] The Marrakesh Accords require Annex I Parties to engage in additional procedures for accounting assigned amounts before the first commitment period. National registries have to be set up since they will serve to track emissions transfers under different units representing greenhouse gasses reductions and must record and account emission reductions. At COP8, Parties agreed to set up technical standards in order to allow the accurate, transparent and efficient exchange of data among national registries, the CDM and the transaction log.

[42] See Article 5, 7 and 8 of the Kyoto Protocol.

[43] See Article 7.4 of the Kyoto Protocol.

[44] Annex I Parties have to submit by 15 April each year, annual national greenhouse gas inventories for the period covering the base year up to the last one year prior to the year of submission.

[45] The same applies to Bulgaria, Romania, Malta and Cyprus.

[46] See decision 11/CP.5 when the Conference of the Parties identified capacity building as an issue that could undermine "the effective participation of countries with Economies in Transition" in both the Convention and the Kyoto Protocol.

support to the EU accession states were agreed[47]. CEECs have raised their efforts in this direction and, together with other Annex I Parties, international organizations and non-governmental organizations, have started to make investments and efforts in the field of capacity building. International projects predominantly concern either activities directed to provide general assistance to Parties to the UNFCCC and the Kyoto Protocol, or actions directly focused on the CEEC region or bilateral initiatives[48]. However, there is still a need for a common strategy, a more coordinated approach, better distribution of tasks and responsibilities, the reinforcement of institutions, the improvement of methodologies, and greater assistance on procedural aspects of ET and JI. In this respect, the development and improvement of capacity building in the region is a logical consequence of the approximation process required by accession.

This is, for instance, the case of the GHG monitoring and reporting systems, which are basic conditions for eligibility to participate in the flexible mechanisms. EU acceding countries are required to strengthen their institutional infrastructures in order to provide adequate data transparency and accountability. Despite the fact that there is still a certain degree of uncertainty concerning which institutions should be responsible for such activities, it is clear that the implementation of the EU rules on greenhouse gas monitoring will lead CEECs to move forward along this line.

The EC monitoring regime is based on Council decision 99/296/EC on a monitoring mechanism for CO_2 and other greenhouse gases[49], establishing a reporting and monitoring mechanism for the fulfillment of EC obligations under the UNFCCC, which has explicitly been part of the *acquis communautaire* since the beginning of the accession negotiations. The EC monitoring mechanism includes reporting and verification requirements regarding national inventories and climate policies, and imposes clear obligations for CEECs, which present serious deficiencies in this sector[50]. Following the developments within the international climate regime, which have brought to the fore the relevance and the difficulties of the monitoring issue, the European Commission has recently adopted decision 280/2004/EC establishing a mechanism for monitoring Community greenhouse gas emissions and for implementing the Kyoto Protocol. This document contains specific references to inventories and reporting obligations related with Joint Implementation and International Emissions Trading, which were lacking in

---

[47] See decision 3/CP.7, Capacity building in countries with economies in transition, FCCC/CP/2001/13/Add.1.

[48] Some of the international organisations involved are UNITAR, EEA, GEG-UNDP, World Bank, OECD/IEA.

[49] See n. 32 above.

[50] On the monitoring deficiencies for environmental policies in CEECs, see European Commission (2001a).

the old monitoring system[51]. The new reporting requirements will cover areas such as the flexible mechanisms including the establishment of registries and will reflect the legal decisions taken at COP 7. The new system, replacing decision 99/296/EC, will therefore facilitate the creation of EU JI initiatives in CEECs and contribute to increased transparency and accountability in these countries. Indeed, monitoring requirements are especially relevant in the context of JI since the Marrakesh Accords offered two options for the implementation of JI projects – track one and track two – depending on the institutional capacity of the host country. Track one projects are those where Parties meet all eligibility requirements; track two projects are those where these rules are not met and a Supervisory Committee (to be set up by COP/MOP 1) must verify the validity of the ERUs generated. On the one hand, track one projects make higher institutional demands in terms of emission monitoring and reporting, like the establishment of a national system for estimating greenhouse gas emission and the development of a registry for recording and monitoring the transfer of emission reductions. On the other hand, track two projects stem from the idea that a monitoring and tracking system is insufficient and additional steps are needed at the JI transaction level to ensure that projects do not sell more emission reductions that they generate. JI investors will generally prefer to opt for track two, and only when compliance is established, will they then switch to track one to benefit from speedier and cheaper project implementation. Thus, the second track increases the transaction costs. Given that for the time being, most of the accession countries are in the process of developing their JI policies and do not yet qualify for track one, the EU would need to invest in capacity building or take the risk that the accession countries can only use the second track[52]. The correct application of the EC monitoring system might foster the development of JI track one by new member states.

The relationship between the new monitoring mechanism and the EATD is clear. This is particularly true for the evaluation of progress[53], where decision 280/2004/EC recalls the importance of periodic reporting to the Council and the Parliament in order to reach Kyoto Protocol targets[54]. The supporting role of the European Environment Agency (EEA) is also

[51] Decision 280/2004/EC of the European Parliament and of the Council of 11 February 2004 concerning a mechanism for monitoring Community greenhouse gas emissions and for implementing the Kyoto Protocol, OJ L 49/1 of 19 February 2004, following European Commission proposal for a monitoring mechanism of Community greenhouse gas emissions and implementation of the Kyoto Protocol, COM(2003)51, Brussels, 05.02.2003.

[52] On this point see Mullins (2000).

[53] Article 5 of decision 280/2004/EC, see n. 51 above.

[54] Member states shall base their periodical assessment following Articles 3 and 6.2 of decision 280/2004/EC as well as Article 21 of the EATD proposal.

mentioned and the recent integration of the CEECs into this institution may enhance the monitoring capacities of these states[55]. But the relationship between the two instruments is even more evident considering national registries. In these cases, decision 280/2004/EC indicates the need to incorporate the Emissions Trading registry system foreseen in the current EATD proposal with the ones covering emissions units like ERUs, CERs and RMUs generated outside of the carbon market. Furthermore, the Central Administrator for national registries designated under the EATD proposal will be the main person responsible for the collection of information on transactions, which take place both outside and within the EU ET scheme[56]. Apart from the requirements of the monitoring mechanism, it should be kept in mind that the EU is currently elaborating a directive on registries[57].

5 The *Acquis Communautaire* Affecting the Development of JI Projects: The Additionality Issue

As already pointed out, JI projects are directly governed by international law: namely the Kyoto Protocol and the Marrakech Accords. Article 6 of the Kyoto Protocol requires Annex I Parties to implement projects reducing emissions in other Annex I Parties. Emission reduction units (ERUs) generated through JI activities can be used by Parties to meet obligations under the Protocol. On the basis of the Buenos Aires Plan of Action agreed at COP 4 and of the key decisions taken at COP 7 in Marrakech, Parties willing to implement JI projects, as well as IET and CDM projects, must meet the eligibility criteria based on methodological and reporting commitments[58]. While JI projects can be performed by the year 2000, the Marrakech Accords stipulate that ERUs cannot be issued before the 1st of

---

55 This is appropriate considering the benefits of cooperation with the EEA in information collection and elaboration of guidelines covering both the collection and evaluation of emission inventories and national programmes.

56 Although it has not yet been decided, the institution that will develop such a function, according to the EATD proposal, will be the European Environmental Agency (EEA). If this is the case, the EEA's competency in the domain of climate change would considerable increase.

57 This provision might also reinforce the institutional and procedural basis so as to allow those countries to opt for JI track I.

58 Annex I Parties are required to be in compliance with their methodological and reporting commitments under the Protocol – submission of national communications and greenhouse gas annual inventories – as well as with the rules for emissions monitoring, government reporting and review of information and an accounting system for transactions under the flexible mechanisms and sinks activities (registries).

January 2008[59]. Trying to foster the use of this mechanism and following a decision taken at COP 1, Parties agreed to launch a "pilot phase of activities implemented jointly (AIJ)" to anticipate the official start of JI. Several AIJ projects have been initiated over the last several years even where credits coming from these projects cannot be issued[60].

CEECs have so far played a relevant role in the implementation of AIJ projects, which is shown by the number of projects developed in the region so far[61]. The success rate of projects in Central and Eastern Europe has been high, although experience shows that the transaction costs of projects can become very large and that a lack of host country institutions can severely hamper the process[62]. On the other hand, it can be argued that most of the AIJ experience was limited to baseline analysis and helped to increase awareness among host countries, whereas commercial, economic and financial assessment of a transaction was limited as no actual emission reductions were conferred under AIJ[63]. One of the most successful AIJ programmes has been the Swedish programme in the Baltic region. Sweden concentrated on a few clearly defined types of projects that were also economically attractive[64]. The Swedish also experimented with different baseline approaches and were among the first to use a countrywide benchmark[65].

After Sweden, the Netherlands were the most active AIJ investors in Central and Eastern Europe and quickly developed plans to go beyond AIJ. In 1999, the Dutch Government initiated the ERU Purchasing Tender (ERUPT) programme. The Dutch Government was of the opinion that private companies could not own ERUs. Thus, an elaborate procedure was set up whereby the Government issued a global tender for ERUs. These were to come from JI projects and a necessary condition was the creation of a framework agreement between the host country and the Netherlands[66].

Article 6 (b) of the Kyoto Protocol requires that the implementation of

---

59 Decision 16/CP.7 Guidelines for the implementation of Article 6 of the Kyoto Protocol, see supra note 3.

60 See decision 5/CP.1, UNFCCC document FCCC/CP/1995/7/Add.1.

61 Two Baltic states – Latvia and Estonia – are ahead in the list of host countries.

62 REC (2001).

63 Fankhauser and Lavric (2003), 12.

64 Michaelowa (1999).

65 Good results were obtained in terms of reporting, transparency and capacity building.

66 These agreements specified that the host country government would transfer ERUs equal to the emission reduction achieved by the private projects during the first commitment period. Furthermore, the necessity to negotiate the framework agreements was an important step in capacity building. The Netherlands supported this by financing JI offices in Bulgaria and Romania.

these projects be based on additionality[67]. According to this criterion, JI projects must be supplementary to existing legal requirements in host countries, i.e. projects that are to be implemented to comply with a national, supranational or international law must not receive credits generated through JI. This also applied to the EU level. Given that additionality will be measured on the "baseline" scenario (what would have occurred in the absence of the project itself), the additionality of a project for CEECs will be marked not only by domestic provisions, but also by the EU requirements as confirmed by the recent directive linking JI/CDM to ET[68]. In this case, the significance of the implementation of the *acquis communautaire* by CEECs for JI projects appears evident and it will be important to focus on the established and forthcoming EU climate regulations especially when considering JI[69].

As for the stringency of baselines, it is clear that the amount of emission reductions generated by JI crucially depends on the baselines. As suggested by certain authors, it would make sense to either apply CDM methodologies or to develop a common approach for all new member states based on the *acquis communautaire*, using these rules to set up baselines[70]. Obviously, these countries might argue that the *acquis* is stricter than a situation without EU accession and thus they may try to obtain more ERUs as they try to be competitive in setting JI conditions. The question is whether there will be a race to the bottom in setting the baselines in order to attract JI projects. If baselines are set up at lower requirements than the *acquis communautaire* in order to generate as many ERUs as possible, that could delay the implementation of important parts of the *acquis* until after 2007[71]. But, as expressed by certain authors, any "leniency in setting entity emissions constraints or project baselines will create domestic distortions because other sectors of the economy will have to do more in order for the country to meet its national target. It is therefore important for governments to be able to set

---

67 Environmental additionality in the context of JI can be defined as the delivery of greenhouse gas emissions reductions that would not have otherwise occurred.

68 The directive of the European Parliament and of the Council amending directive 2003/87/EC establishing a scheme for greenhouse gas emission allowance trading within the Community, in respect of the Kyoto Protocol's project mechanisms was formally approved by the EU ministers at the Council meeting of 18 October 2004, but still not published on the Official Journal at the time if writing.

69 It is important to stress that not all the implementation of the *acquis* involves hard tasks for CEECs. To a large extent, the adoption of the *acquis* will provide opportunities for CEECs in the context of JI. Such is the case with the EU policy on Renewable Energy. See for instance, BMU (2001).

70 Fernández Armenteros and Michaelowa (2002).

71 Lowering baselines serves as a disincentive for investors to engage in necessary high/tech projects, if greater credits are available for low/tech investments.

realistic and fair emission targets and project baselines"[72]. Finally one should not forget that acceding countries will need to make up the ERUs according to the "additionality requirement" and it is unlikely that if the second track were used, the JI supervisory committee would accept baselines softer than those set up by the *acquis communautaire*.

In terms of their potential for low-cost JI, Ukraine, Bulgaria, Russia and Romania are identified as the countries with probably the highest scope for cheap emission reductions, whereas Slovenia, Croatia and Hungary are the countries with the lowest JI potential[73]. It seems then that the countries with a limited JI potential have, quite rationally, refrained from building up expensive JI capacity. This is because they do not expect to participate much in the market. At the other end of the spectrum, the countries with a good JI potential have every interest in developing JI capacity and take advantage of this new opportunity. However, these are also the countries with the most difficult business climate. It appears that the institutional and political constraints, which slow down improvements in the business environment, also hold back the development of JI capacity[74].

The fact that the international regulatory framework governing JI still has to be established adds uncertainty as to the way these rules will influence the attractiveness of these projects through the definition of baselines. It is, in any case, true that a decision on the issue is expected soon, as once the Kyoto Protocol enters into force, the implementation of JI is foreseen to commence from 2008. It appears clear that even if the scope of this paper is to consider the implementation of JI and ET in Central and Eastern Europe focusing on the relation between old and new member states, it is impossible to separate completely the international and the European regime[75].

6 The *Acquis Communautaire* Affecting the Development of JI Projects: the IPPC Directive

The IPPC Directive is certainly a piece of legislation, which will have a big impact on the implementation of JI projects. The reason is that one of its objectives rests on energy efficiency and experience shows that many of the projects implemented under JI in the CEECs have been concentrated on

---

[72] Mullins (2000), 29. See also on general experiences of JI projects Beuerman, Lagrock and Herman (2000).

[73] Fankhauser and Lavric (2003), 7.

[74] Fankhauser and Lavric (2003), 22

[75] The same common position on the EATD, see supra note 32, mentions in its preamble that member states are free to participate in other domestic and/or international ET schemes (point 24). Still, Article 30 establishes a strong link with the international system with regards to the review phase

energy efficiency[76]. Yet, it should be clarified that many of the challenges that the IPPC raises regarding the implementation of climate change flexible instruments are not unique to new member states but rather affect all member states.

Regarding JI, the *acquis communautaire* derived from the IPPC and its BAT (Best Available Techniques) requirements would at first sight imply that, should new member states be interested in JI co-operation, their project baselines should be derived from the emission scenario including those requirements, which will lower the emission reduction potential of the projects. In other words, since the baselines for JI projects should be based according to the environmental standards of EU legislation – which are more stringent than those currently existing in acceding countries – JI projects would yield less emissions reductions than if those baselines were based on less stringent legal requirements. The question is essential as to the "additionality criteria" set out in Article 6 of the Kyoto Protocol.

Important transitional arrangements have been agreed between the EU and acceding countries on the IPPC and specific listed installations will not yet comply with BAT in Latvia (end 2010), Poland (end 2010), Slovenia (end 2011) and Slovakia (end 2011), whereas the current member states have to implement the Directive by the end of 2007[77]. That would mean that these CEECs would have lower IPPC standards at the beginning of the first commitment period (2008–2012) and thus potential lower baselines. It should, however, be clarified that these transitional agreements apply to pre-1997 installations (existing installations), whilst all newer installations had to comply with the directive requirements by the date of accession.

A full understanding of the influence of the IPPC in the development of JI by CEECs requires a brief explanation of the dynamics of IPPC itself. The IPPC Directive requires BAT to control polluting releases to air, land and water. Article 3 of the Directive requires that regulated processes use energy efficiently and that energy efficiency be taken into account when determining BAT. This element appears relevant in the context of JI projects based on energy efficiency and for the establishment of baselines. According to the IPPC, each plant should dispose of a permit including emission limits for pollutants that are likely to be emitted from the installation in taking into account different factors, such as energy efficiency. It is important to note

---

[76] The IPPC Directive is also intimately connected with the ET scheme advanced by the EU. Likewise, the UK scheme uses the IPPC Directive to select facilities and operators. Certain authors argue that the IPPC Directive might not be appropriate for the setting of an ET programme, since the original purpose of the directive was not to address greenhouse gases and climate objectives. To this regard see, Cozijnsen (2001), 106.

[77] See European Commission (2003). The transitional agreements include transitional targets according to which acceding countries will have to achieve environmental objectives progressively.

that this Directive does not contain fixed reduction targets. Rather, permits are negotiated between the administrative authority and the polluter on the basis of BAT, taking account of economic, geographical and environmental conditions. That means that the permits will vary from operator to operator and from country to country, depending on the way administrative authorities implement the concept of BAT[78]. Moreover, it should be noted that Article 9.8 of the IPPC Directive allows for the application of General Binding Rules (GBRs) in place of provisions set out in the Directive. But at the same time, Article 9.4 requires that the local aspects of site specific permitting – technical characteristics, geographical location and local environmental conditions – be still maintained. In other terms, the use of GBRs should provide an equivalent level of environmental protection to that derived from the determination of individual permits. It can be anticipated that GBRs will likely be defined in different ways by different member states and, in the same way that a strict interpretation of BAT by new member states will reduce the scope for JI projects, a strict definition of GBRs will provide less space for JI, or *a contrario*, a lenient GBR will give more space to JI[79]. The ambiguity of the IPPC Directive will lead to different solutions in different countries. Consequently, since CEECs will be competent to define BAT, this will affect the setting of baselines. On the other hand, interpreting and defining BAT for environmental standards is a possible topic of negotiation between EU-15 and new member states. A result of this negotiation could be that for one candidate country the BAT will become less strict that for another country. However, this space for manoeuvering will be narrowed by the dynamics of the IPPC Directive. Indeed, IPPC expert groups compile reference documents (BREFs) for the determination of BAT, which the new member states – as the rest of the EU member states – will have to use as an input into their country specific BAT, and in the individual, plant-specific operational conditions. In other words, BAT requirements will become more and more harmonized. Although BREFs documents are not legally binding, their input into the BAT definitions is more and more important given that they are periodically updated with the latest information about BAT.

Those CEECs with transitional arrangements concerning the IPPC Directive will have a certain margin of manoeuvrability to establish baselines in such a way as to attract JI investments, since for some of those countries the full implementation of the Directive only begins after the start of the first commitment period. Nevertheless, transitional measures in the context of accession should be tested against the principles set up by the

---

[78] See, for an analysis of the legal requirements of the IPPC Directive, Emott (1999).

[79] On the interpretation of the concept of "General Binding Rules" and its use by member states see, IMPEL-Network (2000).

Commission. That means that CEECs, presenting plans containing intermediate legal targets (that is to say, transitional BAT requirements), should ensure full compliance with the *acquis communautaire* according to the calendar fixed by the Commission regarding those transitional measures. That would imply that a new member state that has abided by a transitional agreement could not establish JI baselines below those intermediate legal targets[80]. By the same token, we must not forget the possibility of having early JI projects. Indeed, although Article 6 of the Kyoto Protocol allows for a transfer of ERUs only from 2008 until 2012, that does not mean that the project will have to be carried out during such a period. On the other hand, investments that are mandatory under the *acquis communautaire* after 2008 will not be considered additional JI projects during the commitment period and, as such, will not be qualified as eligible JI projects. But in all those cases in which there will be transitional agreements – for instance until 2011 – early JI projects will be extremely relevant since acceding countries could carry out additional early JI projects until 2011 and thus, will be accounted for in the framework of the first commitment period. Interpreting *a contrario*, a project starting now in a new member state will be deemed non-additional if, by 2008, EU environmental legislation prescribes the baselines used by the project.

7 Implementation of JI: Final Considerations

Focusing on internal market and competition issues, the enlargement process considers in large part the adoption of antitrust and state aid legislation as the necessary means to effectively enforce a competition regime similar to that of the EU[81]. When implementing JI projects and EATD, CEECs should ensure the freedom of establishment as set out in Article 43 of the EC Treaty as well as, the application of the entire part of the *acquis communautaire* on competition. As far as JI is concerned, depending on the way JI will be implemented by CEECs, state aid issues might emerge[82]. The whole EC state aid regime and, in particular, the Commission Guidelines on Environmental

[80] It should also be considered that other directives that might be relevant for JI projects also have important transitional periods. Such is the case of the Large Combustion Plants Directive 88/609/EEC – Czech Republic 2007, Estonia 2015, Lithuania 2015, Poland and Slovakia 2007. The same reasoning applies to other areas which might be relevant for JI projects and which include important transitional agreements, such as in regard to waste management, where Poland obtained a transitional agreement for the Landfill Waste Directive delaying entry into force until 2012 (2009, for current member states). See European Commission (2003a).

[81] See Atanasiu (2001).

[82] See Chapter 8 in this volume.

State Aid should be considered and will be directly applicable[83]. The Environmental Guidelines distinguish investment and operating aid and lay down special provisions for renewable energies and combined heat and power generation that are particularly relevant with respect to JI projects. Although the Guidelines do not contain any particular clauses on flexible mechanisms, the Commission would likely exempt eventual state initiatives related to JI projects, which might fall under the category of state aid. Financial transfers from the investor to an enterprise in the hosting country may be considered as a prohibited subsidy. From the CEECs point of view, certain practices related to JI might require the application of the state aid regime, for instance, in the case of credit sharing between JI investing and host Parties. By the same token, certain developments of JI might oblige CEECs regulators to consider procurement rules.

Three major conclusions can be drawn about JI:

1. The differentiation between what is additional and what would otherwise occur is difficult to work with, especially in the context of the EU (pre-accession) process. As new member states are obliged to implement the *acquis communautaire*, the additionality requirement is much more stringent than in the cases of other EITs that do not have to comply with the EU accession rules. This may reduce the applicability of JI in the CEECs and make other EITs more attractive for the development of these projects. In other words, a substantial greenhouse gas emissions reduction in the new member states should occur in any case as a result of the EU accession process and therefore the additionality of many AIJ/JI projects will have to be examined case by case *vis-à-vis* the obligation of the CEECs to adopt the *acquis communautaire* on environment and energy.

2. The conditions of transparency and institutional stability will be more basic in order to attract investors than by lowering baselines standards. The adoption and implementation of the *acquis communautaire* would entail the improvement of domestic legislation and the need to increase the legal certainty for investors. Basic issues for the development of JI projects from an investor point of view, such as contract enforcement, will be also strengthened by the incorporation of the *acquis communautaire*.

3. It should not be ignored that although the adoption of the *acquis communautaire* by CEECs implies financial costs, the benefits that the EU environmental regulations will provide to new member states can be measured not only from an environmental quality perspective but also in economic terms. This is also true since quite often the hidden effects

[83] Community Guidelines on State Aid for Environmental Protection, Official Journal of the European Communities C37, 03.02.2001.

costs to the economy caused by lower environmental standards through a loss of output and inefficient production is not taken properly into account[84]. Thus, the financial gains of the adoption of the *acquis communautaire* might be superior to those derived from poor implementation of JI projects in the Region.

8 The EU Emission Allowance Trading Directive (EATD)

Before addressing the EU Emission Allowance Trading Directive and the issue of which existing and new EU member states will have to comply by January 2005, a few words are needed to describe the Kyoto Protocol's International Emissions Trading (IET) provisions, which are envisaged to take effect from January 2008.

Article 17 of the Kyoto Protocol institutes International Emissions Trading by providing the possibility for Annex I Parties to acquire emission units from other Annex I Parties[85]. Each Party must hold a minimum amount of units in its national registry. This is called the commitment period reserve. Although IET is a mechanism where units are traded among states, the fact that "Parties may also authorize legal entities to participate"[86] in the scheme increases its relevance to our analysis, especially because the legal terminology used is not very clear.

EU greenhouse gas emission allowance trading is the first example of regional emissions trading and constitutes the basis on which all EU member states will make trades as of 1 January 2005[87]. The scheme has no direct link with the IET and it is simply an internal implementation measure.

The initial EU scheme proposed by the Commission[88] has been slightly modified by the Parliament in first reading[89] and by the Council, which has adopted a common position in March 2003[90]. The compromise package[91] has

[84] On the different economic and monetary benefits of implementation of the *acquis communautaire*, see European Commission (2001b).

[85] These units may be in the form of assigned amount units (AAUs), removal units (RMUs), JI Emission Reduction Unites (ERUs) and CDM Certified Emission Reductions (CERs).

[86] See UNFCCC Decision 19/CP.7, Modalities for the accounting of assigned amounts under Article 7, paragraph 4 of the Kyoto Protocol, FCCC/CP/2001/13/Add.2, p. 55–72.

[87] For an accurate and exhaustive analysis of the EATD, see Chapter 9 in this volume.

[88] European Commission Proposal for a Directive of the European Parliament and of the Council establishing a scheme for greenhouse gas emission allowance trading within the Community and amending Council Directive 96/61/EC, COM(2001) 581 final, Brussels, 23.10.2001.

[89] European Parliament Report A5-0303/2002, P5.TA(2002)0461, 13 September 2002.

[90] The agreement was reached in Brussels at the 2473rd EU Council Environment, on 9 December 2002. See Council Common Position Common position adopted by the Council on 18 March 2003 with a view to the adoption of Directive of the European Parliament and of

been amended and recently adopted after second reading by the Parliament[92] leading to the establishment of "the largest emissions trading scheme in the world" (M. Wällstrom) and "removing any doubt that the scheme will become reality from 2005"[93].

The EU system is a domestic, voluntary and entity-based system compatible with the international market created under the Kyoto Protocol. Under the system, companies which have a CO_2 emission level below their assigned pollution amounts may sell emission rights to other entities that are in danger of exceeding their quotas[94]. Operators that are not in compliance with the system – those that do not surrender sufficient allowances – will have financial penalties imposed on them.

The first period of Emissions Trading will start in 2005, three years before the beginning of the IET under the Kyoto Protocol. During this learning-by-doing phase, the EU member states will have the possibility to test the functioning of the market itself in order to be ready for the implementation of the international system in 2008[95].

The unanimous agreement reached at the EU Council of Environment Ministers was the result of two years of high level negotiations and discussions, whose major points are briefly explained below:

- Allocation of permits:[96] the EATD final text introduces voluntary auctioning for the two commitments periods, respectively 5% and 10% of the total allocation, while the rest of allowances are distributed through

---

cont.
the Council establishing a scheme for greenhouse gas emission allowance trading within the Community and amending Council Directive 96/61/EC, 2001/0245 (COD).

91 Based on the Commission's amended proposal, COM(2002)680.

92 The Parliament proposed in a modified version, the key amendments adopted in first reading and in particular the inclusion of additional gases and sectors in the system, the allocation method, credits, caps for emissions allowances, opt-in and opt-out. Text adopted on 2 July 2003, Plenary session of the European Parliament, Strasbourg, P5_TA-PROV(2003)0319 Greenhouse gas emission allowance trading ***II *(A5-0207/2003) – Rapporteur: Jorge Moreira da Silva.*

93 European Commission (2003b) and Environment Daily (ENDS) (2003).

94 Large plants in key sectors like power generation and heating industries, oil refineries, coke ovens, ore smelters, steel works and cement, glass, tile, ceramics, pulp and paper factories will have CO_2 caps.

95 For a better understanding of the initial EATD proposal, see Lefevere (2002) and Zapfel and Vainio (2002).

96 See Explanatory Statement, point 5 (a), n. 92 above. Emissions allowances in a trading system can be allocated in three different ways: auctioning, when national authorities sell permits to Parties; grandfathering, when permits are allocated free of charge according to past emissions levels or to fuel consumption levels; output basis, when a method similar to grandfathering is used based on current or very recent activity – emissions are considered per unit of output. See Article 10 of the EATD.

grandfathering. The allocation of allowances is a key issue and the European Commission has released an official non-paper with the intention to draw some guidelines for the participants[97].

- Sanctions: the fine for exceeding the maximum emission levels has been fixed at 40 euros per tonne of CO_2 equivalent emissions by an installation in the first phase, and at 100 euros for the second period[98].
- Opting in:[99] from 2008, member states have the possibility to apply to the Commission for the inclusion of additional activities and gases not listed in Annex I of the EATD. Moreover, the Commission may propose by December 2004 "to include other relevant sectors ... chemicals, aluminum and transport sector, activities and emissions of other greenhouse gases".
- Opting out:[100] member states can request to exempt individual installations from the system, and not activities and sectors as requested by the Council in first instance. The scheme will be dependent on Commission approval. Exempted installations will have to reduce emissions and to respond to the verification requirements in the same manner as the ones covered by the scheme[101].
- Pooling system:[102] based on a German proposal, this tool groups together installations within an industrial sector for distribution of allowances, instead of allocating them to individual companies. Installations within an industrial sector that agree to form a "pool" will be free to decide whether to participate in the system or not.
- Credits from JI/CDM:[103] on the possibility to include in the scheme emissions credits from CDM and JI projects, no mention of the new directive linking the mechanisms is made in the text of the EATD.

Although the EU system has been designed to be as compatible as possible with the international regime and mindful that implementation within the EC is irrelevant for the IET in the sense that AAUs cannot be

[97] On this point, see European Commission (2003c) For a better understanding of the allocation system procedures, see also Ministry of Economic Affairs, The Netherlands (2002) and Harrison and Radov (2002).

[98] Article 16 of the EATD.

[99] Article 24 and 30.1, 30.2 of the EATD.

[100] Article 27 of the EATD.

[101] A negative outcome produced by this decision may be the opting out of entire sectors – the German and UK energy sectors for instance – which could have significant negative influences on market prices and liquidity.

[102] Article 28 of the EATD.

[103] Preamble, recital 18 and Article 30.3 of the EATD.

gained through the EATD, some differences between the two systems are relevant and need to be stressed. While the EATD establishes a market among "installations"[104], the IET provides for an exchange of units among states. CEECs, like any other Annex I Party, can participate in such a market and eventually sell part of their emission credits from 2008 onwards. Nonetheless, the Marrakesh Accords establish the possibility for legal entities to participate in the market through the authorization from Parties. It is therefore not completely clear yet what is intended for legal entities. Emissions trading among states goes therefore beyond the scope of the EU scheme and while companies in Central and Eastern Europe will be covered by the EU system, no provisions prevent existing EU member states from concluding a deal with some CEECs under the framework of IET in order to comply with their EU bubble obligations. The EU scheme in itself does not remove power for existing member states to implement IET with other CEECs.

The other main difference between the two trading regimes is timing: while the IET will be operational from 2008 – if the Kyoto Protocol enters into force before this date – the EATD starts from 2005 and several modifications could occur before the start of the second "Kyoto" phase.

9 CEECs and the EATD

CEECs have important roles in the carbon market, as they will have considerable emissions surplus according to the targets agreed under the Kyoto Protocol and they are likely to become net sellers of allowances. The exact amount of such surpluses is still unknown as the EU accession requirements in the environmental field could in fact generate further reductions in greenhouse gas emissions thanks to energy efficiency rules or other EC measures, giving these countries even more emission credits to trade. In any case, according to various different studies, the Kyoto commitment period surplus of CEECs will be enough to cover the EU demand for emission allowances[105].

From a practical point of view, all CEECs, with the exception of Slovenia, will probably be far below their Kyoto targets and will not use emissions trading as a measure to comply with their international obligations, but as an instrument to gain financial benefits for both public and private sectors. This difference is quite important, as the type of system

---

[104] "Installation means a stationary technical unit where one or more activities listed in Annex I are carried out ...", Article 3 (e) of the EATD.

[105] A recent study from the European Commission (ECCP report, June 2001) estimated the EU emissions deficit as 336 Mt.

cap will be decided by each competent authority on the basis of its national perspective and overall goals regarding climate change[106].

While representatives from new member states were only indirectly involved in the negotiations concerning the establishment of the EU trading scheme, after the signature of the Accession Treaties, they had more chances to intervene and contribute to the preparation of the EU legislation. CEECs representatives were able to participate in the discussion of the ET scheme as observers in the Council meetings and in the 2nd reading of the European Parliament regarding the Commission Proposal. At the second meeting of the "Interim Committee" recently created by the EU, CEECs representatives also voiced concerns over the forthcoming EATD[107].

If, from the EU legal point of view, there is no doubt that CEECs will be integrated in the scheme, it is still unclear when and to what extent such provision is going to be implemented by acceding countries and whether some of them will ask for transitional arrangements. Probably, they will accede with just a slight delay, giving birth to a European greenhouse gas emission allowances trading system composed of twenty-five countries. In this sense, the EATD as *acquis communautaire* is really a moving target and the EU enlargement simply means a bigger market.

Timing might also be a considerable problem that new member states, as well as EU-15, will face in implementing the EATD by 2005. Considering the difficulties encountered in the elaboration and submission of their national allocation plans by the agreed deadlines (31 March and 31 May 2004 respectively for EU-15 and new member states), and considering that EU accession is effective only since 1 May 2004, it is hard to see how these countries will (1) allocate the allowances at the latest by the end of September 2004; and, (2) at the same time be able to issue a greenhouse permit. Furthermore, a certain delay in adopting monitoring and allocation guidelines – respectively on 30 September and 31 December 2003 – by the Commission has also to be considered. If the implementation of these tasks is tight for EU countries, they seem even more rigid for CEECs, which were required to comply with these and many other deadlines at the latest by the date of their inclusion in the EU.

---

[106] Emissions can be counted based on current or historic data, as well as on a probable future emissions level.

[107] The Committee, established to allow the acceding states to comment of EU draft legislation concerning the signature of the Accession Treaties, is composed of representatives of the EU, the European Commission, Poland, the Czech Republic, Hungary, Slovakia, Estonia, Latvia, Lithuania, Slovenia, Cyprus and Malta. The committee was set on December 2002 and acceding states will be able to participate in EU meetings as "active observers", until they become full members on 1 May 2004. At the second meeting – held on 13 March 2003 – the discussion on the European Commission's draft EATD was on the agenda.

What has to be fixed *a priori* by the national authorities is the emissions cap on the basis of which the allocation of units will be settled. This is a key decision, which has also important political consequences, especially when the national commitment targets would give space for additional economic growth in the country[108]. As indicated above, CEECs are currently below their greenhouse gas reduction quotas and therefore apparently have no need to set up such a system. Political and public support for emissions trading could consequently suffer from this situation. Beyond the fact that such a measure is a completely new type of instrument, additional investments will be required and some industry sectors may not see the need for this. Some incentives to bring more companies in the market could be introduced. A tricky issue when considering that the allocation of permits is the distribution of the emissions surplus. From the point of view of new member states, the allocation of this surplus to the national state or to private companies could give both the chance to sell units and to generate economic revenues; thus, representing an incentive for private sector participation in the mechanisms[109]. The allocation method and the systems for the distribution of the surplus in CEECs will have to comply with EU competition law and state aid rules in order to avoid distortions in the Community market. A competition issue could arise as to the extent of permits allocated for free, in case an excessive amount of allowances were distributed to operators. The allocation will probably have to be issued according to business as usual emissions levels, without discrimination and without favours between companies and sectors as indicated in Annex III of the EU scheme[110]. In any case, the European Commission will lead the process on the allocation of permits: first of all it has the power to accept or refuse the national allocation plans that member states submit in early 2004; secondly, it is obliged to provide official guidelines for the preparation of the plans.

The use of the EATD will allow the CEECs to be in compliance with EC law, to meet environmental effectiveness and economic efficiency criteria at

[108] For example, see the case of the Czech Republic which has a Kyoto Protocol commitment of 8% and a domestic one of 20%, thus still leaving space for 3.5% industrial growth.

[109] Jilkova and Chmelik (2001), 13.

[110] The question of the quantity of allowances is very tricky. While the initial ET scheme proposal stated that "[T]he plan shall not discriminate between companies or sectors in such a way as to unduly favour certain undertakings or activities, nor shall any installation be allocated more allowances than it is likely to need" – Annex III, point 5. The Council common position changed this point to "[T]he plan shall not discriminate between companies or sectors in such a way as to unduly favour certain undertakings or activities in accordance with the requirements of the Treaty, in particular Articles 87 and 88 thereof". The Parliament reintroduced a provision during its second reading stating that "the total quantity of allowances to be allocated shall not be more than is likely to be needed for the strict application of the criteria of this Annex", Annex III, Point 1.

industrial levels and to gain valuable experience in order to be ready to implement IET as of 2008.

Regarding the possibility for EU new member states to ask for and negotiate possible exemptions or transition periods with regards to the EATD, only Lithuania and Latvia seemed close to seeking this option because of a certain lack of monitoring and reporting capacity at government as well as industry levels. At the second meeting of the Interim Committee mentioned above, CEECs were asking for "further clarification" and time to examine the directive proposal stressing the fact that the economic and/or environmental policy reasons to implement the provision especially in the beginning phase are still lacking within the region[111]. At this point in time, no transition periods have been asked for and new member states seem able to meet the agreed deadlines accordingly.

Finally, the current lack of clarity at the government level in CEECs with regards to the implementation of the EATD and to the size of the allocation is contributing to uncertainty among installations. What if national governments decide to keep part of the national emissions surpluses for their own use in the IET? What if they would decide to go for very stringent allocation plans?

An open question is presented in regard to the modalities of the participation of the enlarged EC in the IET. Indeed, the EATD has been designed to ensure the fulfilment of the EC reduction commitments as they have been negotiated on the burden sharing agreement. Which emissions cap will be put on the EU system? Will the contributions of allowances by CEECs be considered? Will the European Commission be less stringent with national allocation plans for CEECs? If we look at Article 4 of the Kyoto Protocol, there is no room for external contribution of allowances in the bubble and the question appears quite complex. New member states will be integrated into the European system, but to what extent emissions credits from these states could be counted within the bubble to meet their international reduction targets is still unclear. Probably, a solution to the problem will be postponed until the definition and presentation of the

[111] Hungary asked the meeting to convene and Latvia and Malta made clear their interest in the forthcoming provision and asked for further information on the way the proposed directive will be implemented. As economic collapse in most accession countries since 1990 has reduced greenhouse gas emissions well below the targets agreed under Kyoto, a further reduction through emissions trading would leave these countries in a difficult situation ahead of the next round of climate negotiations. Hungarian representatives claimed that emissions growth should have been allowed for acceding countries, as it was decided for some EU countries. However, it is important to stress that stakeholders in these countries are keener to see candidate countries joining the EU scheme from the very beginning of the scheme, considering that this fact will give the chance to extend the size and liquidity of the single markets.

national allocation plans, when the European Commission will have some space for manoeuvre and the global emissions cap will be defined. Indeed, looking at the Kyoto Protocol second commitment period, CEECs with a relatively large emissions surplus might be an attractive Party to be included in the "new EU bubble", as they would widen the group of member states and reduce the burden for those EU-15 with emissions deficits.

10 The EATD and the *Acquis Communautaire*

We proceed now with an evaluation of those pieces of the *acquis communautaire*, which will be of influence in the implementation of the EATD.

Once again, it is obligatory to start with the assessment of the IPPC Directive, to which the EATD is anchored. The linkages between these two instruments, which are even more visible than when compared to JI, are in regard to the coverage of installations, the coverage of the greenhouse gas emissions and the permitting procedures[112].

As it is largely known, the EATD system modifies the IPPC Directive in a way that an installation covered by the emissions trading scheme should not have an emission limit value for direct emissions of carbon dioxide and other greenhouse gases, "unless it is necessary to ensure that no significant local pollution is caused"[113]. However, regarding energy efficiency, whereas the Commission proposal established that the IPPC provides for a common framework, the final text declares in Article 26.2 that "member states may choose not to impose requirements relating to energy efficiency in respect of combustion units or other units emitting carbon dioxide on the site" and in Article 2 deletes the energy efficiency requirements required under the IPPC Directive. That would mean that member states could still choose to impose requirements relating to energy efficiency in respect of combustion units or other units emitting carbon dioxide on the site, but that this is not explicitly required. As regards new member states, it is likely that on the one hand they will be competent to determine the stringency of carbon dioxide abatement efforts that activities might achieve, but they will probably not tend to include energy efficiency requirements. It is also important to recall that Annex III of the Directive does not make an explicit reference to BAT as a criterion in which member states should base their national allocation plans. Nevertheless, point 3 of Annex III includes "technological potential" as one of the criteria to be followed by states when designing their own allocation

[112] See Chapter 9 in this volume, 47 *et seq*. For an analysis of the linkages between ET and the IPPC Directive see also Smith and Sorrel (1998).

[113] See Article 26 of the EATD.

plans. To which extent such a term corresponds or relates to BAT remains to be seen until member states develop their own national allocation plans and the Commission has the opportunity to assess them.

The coverage of greenhouse gas emissions is not exactly the same in the two instruments: emission limit values covered by the EATD are excluded from the scope of the IPPC permit process; but, this does not cause any particular problems in the case of CEECs.

With regards to the relation between the two permitting procedures, from a procedural point of view, the fact that the EATD hinges on the IPPC Directive raises considerable challenges for acceding countries. First of all, such a link implies the necessity of carrying out the correct implementation of the IPPC Directive in order to facilitate the implementation of the ET system. The significance of the greenhouse gas permit scheme – the EATD requires a combination of permits between the two instruments – lies mainly in the fact that it contains specific demands as far as measurements, reports, and verification requirements are concerned. Also, CEECs have to co-ordinate the conditions and procedures for the permits under both schemes, so a greenhouse gas permit could be issued through a single procedure with a permit under the IPPC Directive. On this point, the most recent progress reports from new member states show the significant deficiencies affecting the co-ordination of permits under the IPPC Directive. In this respect, it is illustrative to consider the European Commission's Regular Report on Poland's Progress towards Accession, in which it is specified that regarding the "issuing of permits, delays have occurred in introducing integrated permits", what will constitute a challenge for local and regional authorities[114]. By the same token, in Slovakia, the permitting system is at present media-based with separate permitting systems for different sectors[115]. For many CEECs the above Reports explicitly refer to the lack of human resources and capacity for implementing the IPPC Directive[116]. For obvious reasons, it seems necessary to ensure the correct implementation of the IPPC Directive by acceding countries – through financial and human support from the EU. This will guarantee easier and less problematic implementation of the EATD.

When considering internal market and competition issues related to the implementation of the EATD, the allocation criteria included in Annex III must be considered. CEECs, like any member states, shall prepare national

[114] European Commission (2002a), 108.
[115] European Commission (2002b), 103.
[116] See Caddy (2000), 215, whereas CEECs have proven to be quite successful in meeting the requirements of approximation, effective implementation is hampered by the severity of the policy problem, the lack of personnel and financial resources, as well as the impact of political and administrative traditions. Such factors impinge upon the credibility and legitimacy of environmental policy and hence the effectiveness of its implementation

allocation plans (NAPs) and should consult the non-paper issued by the Commission on 1 April 2003[117] and other forthcoming guidelines on how to develop them. Indeed, how and according to which criteria acceding countries will allocate their emissions, will determine their position in the internal EATD. If CEECs decide to distribute allowances according to their historical emissions – as the final text as well as the Commission's non-paper allow – they could offer them for sale to the rest of Europe. Instead, if the allocation follows the BAT criteria – as it might be interpreted reading criteria 3 in Annex III – they might be in the unlikely position of having to buy allowances. It can be deemed that in the end, the appreciation of the EU Commission regarding the national plans of new member states will be to a certain degree more lenient than with other countries. Indeed, the allocation criteria continue to be vague and consequently acceding countries have maintained flexibility in their interpretations. However, many questions have come up. For instance, criteria 3 listed in Annex III of the final text states that "quantities of allowances to be allocated shall be consistent with the potential including the technical potential of activities covered by this scheme to reduce emissions". The term "technical potential" will have different meanings and it might even be considered as a kind of BAT requirement or at least imply some form of technological consideration. Based on this interpretation, some could consider it as implying a link with the IPPC. However, "technical potential" will not have the same meaning for CEECs as for existing member states, and accordingly, the views of the EU Commission on the way CEECs may implement the criteria included in Annex III – such as the technological potential – will vary depending on the country at issue. Furthermore, other criteria included in Annex III of the Directive will not be easily applicable to new member states. This is the case for instance of criteria 1 which states that "the total quantity of allowances to be allocated ... shall be consistent with the Member State's obligation to limit its emissions pursuant to Decision 2002/358/EC and the Kyoto Protocol". The text explicitly refers to the EU burden sharing agreement and states that as new member states are not bound by the accord, such criteria cannot be applied to CEECs adding even more uncertainty to the issue.

All in all, uncertainty and misunderstanding are common to all CEECs regarding the influence of competition policy on the design of flexible mechanisms. That was even emphasized during the meetings of the ECCP Working Group on Flexible Mechanisms[118] during which the urgency to decrease ambiguities regarding the definition of the EC competition rules applicable to JI was rightly stressed. As mentioned above, many uncertainties continue also to prevail regarding the allocation criteria. Hence,

---

[117] European Commission (2003c).
[118] The group met four times in 2002.

a clarification by the EU Commission of these points would provide better comprehension of these issues by new member states and, in turn, eliminate much of the reluctance that some of these countries have against the current status of the EATD. The final agreed text will give more certainty to the issue and CEECs will start to establish an adequate trading infrastructure to improve data collection and monitoring, reporting and verification facilities. Either from a domestic point of view when the national authorities have to set their own targets, or from an international and/or European perspective, data verification and monitoring are in fact two key points for participation in the ET scheme. An acceptable practice for monitoring in existing EU member states already exists, as a similar monitoring system is issued by the IPPC Directive, which requires member states to report annually on the emissions levels of the sources covered by this measure[119].

11 Slovakia, Czech Republic and Poland: Three Practical Cases

In order to avoid major incompatibilities in the implementation of the EATD in the period prior to the second phase of the European scheme, CEEC authorities have been designing their own domestic scheme and/or pilot projects trying to mimic the structure of the EU system as much as possible, even if it is true that the EU scheme has been defined only recently[120]. What is certain is that the EATD contains already some prescriptions on national initiatives;[121] specifically, an Article linking the EU system with domestic schemes[122]. Looking at the few initiatives undertaken so far in the region – Slovakian and the Czech domestic trading schemes may be operational in the forthcoming years with Poland also making progress in the preparation of an emissions market – CEECs will quite surely opt for a downstream system

[119] It is the decision on the implementation of a European Pollutant Emission Register 2000/479/EC which indicates how to proceed with this reporting.

[120] Legal issues and political concerns, capacity building and technical problems represent big challenges for national governments, which have to design an ET scheme. They will have to decide between a voluntary and a mandatory scheme, which entities will be entitled to participate in the system and how to allocate allowances. They will have to set up national inventory and registry systems, establish a cap for the system, decide which sectors, sources and gases will be included in the system and defining the type of unit to be traded. Finally, they will have to establish also adequate rules and guidelines for monitoring, reporting and verification as well as penalties in case of non-compliance. See OECD/IETA (2002), 15.

[121] The EU ET system could potentially cover 28 countries – EU member states together with candidate countries and European Economic Area states. It is important to remember that domestic ET is not regulated by the Kyoto Protocol.

[122] Article 25 of the EATD.

including, at least for the first phase, only CO_2 emissions. These three countries, at least for the first commitment period, are supposed to be in compliance with their international obligations according to the business-as-usual scenario and current emissions inventories. They will very probably appear as net sellers in the EU scheme and emissions trading can be seen as a new instrument to attract foreign investments and new technologies.

Slovakia is one of the few countries within Central and Eastern Europe that, despite the active participation in some AIJ projects, has so far preferred to implement ET instead of JI projects.

A cap and trade SO_2 market is already operational and the government is thinking of using the same model for the introduction of a CO_2 scheme[123]. The legal basis for the implementation of such a scheme is the Air Act of 2002[124]. The CO_2 market proposed by the Ministry of Environment will be designed to be as compatible as possible with the existing one – a downstream system with permits allocated through grandfathering – and also with the EATD proposal itself. A pilot phase for the period 2005–2007 (and to become fully operational from 2008) has been proposed and the Ministry has just started to translate the provisional text into legal text[125]. Regarding participation in the scheme, all sectors which are contributing to the national CO_2 emissions level should be included and among these sectors, installations exceeding a given size threshold should be permitted to trade[126]. Some sectors experiencing significant difficulties concerning monitoring and verification of data could be excluded from the market[127], while the Slovak cement industry has expressed several times that it favours the EU proposal. Thanks to the requirements of the Air Act, a monitoring, reporting and verification (MRV) system run by the Ministry of Environment already exists, although it needs to be improved.

The Czech Government is currently involved in developing both JI and ET initiatives. Although they do not yet have any experience in ET,[128] this

---

[123] CO_2 emissions covers 80% of national greenhouse gas emissions.

[124] The original Air Act, adopted in 1999, was recently replaced.

[125] "Air Act", Act No. 478/2002 on air protection and amending Act No. 401/1998 Coll. on air pollution charges.

[126] Given a threshold of 5 MW – with the exception of 20 MW for the heating sector – the market participants will include 358 sources, belonging to 12 sectors – covering 70% of the domestic CO_2 emissions and differing in part from the EU proposal which includes fewer sectors. See Williams, Kolar, Levina and Li (2002).

[127] This could be the case of the chemical, waste incineration and commercial sectors, even if the chemical sector is a very large contributor to Slovakian greenhouse gas emissions. Thus, the national government should improve monitoring procedures.

[128] Research on Emissions Trading for the reduction of SO_2 was commissioned in 1996 and published in 1997, but adequate political attention required for the effective implementation of such measures was not included in this initiative. The study considered the district of Sokolov and was supported by the US Agency for International Development and the Harvard

proposal is being issued together with other climate friendly measures in the forthcoming national climate change strategy[129].

The Czech Government started the preparatory phase for the implementation of the ET scheme and an Inter-ministerial Working Group covering all climate issues was set up in 2001. A special working group dealing only with ET is also forthcoming. The ET system designed by the Climate Change Group for the Czech Republic is a downstream cap and trade system where allowances will be distributed to individual companies through grandfathering. Participation should be voluntary for the pre-commitment period of 2005–2007, with the possibility to shift to a mandatory system from 2008 onwards. The voluntary system will also need a system of incentives in order to attract industries to join the scheme: these could be in the form of the reduction of air pollution charges or a sort of facilitative procedure to receive access to some national energy efficiency and renewable energy programmes, or even the possibility of direct incentives through the State Environmental Fund. The gas coverage proposed for the scheme concerns at least in the first phase, only CO_2 emissions from fuel combustion and will therefore be limited to a specific number of eligible sources[130].

Regarding a monitoring system, an emission register already exists in the Czech Republic. The system, which will need some improvements, is called the REZZO database and is based on the self-reporting of the pollution by industrial sources[131]. More research and capacity building on greenhouse gas inventories is needed[132].

As in many other CEECs, the Czech Republic will not have a problem to meet its Kyoto Protocol targets and industries will be keen to participate in the ET system only if they see margins for benefits.

In Poland, where the energy sector is the sector that contributes the most carbon dioxide emissions, both JI and ET initiatives are welcomed at the government level. An adequate structure for the implementation of the

---

cont.
Institute for International Development. The project considered a single district and positive results were registered on the side of the economic efficiency of the ET scheme implemented.

[129] OECD/IEA (2001), 21.

[130] The reason for this choice is that the existing reporting system can only assure data estimation and verification of CO_2 emissions from fuel combustion. See Malkova (2001), 33.

[131] The system is run by two institutes controlled by the Ministry of Environment – the Czech Hydrometeorological Institute and the Czech Inspection Office. Regarding the first institute, one could assume an important role in the domestic ET scheme through its role of taking the coordination of the national registry for allowances in order to organise the permit transfers.

[132] Even at a government level, more capacity building is needed and several working groups on the issue have already been established with the participation of the business sector. Some research is also being developed, especially on the coverage of firms to involve in the scheme and on the quality of data.

Kyoto Protocol mechanisms has already been established and no particular priority has been given to one instrument or another.

Academics and environmental policy makers have been involved in the organisation and implementation of three initiatives aiming to diffuse awareness on such a system among stakeholders and decision makers[133] and several workshops and meetings aiming at the promotion of ET practices among stakeholders have been organised in Poland in recent years.[134]. Still, the results of these initiatives are not very convincing. Introducing an ET scheme in Poland involves considerable challenges. First of all, climate change does not constitute a major priority within the activity of the Polish Ministry of Environment[135]; secondly, resistance is coming from politicians and decision makers which still prefer other instruments already established and which present faster and more visible positive results on the national environment than a new system where there is not yet sufficient experience. The weak support for ET is further caused by the repetitive failures at the political level to find an agreement in order to establish some kind of legal basis for such a system. At the moment, the legal basis for the introduction of environmentally friendly measures is lacking. Neither the Environmental Protection Act of 1999, nor the current Act on Environmental Protection[136] refers to ET. The recent National Environmental Policy Action Plan confirmed that an ET system should be established by 2005.

Despite negative remarks, both stakeholders and environmentalists still believe that a Polish ET market could be in place by 2005. A recent political decision taken at the ministerial level is the best evidence of this. In 2002, the Polish Ministry of Environment and some stakeholder representatives decided to start a pilot project financed by the industrial sector aiming at the

133 A pilot project including a steel mill, a power plant and local, small-scale heat producers aims at showing how to implement such a system using a computer trading simulation among experts and practitioners from the power and combined heat and power sectors. This first two projects – Chorzów project, 1991; Opole project, 1996 – were commissioned by the Ministry of Environment and focused on sulphur dioxide emissions (SO_2) – finally a workshop where Emissions Trading issues and practices were analysed – Jadwisin workshop, 1996.

134 One Polish economic sector has been so far quite active: the national energy sector was in fact involved in some trading simulation organised initially by EURELECTRIC and then by the Baltic Sea Region Energy Cooperation (BASREC). See also the Greenhouse gas and Electricity Trading Simulations (GETS) in 1999, 2000 and 2002 and the BASREC in 2002.

135 The Ministry of Environment is currently giving greater priority to the transposition indicated in the *acquis communautaire*, while new policy proposals cover a second level. Also for this reason, it is important that the EU proposal on ET is adopted as soon as possible and that positive further steps are made during the preparation of the Accession Treaties when ET will be discussed and possibly inserted into them.

136 See Dz. U. 2001.62.627. This provision deals with compensation procedures. Although they have been used for many years to regulate some emissions sources, these procedures do not constitute adequate legal bases for ET.

introduction of a downstream national emissions trading system to enter into force in 2004. The project, which is going to be designed to be as compatible as possible with the EU scheme, has just begun and a preliminary assessment of the Polish situation (which should form the basis for a tender by the end of 2003) has been concluded. Presently, it is not possible to say whether Poland will ask for any transition period or not.

The scheme will initially cover CO_2, SO_2, NO_x and PM emissions including the power sector – electricity and heat, iron and steel, cement and oil refineries. The other sectors, such as the chemical sector, are excluded because of administrative reasons or incompatibility with the national monitoring and verification system[137]. Concerning the sources to be involved in the scheme, the choice will be between units, installations and companies[138]. The Polish system will likely set out an emissions level for each installation stating when trading would be permitted. All major companies belonging to the same sector would likely be included in order to avoid possible distortions within the sectors[139]. Concerning the allocation system, an approach based on grandfathering with a small part of allowances being auctioned, is envisaged for the Polish system[140]. The Polish project aims also to establish the missing legal basis discussed above. It is not yet determined whether it will be an amendment to the existing Act on Environmental Protection or a new law.

12 What About the Other EU New Member States?

In Slovenia, where there are not many opportunities to host JI projects, the implementation of an ET scheme is seen as a tool to meet international obligations, as the Kyoto targets are still not easy to meet. A step in the direction of an ET system is considered the amendment of the CO_2 tax provision, a regulation that introduces the concept of permits for emissions in several sectors.

---

[137] Kittel and Blachowicz (2002), 6 *et seq*.

[138] Again, the Polish scheme should look at the EU one, which, following the IPPC and the decision for the monitoring mechanisms terminology – both of these instruments require installations to provide data on the emissions coming from all their sources – is considering to regulate emissions on the installation level.

[139] Hauff (2000), 54.

[140] A combination between grandfathering for historic emissions based on the average from 1998 to 2000 and auctioning for the difference between the historical level and the hypothetical growth level in expected emissions would be by far preferable as they would be more efficient from an economic point of view. New potential emissions could be sold through auctioning and the revenues could be used by national governments for more investments or for reducing taxes, even where EC state aid law applies.

There is still no concrete strategy on ET in the Baltic States either. Their good relationship with and dependence on countries like Sweden and Finland is the reason of a general preference for JI projects instead of ET. Moreover, the small size of their potential markets and the monopolistic structure of their economies represent additional obstacles. A solution could have been either the creation of a single market grouping together the Baltic States or the establishment of a connection with the countries mentioned above[141].

The situation is still vague in Hungary, Romania and Bulgaria as well[142]. However, these countries generally are keener to implement JI projects[143]. In Hungary, the Ministry of Environment set up in 2001 two expert groups in order to prepare a registry system for counting emissions units. From this initiative, some private investors have already started plans for the elaboration of a regional ET scheme for the area of Budapest[144].

In Bulgaria, the National Climate Change Action Plan, adopted in 2000, expressly promotes the country's participation in JI projects and IET. However, the national government has not yet started any preparations for the introduction of a domestic ET scheme.

Romania, one of the firsts industrial countries to ratify the Kyoto Protocol, is very keen to use the flexible mechanisms. Regarding JI, the country has been so far very active. Although workshops aimed to encourage better understanding of the mechanism were organised by the OECD and IETA, several years will be needed for the implementation of ET[145]. A precondition is the conclusion of the process concerning the privatisation of the energy sector.

Finally, CEECs are showing a growing general interest in ET programmes. Despite this emerging interest, political and business support is lacking. Furthermore, their internal administrative structures have not been fully co-ordinated to design complex systems for ET, in which several ministries – Energy, Environment, Economy – and different actors should be involved. The definition of the EATD system is of course a positive outcome in terms of legal certainty. Institutional and financial support from the EU

---

[141] Bodnar, Hannes, Hauff and Lipcsey (2002), 77.

[142] The situation in Bulgaria and Romania is slightly different, as these two countries are not joining the EU in the first wave of accessions. Thus, they have to comply with the EU *acquis communautaire* and with the ET Directive probably by 2007, which is the year foreseen for the integration of these two countries into the EU. Moreover, this delay is an attractive force for JI projects.

[143] Bodnar, Hannes, Hauff and Lipcsey (2002), 73.

[144] Id., 77.

[145] The government is already involved in the ERUPT programme with the Netherlands and has begun negotiations with Japan, Switzerland, Norway and France for more JI projects.

will create a more favourable environment for the development of ET in those countries[146].

The introduction of such a market, independently from the EATD, in the short term does not therefore appear to be feasible with Hungary, Bulgaria, Slovenia and the Baltic states probably able to set up a domestic system for 2008. In Romania, an ET scheme will be probably ready for the start of the second Kyoto commitment period in 2012.

13 JT and ET: A Comparative Assessment

The analysis carried out in the previous sections allows us to draw certain conclusions regarding the possibilities and challenges for both JI and ET activities in new member states. A comparative assessment of both instruments would also serve to advance some of the considerations to which EU policy makers should be more attentive in developing special strategies regarding the support to be transmitted to CEECs in their first steps in the design of the new instruments.

When trying to highlight the differences between JI and ET, it is important to recall once again that we are talking about two mechanisms whose implementation are covered by two different legal regimes, at least for the first phase: the Kyoto Protocol and the Marrakesh Accords for JI; and EC law for ET at least until 2008. Furthermore, while ET is a cap and trade system based on the *ex ante* allocation of emission allowances, JI is a "baseline and credit" instrument where the verification of the difference between baseline and actual emissions is done *ex post*. Two other main differences between these two instruments arise from the procedural point of view: timing, and the different type of units to be traded.

JI appears as a policy enlargement instrument and initial JI activities may emerge as one of the most promising tools in the context of the enlargement process. Yet, the development of JI projects in CEECs will reveal the different and, to a certain extent, competing interests of participating actors, namely, the European Union, the member states (investors) and CEECs (host countries)[147]. The EU itself has on the one hand, an interest in achieving its commitment targets under the EU bubble and on the other hand, aims at

[146] This is also why the European Commission has recently established, with the support of the Center for Clean Air Policy, a series of workshops at the ministerial level to take place within CEECs. The aim of this initiative is to help acceding countries prepare for participation in the EATD, throughtrading simulation and the production of capacity needs assessment reports.

[147] See Stronzik (2001), 131 on the issue of conflicting interests between host and investor countries. See also Khovanskaia (2001), 3. "There is a critical trade-off between procedures, which are more attractive to investors, but offer less environmental security".

pushing CEECs to adopt the *acquis communautaire* and its parallel increase of environmental and development standards. Despite these theoretically conflicting interests, it is obvious that the advantages of the EU enlargement process will bring about positive results to all actors. In this respect, JI contains particularities (such as increases to environmental and development standards as well as financial transfers) that make it a particularly good tool for furthering the enlargement process. As such, the EU should take advantage of JI's properties so as to squeeze its "plus points". From a political point of view, early JI activities might become an asset for EU enlargement. JI within the context of the European integration and accession process should be used to strengthen the accession process on two sides. By advancing the implementation of EU regulations and contributing to the achievement of EU environmental goals, early JI activities firstly increase the degree of certainty about enlargement by CEECs and also enhance the confidence of EU-15 in the implementation of the *acquis communautaire* by CEECs. Equally, given their common character of being public/private projects, JI activities can foster private/public partnerships.

Regarding financial constraints, there is no doubt that the monetary needs derived from the implementation of the *acquis communautaire* are added to the financial needs arising from putting in place regulatory and administrative frameworks for the implementation of JI projects in CEECs[148]. In this respect, certain countries have manifested their preference for JI over ET because of the environmental goals of the project mechanism[149]. Thus, because of its transfer finance nature, JI might serve to overcome those financial constraints. Given that the Commission's Communication on Accession Strategies for the Environment (1998) required new member states to develop investment strategies for the implementation of the *acquis communautaire*, along with their legislative approximation strategies, JI should be integrated by CEECs in their EU accession strategies as a way not only to meet *acquis* requirements, but also as mechanisms for facing financial constraints. Hence, by setting an appropriate national JI model, CEECs might use JI investments for attaining economic development and environmental objects related to EU accession.

In terms of the choice between these two mechanisms, EU enlargement will need additional considerations. The application of the *acquis communautaire* with all its environmental measures among CEECs, could

---

[148] For an overview of the environmental and financial needs of candidate countries, see European Commission (2001c).

[149] "Latvia will focus on JI instead of ET because of environmental benefits. By definition, JI necessitates an investment project that will help alleviate an environmental problem. Emissions trading only results in revenue for the government in exchange for emissions allowances. Thus, there is no environmental effect unless this revenue is tied to environmentally friendly expenditures" Jurjans (2001), 111.

lower the interest for JI and make these projects less attractive to investors. On the other hand, CEECs will have to comply with the EU legislation and will be therefore constrained to adopt the EATD probably as soon as 2005. Thus, the scope for JI would be reduced by: (1) strict environmental standards under the *acquis*, which by consequence would incite participation in ET; and (2) the implementation of the EATD itself as those activities and areas falling under this mandatory scheme will be excluded from the JI sphere[150]. The recent agreement on the EATD gives the possibility to member states to include installations in the ET system that carry out activities listed in Annex I at levels below the specific capacity limits[151]. Thus, the relevant question regarding the use of this option by new member states is whether in these circumstances JI possibilities would be reduced by the implementation of the EATD. But the final text also recognizes that CEECs – as well as the rest of EU member states – could, under Article 27.1, ask for a temporary exclusion of certain installations and activities from the system until 31 December 2007. In this case, the question would be whether those excluded activities and installations could be the subject of a JI project and under which conditions. For instance, could an acceding country whose activities are excluded from the scope of ET, incorporate such activities under the JI program? In this case, would the stringency of baselines be the same as it would be according to the allocation criteria included in the EATD?

While ET has the advantage of lower transaction costs and a relative quick and easy means to proceed once it is introduced, although there is a stringent need to speed up with the preparation of domestic schemes in both the old and new member states, there is still a lot of uncertainty on the market price due to the vagueness on the number of participants in the scheme. The disadvantages of JI implementation are that the transfer of units will be valid only for 2008–2012, there will be high administrative costs and very often there will not be a very well-organised structure at the national level[152].

From an environmental point of view, while JI activities produce positive results for the environment by the effective reduction of emissions, ET is often accompanied by sceptical concerns regarding its effective environmental integrity/effectiveness in reducing other environmental problems. Nevertheless, the environmental integrity of ET should be guaranteed by the correct implementation of the Directive. ET will provide some economic profits eventually to be reinvested in environmental friendly projects, while JI will facilitate the transmission of information,

150 OECD/IETA (2002), 11.
151 Article 23.1 of the EATD.
152 Bodnar, Hannes, Hauff and Lipcsey (2001), 75.

understanding and technology, as well as facilitate funding for projects and additional benefits for other policies[153].

Some observers have argued that the more developed CEECs would prefer full integration into the EATD, whereas those countries lagging behind would prefer to implement JI projects as they bring more direct investment[154]. Certainly, full integration in the EATD is quite complex and has to be complemented by a clearing procedure to balance national accounts at least to the end of the first commitment period. ET is therefore likely to become a more relevant mechanism for some new member states as the emission reductions due to the accession process increase the scope for ET. Despite these facts, many CEECs – Poland and the Czech Republic for instance – have declared their interest to participate actively in both JI and ET.

It is generally expected that some of the EU-15 member states will have to face large emission deficits during the first commitment period, which they will have to compensate through the acquisition of assigned amounts from other countries. In this regard, a sort of competition as to the sale of credits could arise among new member states and other EITs. Although we are assessing JI and ET *vis-à-vis* CEECs, both mechanisms can be used as well by any Annex I Party, including Bulgaria, Romania, Ukraine and Russia. These countries will compete with other EITs regarding the supply of emissions units/reductions under JI and ET. Moreover, in terms of JI, these countries which are not involved in the immediate EU accession process will not need to consider any requirements derived from the *acquis communautaire* legislation, when setting baselines. By the same token, CDM projects will also influence the JI market: the regulatory structure already established and the possibility to earn credits before 2008 makes CDM a more attractive investment[155].

Despite the fact that JI policies and activities are more developed and more readily receive approval than ET in the EU member states, it can be concluded that there is not yet a strategic approach to the combination of policies and measures to curb climate change and no balance in the use of the Kyoto Protocol flexible mechanisms within CEECs. However, the debate over the possibility to set up ET schemes has already begun in the region. Early action on ET will facilitate the work for the implementation of the EU

---

153 Missfeldt and Villavicenco (2002), 5 *et seq*.

154 Van der Gaast (2001).

155 For countries like Romania and Bulgaria, advanced environmental technologies are often not yet economically and technically viable due to insufficiently developed products and capital markets and a lack of institutional capacity to install and maintain those techniques. For those countries, JI could represent a significant tool in attaining EU standards by the time the candidate becomes a EU member state.

scheme and could foster the involvement of a wider number of industries once the link with the EU scheme is established. In any event, the involvement of stakeholders has been generally low so far and most countries have been waiting for the final developments of the EATD scheme proposal.

The recent developments in terms of EU climate policy, namely, the approval of the EATD and the linking directive in particular, can only facilitate the improvement of the climate policies of CEECs and increase the interest and involvement of policy-makers, stakeholders and competent authorities.

References

Atanasiu, I. (2001). 'State Aid in Central and Eastern Europe'. *World Competition*, 24/2: 257–283.

Beuerman, C., Lagrock, T. and Hermann, E. (2000). *Evaluation of (non-sink) AIJ-Projects in Developing Countries*. Wuppertal: Wuppertal Institute, 100.

Bodnar, P., Hannes, B., Hauff, J. and Lipcsey, G. (2002). 'CO_2-Emissionshandel in Mittel- und Osteuropa – eine Chance für Unternehmen des Energiesektors'. *Zeitschrift für Energie Wirtschaft (ZfE)*, 26/1.

Caddy, J (2000). 'Implementation of EU Environmental Policy in Central European Applicant States: The Case of EIA', in Knill, C. and Lenschow, A. (eds.), *Implementing EU Environmental Policy: New Directions and Old Problems*. Manchester: Manchester University Press, 196–221.

Cozijnsen, J. (2001). *The Development of Post-Kyoto Emissions Trading Schemes in Europe: An Analysis in the Context of the Kyoto Process, in Greenhouse Gas Market Perspectives, Trade and Investment Implications of the Climate Change Regime*. United Nations Conference on Trade and Development (UNCTAD). http://www.unctad.org.

Emott, N. (1999). 'An Overview of the IPPC Directive and its Development', in *Integrated Pollution Prevention and Control, The EC Directive from a Comparative Legal and Economic Perspective*. London: Kluwer Law International, 23–41.

Environment Daily (ENDS) (2003). 'EU Climate Emissions Trading Breakthrough'. *Environment Daily (ENDS) 1470*. http://www.environmentdaily.com.

European Commission (1997). *Guide to the Approximation of European Union Environmental Legislation*. Brussels: European Commission. *Staff Working Paper SEC (97) 1608*.

European Commission, (2001a). *Report on Administrative Capacity for Implementation and Enforcement of EU Environmental Policy in the 13 Candidate Countries*. Brussels: European Commission.

European Commission (2001b). *The Benefits of Compliance with the Environmental Acquis for the Candidate Countries*. Brussels: European Commission.

European Commission (2001c). *The Challenge of Environmental Financing in the Candidate Countries*. COM (2001) 304 final, 8.6.2001. Brussels: European Commission.

European Commission (2002a). *Regular Report on Poland's Progress towards Accession*. Brussels: European Commission. COM (2002) 700 final, SEC (2002) 1408.

European Commission (2002b). *Regular Report on Slovakia's Progress towards Accession.* Brussels: European Commission. COM (2002) 700 final, SEC (2002) 1410.

European Commission (2003a). *Report on the results of the negotiations on the accession of Cyprus, Malta, Hungary, Poland, the Slovak Republic, Latvia, Estonia, Lithuania, the Czech Republic and Slovenia to the European Union prepared by the Commission's departments.* Brussels: European Commission. http://europa.eu.int/comm/enlargement/negotiations/pdf/negotiations_report_to_ep.pdf.

European Commission (2003b). *Commission Welcomes Final Adoption by Council of EU Emissions trading Directive.* Brussels: European Commission. Press Conference IP/03/1073.

European Commission (2003c). *The EU Emissions Trading Scheme: How to Develop a National Allocation Plan.* Brussels: European Commission.

European Parliament (2000). *Report of the European Parliament, Environmental Aspects of the Enlargement Negotiations of the Committee on the Environment, Public Health and Consumer Policy.* Brussels: European Parliament.

Fankhauser, S. and Lavric, L. (2003). *The investment Climate for Climate Investment: Joint Implementation in transition countries.* EBRD (European Bank for Reconstruction and Development), 77.

Fernández Armenteros, M. and Michaelowa, A. (2002). 'Joint Implementation and EU Accession Countries'. Hamburg: HWWA (Hamburg Institute of International Economics). *Discussion Paper 173.*

Harrison, D. and Radov, D. (2002). *Evaluation of Alternative Initial Allocation Mechanisms in a EU Greenhouse Gas Emissions Allowance Trading Scheme.* National Economic Research Associates.

Hauff, J. (2000). *The feasibility of domestic CO_2 Emissions Trading in Poland.* Roskilde: Risø National Laboratory.

IMPEL-Network (2000). *The Application of General Binding Rules in the Implementation of the IPPC Directive.* http://europa.eu.int/comm/environment/impel.

Jilkova, J. and Chmelik, T. (2001). *Domestic Emissions Trading System in the Czech Republic: Options for an Implementation Framework.* Prague: Institute for Economic and Environmental Policy, University of Economics.

Jurjans, A. (2001). 'Country Report: Latvia', in Wuppertal Institute for Climate (ed.), *Emissions Trading and Joint Implementation as a Chance for the CEECs.* Berlin: Federal Ministry for the Environment, Nature Conservation and Nuclear Safety, 111–119.

Kittel, M. and Blachowicz, A. (2002). *Emissions Trading As The Key Option For CO_2 Mitigation.* Washington: Center for Clean Air Policy (CCAP).

Khovanskaia, M. (2001). *JI and Business Involvement in CEE.* Workshop of JI experts: 18–19 April 2001. Budapest: Regional Environmental Centre (REC).

Joint Implementation Quarterly (JIQ) (2002). *ERUPT-2Waste Management in Slovakia.* http:///www.jiq.nl.

Lefevere, J. (2002). *Greenhouse Gas Emission Allowance Trading in the EU: A Background.* London: Foundation for International Environmental Law and Development (FIELD).

Malkova, J. (2001). *Domestic ET System in the Czech Republic: Options for an Implementation Framework.* Prague: ECON.

Massai, L. (2002). 'Central and Eastern European Countries and Climate Change Regime'. *ELNI (Environmental Law Network International) Review*, 1/03:25–35.

Mayhew, A. (2000). 'Enlargement of the European Union: An Analysis of the Negotiations with the Central and Eastern European Candidate Countries'. Sussex, UK: Sussex European Institute, *Working Paper 39*.

Michaelowa, A. (1999). *Review of Reports on Activities Implemented Jointly (AIJ) under the Pilot Phase with a Specific Focus on Baseline and Additionality Issues: Lessons Learned and Recommendations Regarding Practical Options*. Bonn: UNFCCC Secretariat.

Ministry of Economic Affairs, The Netherlands (2002). *Allocation of CO_2 Emission Allowances – Distribution of Emission Allowances in a European Emissions Trading Scheme*. The Hague: Ministry of Economic Affairs, The Netherlands.

Missfeldt, F. and Villavicenco, A. (2002). 'How can Economies in Transition Pursue Emissions Trading or Joint Implementation?'. Milano: Fondazione Eni Enrico Mattei. *Working paper*, 59.

Mullins, F. (2000). *Capacity Needs of Central and Eastern Europe: An Assessment of National Systems for Reporting Participation in the Mechanisms in Six CEE Countries*. London: Environmental Resources Management.

OECD/IEA/IETA (Organisation for Economic Co-operation and Development Environment Directorate, International Energy Agency and International Emissions Trading Association) (2001). *Designing Inventory, Registry And Trading Systems In Countries With Economies In Transition*. Workshop Report: COM/ENV/EPOC/IEA/SLT(2001)14.

OECD/IEA/IETA (Organisation for Economic Co-operation and Development Environment Directorate, International Energy Agency and International Emissions Trading Association) (2002). *National Systems for Flexible Mechanisms: Implementation Issues in Countries with Economies in Transition*. Workshop Report: COM/ENV/EPOC/IEA/SLT(2002)4.

PointCarbon (2002). *Carbon Market Europe 14 March 2003*. http://www.pointcarbon.com.

Regional Environment Center and World Resources Institute (REC/WRI) (2001). *Capacity for Climate Protection in Central and Eastern Europe, Activities Implemented Jointly*, Budapest: Regional Environmental Centre (REC).

Smith, A. and Sorrel, S. (1998). 'Interaction Between Environmental Policy Instruments: Carbon Emissions Trading and Integrated Pollution Prevention and Control'. Sussex: University of Sussex – Science Policy Research Unit. *Working Paper 27*.

Sofia Initiative on Economic Instruments (1999). *Source Book on Economic Instruments for Environmental Policy in Central and Eastern Europe*. Klarer, J., McNicholas, J. and Knaus, E. (eds). Szentendre: Hungary, Regional Environmental Centre (REC).

Stronzik, M. (2001). 'Joint Implementation-Investors and Hosts- Reconciliation of Differing Interests', in Wuppertal Institute for Climate (ed.), *Emissions Trading and Joint Implementation as a Chance for the CEECs*. Berlin: Federal Ministry for the Environment, Nature Conservation and Nuclear Safety, 131–133.

Szécáks, G. (2001). 'Country Report: Hungary', in Wuppertal Institute for Climate (ed.), *Emissions Trading and Joint Implementation as a Chance for the CEECs*. Berlin: Federal Ministry for the Environment, Nature Conservation and Nuclear Safety, 109–110.

Van der Gaast, W. (2001). *Project-based Mechanisms: Past Experiences and the Future*. Workshop presentation. http://www.hwwa.de/climate.

Williams, E., Kolar, S., Levina, E. and Li, J. (2002). *Developing a CO_2 Emissions Trading Design for Slovakia*. Washington: Center for Clean Air Policy.

Zapfel, P. and Vainio, M. (2002). 'Pathway to European Greenhouse Gas Emissions Trading History and Misconceptions'. Milano: Fondazione Eni Enrico Mattei. *Working Paper*, 85.